FOUNDATIONS OF
GEOMETRY

FOUNDATIONS OF
GEOMETRY

C. R. WYLIE, Jr.

DOVER PUBLICATIONS, INC.
Mineola, New York

Bibliographical Note

This Dover edition, first published in 2009, is an unabridged republication of the work originally published by the McGraw-Hill Book Company, Inc., New York, in 1964.

Library of Congress Cataloging-in-Publication Data

Wylie, Clarence Raymond, 1911–.
    Foundations of geometry / C. R. Wylie. — Dover ed.
      p. cm.
    Originally published: New York : McGraw-Hill, 1964.
    Includes index.
    ISBN-13: 978-0-486-47214-0
    ISBN-10: 0-486-47214-0
    1. Geometry—Foundations. I. Title.

QA681.W9 2009
516—dc22

2008056113

www.doverpublications.com

This book has been written primarily for students preparing to become teachers of secondary school mathematics, although it should also be of interest to practicing teachers and to undergraduates majoring in mathematics who wish to review or to extend their background in geometry. Its purpose is to present a careful axiomatic development of certain important parts of elementary euclidean and non-euclidean geometry, and, in so doing, to acquaint the student with the axiomatic method as a general pattern of thought. To some, the word "Foundations" in the title of a book suggests a boring preoccupation with details or an intensive belaboring of the obvious. However, to the prospective reader of this book, I would express both the conviction that its concern with details, while extensive, is not significantly greater than that found in the new high school programs, and the hope that such concern is adequately motivated and properly balanced by exciting glimpses of what lies beyond the bounds of traditional euclidean geometry.

The book begins with a chapter on the axiomatic method and its major features, independent of its use in geometry. Then, in Chapter 2, postulates for three-dimensional euclidean geometry are introduced and the principal results up to, but not including, the measurement of volume are carefully developed. Chapter 3, though it contains nothing original, is perhaps the most novel in the book. It is devoted to an axiomatic development of the simpler aspects of four-dimensional euclidean geometry, and is intended to give the student practice in the axiomatic method in a setting, essentially as simple as the geometry of three dimensions, but in which he has no familiarity with the main results to guide or to hinder him. Chapter 4 provides an introduction to plane hyperbolic geometry and carries the development as far as the measurement of area. Finally, in Chapter 5 the question of the consistency of hyperbolic geometry is considered; and by describing in detail a model in the euclidean plane in which each of the postulates of hyperbolic geometry can be verified, it is shown that hyperbolic geometry is relatively consistent.

Those who are familiar with the geometry texts prepared by the School Mathematics Study Group will find a striking resemblance between the postulates employed in those books and the postulates adopted here. It was my great good fortune to participate in the writing of the School Mathematics Study Group's "Geometry with Coordinates," and the present book owes much to the stimulating contacts I had with my colleagues in this project. Of course, the School Mathematics Study Group has no responsibility for what I have written here, and my obvious indebtedness to it does not imply any indorsement of my work. However, it is a pleasant obligation on my part to acknowledge the kindness of the Group and its Director, Professor E. G. Begle, in permitting me to employ its wording of a number of postulates and theorems and to use, without change, occasional passages which I wrote during the preparation of "Geometry with Coordinates."

The author of any textbook owes much to his own teachers, colleagues, and students; and to all who have assisted me, consciously or unconsciously, in the preparation of this book, I express my appreciation. Finally, it is a pleasure to acknowledge the welcome assistance of Miss Maxine Winterton, who typed the manuscript, and of my wife, Ellen, and my secretary, Mrs. Patricia Everts, who shared the task of reading the proof.

C. R. Wylie, Jr.

CONTENTS

1

THE
AXIOMATIC
METHOD

1.1 Introduction. The origin of the discipline we call geometry is clearly evident in the name itself, which derives from the Greek words *ge*, meaning *the earth*, and *metrein*, meaning *to measure*. In the beginning geometry was, indeed, the art (not science) of earth measurement and consisted of a disorganized collection of rules for computing simple areas and volumes and carrying out a few elementary constructions. Such results were the fruits of long centuries of trial and error by the Babylonians and the Egyptians who, in the dawn of civilization, had to develop practical procedures for such things as land surveying, the construction of granaries and canals, and the erection of tombs and temples. Some of this information was correct and some incorrect though useful as an approximation, but all had at best only the sanction of plausibility. In other words, for the first two thousand years or so of its existence, geometry was a body of empirical knowledge obtained inductively from a consideration of many special cases and completely unsupported by anything resembling logical proof.

Then in the millennium immediately preceding the Christian era, geometry underwent a remarkable change. The Greeks, inclined by temperament toward philosophy and abstraction and blessed with security and leisure to follow these inclinations, took the geometry of the Egyptians and recast it in the form of a deductive science. Beginning with Thales (Thā′-lēz, 640–546 B.C.), this transformation culminated in the work of Euclid (365?–275? B.C.), whose "Elements" presented the sum total of current geometrical knowledge, not as a disjointed collection of empirical results, but as a well-organized chain of theorems following inevitably by the laws of logic from a few simple

1

initial assumptions. Euclid's "Elements" remains the most famous and most important textbook ever written. Not only did it permanently establish the character of geometry as a deductive science, it also exemplified a pattern of logical organization so effective and so elegant that today most of mathematics is constructed according to the same plan. In major areas of other disciplines, such as biology, chemistry, economics, physics, and psychology, the goal of scholars is still to achieve a comparable logical structure.

The abstract logical plan which Euclid conceived and so ably illustrated in his "Elements" we now refer to as the **axiomatic method,** and any particular instance of it we call an **axiomatic system.** The impact of the axiomatic method upon mathematics, and upon other sciences as well, has been so profound that not only the scholar who must be prepared to use it in his own work but also the intelligent layman who would achieve some understanding of the nature of scientific thought must be familiar with it. Accordingly, we shall devote the balance of this chapter to a discussion of the axiomatic method and its major features.

1.2 Inductive and Deductive Reasoning. A careless reading of the preceding section might well leave one with the impression that inductive reasoning is primitive and unscientific and that deductive reasoning is the only mode of thought appropriate to genuine scientific inquiry. Nothing could be further from the truth. Although this book is devoted almost exclusively to instances of deductive reasoning, we should understand from the outset the nature of induction and the important role which it plays in every science, including mathematics.

Induction can be described briefly as the process of inferring general properties, relations, or laws from particular instances which have been observed. Deduction, on the other hand, is the process of reasoning to particular conclusions from general principles that have been accepted as the starting point of an argument. Both induction and deduction have their merits and their defects, and neither by itself is sufficient to support genuine scientific progress.

Induction has at least two obvious weaknesses. In the first place, no matter how many completely correct observations have been made, unless every possible instance has been examined, no generalization can be made with certainty, because any of the uninvestigated cases may contradict it. Second, the assumption that the observations actually made have been made with perfect accuracy is often false, so that in many cases there is not exact information on which to base a generalization.

To illustrate, if we evaluate the expression $n^2 - n + 17$ for the first few positive integers we obtain the following table:

n	$n^2 - n + 17$
1	17
2	19
3	23
4	29
5	37
6	47
7	59
8	73
9	89
10	107

Here, we are in the fortunate position of having a set of completely accurate observations from which to generalize, and by induction we may be led to any of several plausible conjectures. For instance, from the particular cases before us, we may draw the almost obvious conclusion that for every positive integer n the expression $n^2 - n + 17$ is a number which is odd. Or, taking a somewhat closer look, we may conclude further that for every positive integer n the expression $n^2 - n + 17$ is a number which ends in 3, 7, or 9. Or, observing that 17, 19, 23, . . . , 107 are all prime numbers, we may infer the still more remarkable property that $n^2 - n + 17$ is a prime number for every positive integer n. Each of these conclusions is strongly suggested by the data. But are they all correct, and if so, how can we be sure? Trying additional values of n *may* answer the question, for if we find for some n that $n^2 - n + 17$ is not odd, or does not end in one of the digits 3, 7, or 9, or is not a prime number, then the corresponding inference is immediately overthrown by that one counterexample. But if we investigate additional cases and find them all consistent with our conjectures, the issue remains in doubt. We may feel that the additional supporting examples increase the probability that our generalizations are correct, but we still must admit the possibility that among the cases not yet examined there may be at least one which will contradict, and hence overthrow, one or more of our conjectures. Specifically, if we extend our table a bit, we find

n	$n^2 - n + 17$
11	127
12	149
13	173
14	199
15	227
16	257
17	289

Through $n = 16$, the entries support each of our conjectures. However, for $n = 17$ we find that $n^2 - n + 17 = 289$ is not a prime but in fact is equal to $(17)^2$. Thus our third conjecture is false, while the other two, though perhaps more plausible because of the additional supporting evidence, remain uncertain.

It is precisely at this point that deduction "comes to the rescue" and "takes over" from induction. When further search for a counterexample seems fruitless and the evidence of a sufficient number of supporting examples convinces a mathematician that a conjecture is probably true, he abandons induction and tries to *prove* the conjecture by deriving or deducing it from more fundamental principles. In particular, though we shall not digress to do so, it is easy to deduce from the principles of arithmetic that for every positive integer n the expression $n^2 - n + 17$ is a number which is odd and which moreover ends in 3, 7, or 9.

On the other hand, deduction has its weaknesses. In the first place, deduction can only provide us with conditional statements of the form "*If* something is true, *then* something else is true." It is essentially unconcerned with whether or not the statements with which an argument begins are true or false. Second, deduction is, in itself, incapable of providing us with either the results which we hope to prove or the initial statements from which we propose to evolve a proof.

It is here that induction steps in and saves deduction from its inherent sterility. It is induction that suggests the theorems which deduction subsequently so painstakingly tries to prove. It is induction, too, which provides us with the insights that we formalize in the principles on which our proofs are based. And insofar as these principles have any claim to truth, it is induction which ultimately supports that claim. Without the body of geometric information accumulated by the Babylonians and the Egyptians and his Greek predecessors, Euclid would have had no material to organize, no results to set in logical order, no initial principles from which to reason. The proofs of theorems are achievements of deductive reasoning, but theorems themselves are the fruits of induction, of intuition, of creative insight habitually examining every special case for suggestions of more general relations.

On its lower levels, induction is merely the tedious cataloging of observations; at its best, it is the imaginative recognition, through a thousand irrelevancies, of the essential nature of a situation. Without induction, deduction can only wait, idly, for something to prove. Without deduction, induction is always unsure of itself, its inferences suspect, its insights, no matter how brilliant, vulnerable to counterexamples and disproof. And in more practical terms, without skill in both induction and deduction, that is, without both intuition and a clear feeling for proof, no student of mathematics is more than marginally prepared for his work.

The cooperative relation between induction and deduction is roughly comparable to the relation between mathematics and the other sciences. Historically, mathematics developed out of man's concern with physical problems. And though mathematics is ultimately a construction of the mind alone and exists only as a magnificent collection of ideas, from earliest times down to the present day it has been stimulated and inspired and enriched by contact with the external world. Its problems are often idealizations of problems first encountered by the physicist or the engineer, or, more recently, by the social scientist. Many of its concepts are abstractions from the common experience of all men. And many areas of mathematics stem originally from the needs of scholars for more powerful analytical tools with which to pursue their investigations of the world around them. Surely, without contact with the external world, mathematics, if it existed at all, would be vastly different from what it is today.

But mathematics repays generously her indebtedness to the other sciences. At the mere suggestion of a new problem, mathematics sets to work developing procedures for its solution, generalizing it, relating it to work already done and to results already known, until finally it gives back to scholars in the original field a well-developed theory for their use.

Oftentimes mathematics outruns completely the demands of the physical problem which may have stimulated it, and careless critics scoff at its "pure" or "abstract" or "useless" character. Such criticism is absurd on two quite different counts. In the first place, all mathematics worthy of the name is pure or abstract or even, in a certain sense, useless, just as poetry and music and painting and sculpture are useless except as they bring satisfaction to those who create and to those who enjoy. It is no more appropriate to criticize mathematics for possessing the attributes of one of the creative arts than it is to criticize the arts themselves, unless it be that perhaps the number of those who find enjoyment in mathematics, though considerable, is less than those who enjoy the more conventional arts. Then in the second place, by a remarkable coincidence which is almost completely responsible for the existence of all present-day science and technology and which has no counterpart in the arts, even the most abstract parts of mathematics have turned out again and again to be highly "practical," as science, becoming ever more sophisticated, finds that it needs mathematical tools of greater and greater refinement.

As we now begin to employ the deductive method in our investigation of one small part of the great field of mathematics, it is important that we do not lose sight of the great significance of the inductive method, which suggested most, if not all, of the results with which we shall be concerned. And it is important, too, that even while we remain thoroughly committed to the conviction that geometry is a branch of pure mathematics, we never forget that it had its origins in the physical world, owes much of its vitality to its contacts with the physical world, and, properly understood, is a tool of magnificent power and effectiveness for the study of the physical world.

EXERCISES

1. Prove that for all integral values of n, $n^2 - n + 17$ is odd and ends in one of the digits 3, 7, or 9.
2. What do you think is the most plausible value for the next number in each of the following sequences?

    (a) 0, 0, 0, 0, . . .   (b) 1, 0, 1, 0, . . .
    (c) 1, 2, 3, 4, . . .   (d) 1, 2, 4, 8, . . .

    Can you prove your conjectures? Can you construct expressions which will yield the given values for $n = 1, 2, 3$, and 4 but in each case will yield the value π when $n = 5$?
3. Evaluate $x^7 - 14x^5 + 49x^3 - 39x$ for $x = 0, \pm1, \pm2, \ldots$ What generalizations occur to you? Are your inferences correct for all values of x? Are they correct for all integral values of x?
4. Determine the sum of the first k odd positive integers for a number of values of k. What generalizations occur to you? Are your inferences correct for all positive integers k?
5. Determine the sum of the cubes of the first k positive integers for a number of values of k. What generalizations occur to you? Are your inferences correct for all positive integers k?
6. Evaluate $n^2 - 39n + 421$ for a number of integral values of n. What generalizations occur to you? Are your inferences correct for all integral values of n?
7. Let a sequence of integers $u_1, u_2, u_3, \ldots, u_r, \ldots$ be defined by the conditions

$$u_1 = 1$$
$$u_2 = 1$$
$$u_{r+2} = u_{n+1} + 2u_n \qquad n = 1, 2, 3, \ldots$$

What general properties of this sequence occur to you? Let Π_r denote the product of the first r numbers in the sequence and evaluate the expression

$$\frac{\Pi_r}{\Pi_k \Pi_{r-k}} \qquad 0 < k < r$$

for various values of r and k. What generalizations occur to you?
8. Perform the following geometric "experiment" several times: Construct a quadrilateral of any shape and determine the midpoints of its sides. What properties of the four midpoints occur to you?

9. Perform the following geometric "experiment" several times: Construct two triangles so related that the lines joining corresponding vertices pass through the same point, and locate the point of intersection of the lines determined by corresponding sides of the two triangles. What properties of the three points of intersection occur to you?

10. Perform the following geometric "experiment" several times: Let l_1 and l_2 be two intersecting lines, and on each choose three points, say A_1, B_1, C_1 and A_2, B_2, C_2, distinct from the intersection of l_1 and l_2. Determine the intersection of each of the following pairs of lines:

    A_1B_2 and A_2B_1.　　A_2B_3 and A_3B_2,　　A_1B_3 and A_3B_1

    What properties of the three points of intersection occur to you?

11. Repeat the "experiment" described in Exercise 10, only this time let the points A_1, B_1, C_1 and A_2, B_2, C_2 be chosen so that the lines A_1A_2, B_1B_2, C_1C_2 all pass through the same point. What properties of the three points of intersection defined in Exercise 10 occur to you in this case?

12. Perform the following geometric "experiment" several times: On any circle choose any six points, P_1, P_2, P_3, P_4, P_5, P_6, and determine the intersection of each of the following pairs of lines:

    P_1P_2 and P_4P_5,　　P_2P_3 and P_5P_6,　　P_3P_4 and P_6P_1

    What properties of the three points of intersection occur to you? Can you think of a generalization of this "experiment" to curves other than circles?

13. Perform the following geometric "experiment" several times: Draw three circles, C_1, C_2, and C_3, so related that each one intersects each of the others at two distinct points. Draw the lines determined by the points of intersection of each pair of circles. What properties of the three lines occur to you? Can you think of a generalization of this "experiment" to curves other than circles?

14. Perform the following geometric "experiment" several times: Construct a triangle of any shape and on each of its sides construct an equilateral triangle having that side as base. For each of the three equilateral triangles, determine the point of intersection of its medians. What properties of these three points occur to you? Can you think of a generalization of this "experiment" to polygons other than triangles?

15. Discuss the following "proof" that 1 is the largest positive integer:

    "If n is any positive integer except 1, it is obvious that n^2 is an integer which is still larger. Hence no positive integer different from 1 can be the largest and so, perforce, 1 must be the largest positive integer."

    What theorem, if any, is established by this argument?

16. Discuss the "moral," if any, to the following anecdote:

    An engineer, a physicist, and a mathematician were once riding together through the sheep country of Montana. Glancing across the plains, the engineer saw a small flock of sheep and remarked, "Well, I see there are some black sheep in Montana." The physicist, looking out and observing that there was but a single black sheep in the little flock, rebuked the engineer, saying, "As scientists, don't you think we

should say simply that there is at least one black sheep in Montana?"
Then the mathematician, having made his survey of the flock, said,
"Gentlemen, it appears to me that all we are entitled to assert is that
there is at least one sheep in Montana which is black on one side."

17. Discuss the following statement of the principle of induction:

"If certain members of a class are observed to have a given property
and if the rest of the members of the class have this property, then all
members of the class have the given property."

Discuss each of the following quotations:

18. "We see that experience plays an indispensable role in the genesis of
geometry; but it would be an error then to conclude that geometry is
even in part an experimental science."

Henri Poincaré, "Science and Hypothesis," p. 79 [1].*

19. "As we emphasize the deductive structure of our science [mathematics]
and of acceptable proof, let us not lose sight of the fact that many of the
most significant results that we prove were arrived at by guess-work, by
intuition, by brilliant insight."

Mina Rees, "The Nature of Mathematics," *Science*, Oct. 5, 1962, p. 11
[2].

20. "I address myself to all interested students of mathematics of all grades
and I say: Certainly let us learn proving, but also *let us learn guessing*."

George Polya, "Induction and Analogy in Mathematics," p. v [3].

1.3 Axiomatic Systems. Before we can begin to organize our
knowledge of geometry into a logical, deductive structure, we must
first become familiar with the general features of the kind of organization we hope to achieve. Briefly, an axiomatic system consists of the
following:

1. A set of undefined terms which forms the basis of the necessary
technical vocabulary
2. A set of unproved initial assumptions
3. The laws of logic
4. The body of theorems, expressing properties of the undefined
objects, which are derived from the axioms by the laws of logic

At first glance it might appear that in any logical discussion every
term should be carefully defined, but a moment's reflection shows that
this is impossible! New terms can be defined only by means of others
already defined and understood. Thus, attempting to define every
term either leads us to some first word, for whose definition no other
words are available, or else leads us in circles in which, in effect, we
define A in terms of B, B in terms of C, and C in terms of A! How

* Bracketed numbers refer to full bibliographic credits listed at the end of the
chapter.

common the latter process is can be seen by looking up almost any word in a dictionary, then looking up the words used in its definition, and so on. Usually in just a few steps one reaches a definition in which the original word reappears! For instance, in one of the standard unabridged dictionaries we find the following chain of definitions purporting to give meaning to the term *magnitude:*

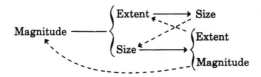

In the development of an axiomatic system the impossibility of defining every term is explicitly recognized and certain technical terms are deliberately left undefined. The vocabulary of the discussion then consists of these undefined terms, other technical terms defined by means of them, and the nontechnical vocabulary of everyday discourse which, of course, we implicitly assume to be available.

Similarly, it might be thought that in any scientific discussion every assertion should be carefully proved, but this, too, is impossible. If we attempt it, we either embark upon an infinite regression, in which we assert that

A is true because B is true.
B is true because C is true.
C is true because D is true.
. . . .

or else we reason in a circle and assert, in effect, that

A is true because B is true.
B is true because C is true.
C is true because A is true.

The impossibility of proving every statement is also explicitly recognized in the construction of an axiomatic system, and certain statements, nowadays referred to interchangeably as **axioms** or **postulates**, are accepted without proof as a necessary starting point for the discussion.

Occasionally it is said that axioms are facts which are taken for granted because their truth is so obvious that it needs no proof. Since the purpose of an axiomatic system is to provide an orderly development in which complicated and difficult results are deduced from simpler and more fundamental ones, the initial assumptions are often so simple that they do, indeed, seem obviously true, and on psychological and pedagogical grounds this is probably desirable. Moreover, in choosing among various sets of axioms which may serve equally well as the starting point of a deductive development, it is eminently proper to choose the one which seems most natural or most plausible. Nevertheless, the ultimate reason for accepting a set of axioms without proof is simply that no other course is possible and has nothing to do with the obviousness or intuitive appeal of the individual axioms. In fact, in Chap. 4, in order to make possible our development of plane hyperbolic geometry we shall have to accept (without proof, of course) a new parallel axiom which not only is not obvious but in fact contradicts all our intuitive notions of parallelism.

It is important to observe that although the objects and relations with which an axiomatic system deals are ultimately undefined, they are certainly not meaningless. In fact, the axioms make statements about them, and by means of the accepted laws of logic additional results, or theorems, are proved to be true about them. Thus, by the processes of deductive reasoning, more and more properties of the objects of the system become established, not just with experimental accuracy, but with the certainty that they follow as logical consequences of the initial assumptions.

It is sometimes said that a mathematician working with an axiomatic system is merely playing a meaningless game with undefined pieces subject to arbitrary rules, and in a very real sense this is true. However, it is not the whole truth, and without further qualification it is only a caricature of the truth. For as we pointed out in the last section, the construction of an axiomatic system is usually motivated by ideas drawn from the external world. The undefined terms are often idealizations suggested by objects in the world of our physical experience, and the axioms are abstract formulations of the fundamental observed properties of these objects. To abandon contact with the "real" world in this fashion may seem foolish to those of a practical turn of mind, but it is highly practical. It allows the instruments of the mind to replace the instruments of the hand and eye in the study of the phenomena of original interest. And if the initial abstraction from the world of experience was made with appropriate care, the results deduced from the axioms by the laws of logic can be transported back into the "real" world either as properties supported now by deduction as well as induction or often as new properties that were previously unknown. Moreover, the fact that an axiomatic system deals with undefined things means, in effect, that as we work with it we are "killing many birds with one stone." For an abstract system, though it may have been motivated or suggested by a specific set of objects or facts, is bound to no particular interpretation. Its results can be applied equally well to any system whose elements can be identified with the undefined objects of the abstract system and verified to have the properties ascribed by the axioms. Thus if the study of such a system be a game, it is certainly a game which affords not only intellectual pleasure and satisfaction to those who play it but which often has intimate and important connections with the external world.

From what we have said thus far, it might seem that the choice of the axioms upon which a logical system is based is completely arbitrary, but this is not the case. There is in fact one essential requirement, namely, that the system be *consistent*. By this we mean that no one of the axioms should contradict any other and that among all the theorems which can be derived from the axioms there should never be two which contradict each other. This is a perfectly natural requirement, of course, for it is clear that any system in which we could deduce that a statement, *p*, was both true and false would certainly be worthless.

In addition to the necessary property of consistency, there are several other properties which an axiomatic system may, but need not, have and which, depending on the purpose the system is designed to serve, may or may not be desirable. The first of these is *independence*, which is the requirement that no one of the axioms be derivable from the others, that is, that there should be no theorems of the system "masquerading" as axioms. The next is *completeness*, which is the requirement that for any statement involving the undefined objects and relations of the system, the axioms be sufficient to determine whether it is true or false. The last, which is very closely related to completeness, is *categoricalness*, which, roughly speaking, is the requirement that all systems obtained by giving specific interpretations to the undefined terms of the abstract system should be essentially the same. In succeeding sections we shall elaborate these definitions and examine each of the properties in some detail.

EXERCISES

Discuss each of the following quotations:

1. "The most precious production in modern philosophy [Spinoza's "Ethics"] is cast into geometric form to make the thought Euclideanly clear; but the result is a laconic obscurity in which every line requires a Talmud of commentary. . . . Descartes had suggested that philosophy could not be exact until it expressed itself in the forms of mathematics; but he had never grappled with his own ideal. Spinoza came to the suggestion with a mind trained in mathematics as the very basis of all rigorous scientific procedure. . . . To our more loosely textured minds the result is an exhausting concentration of both matter and form; and we are tempted

to console ourselves by denouncing this philosophic geometry as an artificial chess game of thought in which axioms, definitions, theorems and proofs are manipulated like kings and bishops, knights and pawns; a logical solitaire invented to solace Spinoza's loneliness. But Spinoza had but one compelling desire—to reduce the intolerable chaos of the world to unity and order."

Will Durant, "The Story of Philosophy," pp. 185–186 [4].

2. "The axiomatic method has provided deep insights into mathematics, disclosing identities where none had been suspected. In the hands of mathematicians of genius this method has been used to strip away exterior details that seem to distinguish two subjects and to disclose an identical structure whose properties can be studied once for all and applied to the separate subjects."

Mina Rees, "The Nature of Mathematics," *Science*, Oct. 5, 1962, p. 10 [2].

3. "The major improvement in the modern viewpoint on axiomatics has been directed toward the elimination of intuition from proofs. It has been found that the only safe way to avoid intuition is to make its use impossible. This is accomplished by conscientiously refusing to know anything at all about the entities with which you are dealing be they called numbers, points, lines, or what have you, beyond what is stated explicitly about them in the axioms. Thus the entities with which a branch of mathematics is concerned, enter, in the first instance, as completely abstract, formless objects. Then a collection of axioms stating certain facts about these abstract objects is announced as the basis of the mathematical structure. These axioms are to be considered not as hints or clues as to the nature of the abstract objects with which you are concerned, but rather as the complete statement of *all* you know about them."

Kershner and Wilcox, "The Anatomy of Mathematics," p. 26 [5].

4. "The emergence of the abstract viewpoint adopted in the Twentieth Century by a large number of mathematicians led finally to a feeling that the subject matter of mathematics was not the study of numbers or space or any elaborations thereon, but simply the determination of consequences of systems of axioms. From this standpoint any system of axioms whatsoever is fair material for investigation. Thus mathematics has come to be, at least in the eyes of many practitioners of the art, something which can be loosely described as the science of axiomatics."

Kershner and Wilcox, "The Anatomy of Mathematics," p. 27 [5].

5. "There seems to be a great danger in the prevailing overemphasis on the deductive-postulational character of mathematics. . . . A serious threat to the very life of science is implied in the assertion that mathematics is nothing but a system of conclusions drawn from definitions and postulates that must be consistent but otherwise can be created by the free will of the mathematician. If this description were accurate, mathematics could not attract any intelligent person. It would be a game with definitions, rules, and syllogisms, without motive or goal."

Courant and Robbins, "What Is Mathematics?" p. xvii [6].

1.4 Consistency. As we pointed out in the last section, a set of axioms is consistent if and only if there are no contradictions among the axioms and the theorems which can be derived from them. The consistency of an axiomatic system thus depends, in principle, upon *all* the theorems, the undiscovered ones as well as those already known, which can be proved in the system. Hence it is clear that it may be very difficult to prove that a particular system is consistent. In fact, in spite of the efforts of many eminent mathematicians to find one, no direct procedure is known for showing that from a given set of axioms it is impossible to deduce two contradictory statements. Instead, the consistency of a set of axioms is ordinarily established by the pragmatic device of exhibiting a specific model whose elements and relations are particular interpretations of the undefined terms of the abstract system for which each of the axioms can be verified. If such a model involves objects and relations from the external world, we say that the abstract system is **absolutely consistent,** for any inconsistency in it would necessarily appear in the model as an inconsistency in some portion of the "real" world, which is impossible. On the other hand, if the model involves, as it often does, objects and relations from some other axiomatic system, say elementary arithmetic or euclidean geometry, we say that the abstract system under consideration is **relatively consistent.** In other words, it is consistent *if* the system from which the model is taken is consistent, for any inconsistency in the first system would appear, via the model, as an inconsistency in the second system. In particular, although neither euclidean geometry nor arithmetic has been shown to be absolutely consistent, four thousand years of experience with these systems provides us with overwhelming inductive evidence that they actually are consistent. Hence we are well persuaded of the consistency of any axiomatic system for which we can construct an arithmetic or geometric model.

Theoretically, an axiomatic system may spring full-blown from a mathematician's imagination, and in such a case (if it ever occurred) the construction of a model to prove its consistency might be very difficult. In almost all cases, however, a new set of axioms arises in one or the other of the following ways:

1. From a study of some particular system which a person wishes to generalize or to which he wishes to give a deductive structure
2. From a desire to explore the consequences of altering or omitting one or more of the axioms in some familiar axiomatic system

In either case, the particular motivating system or some variant of it usually provides the necessary consistency model.

To illustrate these ideas, let us consider a simple axiomatic system involving two types of undefined objects which, with complete neutrality, we shall call simply x's and y's, an undefined relation called "belonging to" which either does or does not exist between a given x and a given y, and the following axioms:

$A.1.$ If x_1 and x_2 are any two* x's, there is at least one y belonging to both x_1 and x_2.

$A.2.$ If x_1 and x_2 are any two x's, there is at most one y belonging to both x_1 and x_2.

$A.3.$ If y_1 and y_2 are any two y's, there is at least one x belonging to both y_1 and y_2.

$A.4.$ If y_1 is any y, there are at least three x's belonging to y_1.

$A.5.$ If y_1 is any y, there is at least one x which does not belong to y_1.

$A.6.$ There exists at least one y.

Since this system comes to us with no apparent motivation, it is not clear just how we should begin our search for a model to establish its consistency. However, the undefined relation called "belonging to" suggests that perhaps we might find a model in which the x's are men, the y's are clubs or lodges, and "belonging to" has its literal significance (with the qualification that if a man belongs to a lodge, the lodge will also be said to belong to the man). With this clue, a little experimentation could easily lead us to a system like that in Model 1,

Model 1

	Ames	Berg	Cody	Dunn	Egan	Ford	Grey
Eagles	B	B	B				
Elks	B			B	B		
Masons	B					B	B
Moose		B		B		B	
Oddfellows		B			B		B
Shriners			B	B			B
Woodmen			B		B	B	

* Throughout this book when we speak of two objects we shall always mean two distinct objects. Thus if we say, for instance, "Consider the two points A and B," we mean that A and B are the names of two distinct points and are not just two different names for the same point. On the other hand, if we say "Consider the points A and B," we permit the possibility that A and B are different names for the same point. Similar conventions apply, of course, to the words *three, four, five.* . . .

in which the fact that a man (that is, an x) belongs to a lodge (that is, a y) is indicated by a B in the square common to the corresponding column and row.

With these interpretations for the undefined x's and y's and the undefined relation "belonging to," it is easy to verify that each of the six axioms is fulfilled. Axioms 4, 5, and 6 are obviously satisfied. In order to verify $A.1$ and $A.2$ we must consider every possible pair of men (21 pairs in all) and verify that there is one and only one lodge to which both men belong. For instance, Berg and Grey both belong to the Oddfellows but to no other lodge, Cody and Ford both belong to the Woodmen but to no other lodge, and so on. Finally, to verify $A.3$ we must consider every possible pair of lodges (21 pairs in all) and verify that in each case there is at least one man who belongs to both lodges. For instance, the Masons and the Oddfellows have Grey as a member in common, the Elks and the Shriners have Dunn as a member in common, and so on.* Thus we have exhibited a concrete representation of our axiomatic system involving objects and relations from the "real" world. If there were any contradictions among the axioms or any of their consequences, this would imply an inconsistency in the model, since each of the axioms is satisfied in the model and therefore any deduction from them is verifiable in the model. But we know that there can be no inconsistency in any model drawn from real life. Hence we know that the given system is absolutely consistent.

Once a single model of an axiomatic system has been found, the consistency of the system is established and no further models need be considered, at least for this purpose. However, as an illustration of the variety of models which may often be found for a given system, let us consider one more interpretation. Let the x's this time be the seven points shown in Fig. 1.1, let the y's be the six segments P_1P_2, P_2P_3, P_1P_3, P_1P_6, P_2P_4, P_3P_5, and the circle $P_4P_5P_6$, and let the relation "x belongs to y" or "y belongs to x" mean that the point x is a point of the segment or circle y. With this interpretation it is again easy to verify that each of the six axioms is satisfied. Axioms 4, 5, and 6 can be checked immediately. To check $A.1$ and $A.2$ we must verify that any two of the seven points lie on one and only one member of the set consisting of the six segments and the circle, and this, of course, can easily be done by inspection. For instance, P_2 and P_6 both lie on the segment P_2P_3 and on no other segment nor on the circle. Similarly, P_4 and P_5 both lie on the circle but on no one of the six segments. Finally, to check $A.3$ it is necessary to verify that any two of the segments or any one of the segments and the circle have at least

---

* It appears that every two lodges also have *at most* one member in common, but since the axioms do not assert this, it is unnecessary to verify that this is the case. Actually, as we shall see in Sec. 1.7, this is one of the theorems of the system.

one point in common.   This, too, can be done by inspection, and so
we have another and apparently quite different model of the given
abstract system.

Model 2

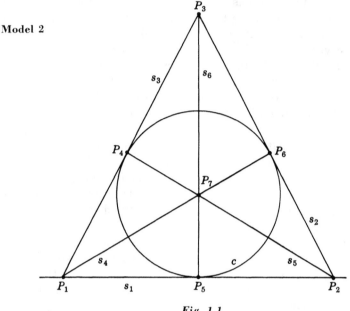

Fig. 1.1

EXERCISES

1.   Consider the system consisting of seven triples of the form $x = (u,v,w)$
where, specifically,

$$x_1 = (1,0,0) \qquad x_2 = (0,1,0) \qquad x_3 = (0,0,1)$$
$$x_4 = (0,1,1) \qquad x_5 = (1,0,1) \qquad x_6 = (1,1,0)$$
$$x_7 = (1,1,1)$$

Let the y's be those sets of x's which satisfy the respective equations

$$u = 0 \qquad\qquad v = 0 \qquad\qquad w = 0$$
$$v + w = 0 \qquad u + w = 0 \qquad u + v = 0$$
$$u + v + w = 0$$

where all calculations are to be carried out modulo 2; that is, the results
of all calculations are to be divided by 2 and only the remainders retained.
List the x's which belong to each y.   List the y's which belong to each x.
Verify that in this system Axioms 1 to 6 are satisfied.

2.   Show that the following statements are inconsistent:
   (a) The population of Smalltown consists exclusively of young married
       couples and their children.
   (b) There are more adults than children.
   (c) Every boy has a sister.

(d) There are more boys than girls.

(e) There are no childless couples.

Is it meaningful to ask which of these statements is inconsistent?

3. Show that if any one of the five statements in Exercise 2 is omitted, the remaining four are consistent.

4. (a) The following is a multiplication problem in which all but four of the digits have been suppressed and it is required to reconstruct the calculation:

$$
\begin{array}{r}
2 \ \cdot \\
\cdot \ \ \cdot \\
\hline
\cdot \ \ 1 \ \ \cdot \\
\cdot \ \ 0 \ \ \cdot \\
\hline
\cdot \ \ \cdot \ \ 3 \ \ \cdot
\end{array}
$$

Prove that this is impossible; that is, prove that the assertions made about the occurrences of the digits 0, 1, 2, and 3 are inconsistent.

(b) Prove that the data become consistent and the calculation can be reconstructed if the digit 3 is replaced by the digit 4.

5. Prove that the following description is inconsistent: "In a far corner of the world there is a city built on ten islands, which is known as the 'City of Bridges.' Bridges lead directly from five of the islands to the mainland; four islands have four bridges leading from them; three islands have three bridges leading from them; two islands have two bridges leading from them; and one of the islands has a single bridge leading from it."

Is the following description consistent? "In a far corner of the world there is a city built on six islands, which is known as the 'City of Bridges.' Bridges lead directly from four of the islands to the mainland; three of the islands have three bridges leading from them; two of the islands have two bridges leading from them; and one of the islands has a single bridge leading from it."

6. A system consists of a set of elements, S, and an operation of combination, denoted by the symbol ∘, such that the result of combining a pair of elements of S is also a member of S. The structure of the system is described by the following axioms:

   $A.1.$ If a and b are any elements of S, then $a \circ b = b \circ a$.

   $A.2.$ If a and b are any elements of S, then $a \circ b = b$.

   $A.3.$ S contains at least two elements.

   Prove that the system is inconsistent.

7. A system consists of a set of elements, S, and a relation, R, which either does or does not hold between any pair of elements in S, taken in a given order. If an element a bears the relation R to an element b, we write $a \, R \, b$. The structure of the system is described by the following axioms:

   $A.1.$ If a and b are distinct elements of S, then either $a \, R \, b$ or $b \, R \, a$.

   $A.2.$ If a and b are elements of S and if $a \, R \, b$, then a and b are distinct.

   $A.3.$ If a, b, and c are elements of S and if $a \, R \, b$ and $b \, R \, c$, then $a \, R \, c$.

   $A.4.$ S contains exactly four elements.

   Prove that the system is consistent. (Hint: Consider four men of different ages and let R be the relation "is older than.") Prove the following theorems about this system:

(a) If $a\,R\,b$, then it is false that $b\,R\,a$.

(b) There is at least one element, a, in S such that $a\,R\,x$ is false for every x in S.

(c) There is at most one element, a, in S such that $a\,R\,x$ is false for every x in S.

8.   A system consists of two types of elements called x's and y's and a relation called "belonging to" which either does or does not exist between a given x and a given y. The structure of the system is defined by the following axioms:

    A.1.   If x_1 is any x, there are exactly four y's which belong to x_1.

    A.2.   If y_1 is any y, there are exactly two x's which belong to y_1.

    A.3.   If x_1 and x_2 are two x's, there is exactly one y which belongs to both x_1 and x_2.

    A.4.   There exists at least one x.

Prove that this system is consistent. Prove the following theorems about the system:

(a) There are exactly five x's.

(b) There are exactly ten y's.

9.   A system consists of a set of elements, S, and an operation of combination denoted by the symbol $\circ$, such that the result of combining a pair of elements of S is a unique member of S. The structure of the system is described by the following axioms:

    A.1.   If a, b, and c are any elements of S, then

$$(a \circ b) \circ c = a \circ (b \circ c)$$

    A.2.   If a and b are any elements of S, then there is a unique element, x, of S such that $a \circ x = b$.

    A.3.   If a and b are any elements of S, then there is a unique element, y, of S such that $y \circ a = b$.

    A.4. S contains at least one element.

Prove that this system is consistent. (Hint: Consider the numbers 1, -1, i, $-i$ combined under multiplication.)

Prove the following theorems about the system:

(a) There exists a unique element, I_r, such that $a \circ I_r = a$ for every element a in S.

(b) There exists a unique element, I_l, such that $I_l \circ a = a$ for every element a in S.

(c) $I_r = I_l$.

(d) For every element a in S there exists a unique element, a^{-1}, such that $a \circ a^{-1} = a^{-1} \circ a = I$, where $I = I_r = I_l$.

10.   Show that if the following axioms are added to those listed in Exercise 9, the resulting system is inconsistent:

    A.5. S contains at least two elements, a and b, such that

$$a \circ b \neq b \circ a$$

    A.6. S contains at most four elements.

1.5 Independence. An axiomatic system is said to be independent if the axioms themselves are independent, that is, if no one of the axioms is a logical consequence of, or deducible from, the others. Unlike consistency, which is an absolute requirement, independence is not a necessary property of an axiomatic system and, in fact, for certain purposes may actually be undesirable.

From a purely aesthetic point of view, independence is desirable because it means that the axiomatic system contains no superfluous assumptions; in other words, nothing that can be proved has been taken for granted, and the development proceeds from the absolute minimum which will still suffice. The system thus possesses a logical economy which a mathematician would describe as "elegant." In nonmathematical terms, the appeal of an independent set of axioms is perhaps best explained by analogy with the game of chess. A good chess player looks with disdain upon a game in which the winner merely crushes his opponent beneath the weight of a great material superiority. On the other hand, he is enthusiastic about a game in which the winner sacrifices piece after piece in some brilliant combination leading to checkmate with the barest minimum of pieces. Somewhat facetiously, we might say that an independent axiomatic system is more "sporting" than one which is not.

But the less we assume, the less we have to work with, and the harder our proofs will be. This is sometimes a major consideration in the axiomatic development of a given subject in the classroom, for with an independent set of axioms, the proof of one or more important early theorems, though possible, may be so long or so difficult as to be completely infeasible. In such cases, mathematical elegance must yield to pedagogical necessity, and one or more of the theorems in question should be included among the axioms, with, of course, an honest statement that they are actually theorems and could be deduced from the other axioms.

It is also possible for a set of axioms to include one or more which are dependent without this fact being recognized, but this in no way impairs the usefulness of the set. For instance, Hilbert's famous set of axioms for euclidean plane geometry contained two which were actually theorems, although this was not realized until years after the axioms were first published.

Like consistency, independence is also established through the use of models. But whereas a single model is sufficient to prove an axiomatic system consistent, to prove it independent we need as many models as there are axioms. If we assume that the system contains n axioms, $A_1, A_2, \ldots, A_n$, the procedure is as follows: First, we observe that if A_1 is a consequence of or can be derived from $A_2, A_3, \ldots, A_n$, it must be true whenever $A_2, A_3, \ldots, A_n$ are true. Then

we attempt to construct a model in which A_2, A_3, . . . , A_n are true
but A_1 is false.   If we can do this, we have proved that A_1 is independent of A_2, A_3, . . . , A_n, for if it were a consequence of these axioms,
it would have to be true in our model, since A_2, A_3, . . . , A_n are true.
Similarly, we prove that a general one of the axioms, A_k, is independent of A_1, . . . , A_{k-1}, A_{k+1}, . . . , A_n by exhibiting a model in which
A_1, . . . , A_{k-1}, A_{k+1}, . . . , A_n are true but A_k is false.

As an illustration of these ideas, we return to the axiomatic system
we discussed in the last section.   As a model to prove that the first
axiom is independent of the other five, we can use Fig. 1.2, in which the

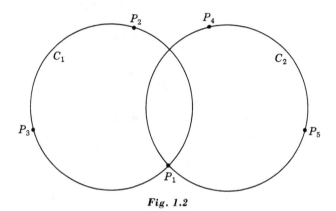

Fig. 1.2

x's are the five points P_1, P_2, P_3, P_4, P_5, the y's are the two circles, and
an x belongs to a y, and vice versa, if and only if the corresponding
point lies on the corresponding circle.   Clearly A_1 is false since,
specifically, neither circle contains both P_3 and P_4.   That the other
axioms are satisfied is almost obvious and easily verified.

To prove the independence of A_2 we can use the model described by
Fig. 1.3, in which the four points P_1, P_2, P_3, P_4 are the x's, the three
circles are the y's, and an x belongs to a y, and vice versa, if and only if
the corresponding point lies on the corresponding circle.   In this
system A_2 is false since there are two circles which contain both P_1
and P_2, for instance.   The other axioms are readily verifiable, however, which completes the proof.

Among the various models which might be used to demonstrate the
independence of A_3, the most obvious is ordinary euclidean plane
geometry, points and lines being, respectively, the x's and y's of the
abstract system and "belongs to" being interpreted naturally to mean
"lies on" or "passes through."   Clearly A_3 is false because although
many pairs of lines do have a point in common, there are some that do
not, namely, pairs of parallel lines.   Obviously, the other axioms are
all satisfied.

The independence of A_4 is apparent at once from the array

$$P \quad Q \quad R$$
$$Q \quad R \quad P$$

in which the three letters are the x's, the three columns are the y's, and an x belongs to a y and vice versa if and only if the corresponding letter lies in the corresponding column. Obviously, each column contains but two letters, that is, only two x's belong to each y, and so

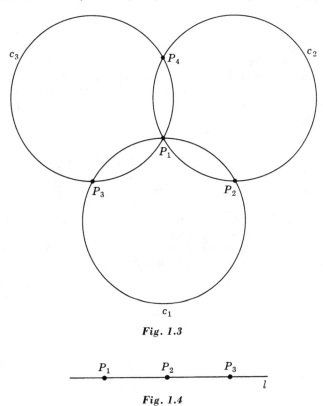

Fig. 1.3

Fig. 1.4

$A.4$ is violated. It is easy to see that the other axioms are all true, however, which shows that $A.4$ is independent of the other axioms.

To prove the independence of $A.5$ we may use the very simple system shown in Fig. 1.4, in which P_1, P_2, P_3 are the x's, the line, l, is the only y, and "belongs to" means "lies on" or "passes through." Since every point of the system lies on l, $A.5$ is clearly violated. Obviously, $A.1$, $A.2$, $A.4$, and $A.6$ are satisfied. That $A.3$ is satisfied follows trivially from the fact that there is only one line, or y, in the system. Hence the hypothesis of $A.3$ is not fulfilled and therefore, from the accepted meaning of conditional statements, $A.3$ is true.

Finally, to show that $A.6$ is independent of the first five axioms we need only consider a system consisting of a single x and no y's. For such a system $A.6$ is clearly false. On the other hand, each of the other axioms is a true statement since each is a conditional statement with a false antecedent. Hence $A.6$ is independent of the other axioms.

EXERCISES

1. Construct models different from those in the text to show that $A.1$, $A.2$, $A.3$, $A.4$, $A.5$, and $A.6$ are independent.
2. In Exercise 2, Sec. 1.4, show that every set of four statements obtained by omitting one of the original statements is independent.
3. The following is a multiplication problem in which all but three of the digits have been suppressed and it is required to reconstruct the calculation:

$$
\begin{array}{r}
2\ \cdot \\
\cdot\ \cdot \\
\hline
\cdot\ 1 \\
\cdot\ \cdot\ \cdot \\
\hline
\cdot\ \cdot\ 4\ \cdot
\end{array}
$$

Show that the three given facts, namely, that the digits 1, 2, and 4 occur in the indicated positions, are independent.
4. The following is a multiplication problem in which all but four of the digits have been suppressed and it is required to reconstruct the calculation:

$$
\begin{array}{r}
4\ \cdot \\
\cdot\ \cdot \\
\hline
\cdot\ 8\ \cdot \\
8\ \cdot \\
\hline
\cdot\ \cdot\ 4\ \cdot
\end{array}
$$

Show that the given facts, namely, that the digits 4 and 8 occur in the indicated positions, are dependent and determine which one or ones are dependent.
5. Prove that the four axioms in Exercise 7, Sec. 1.4, are independent.

1.6 Completeness and Categoricalness. Any statement involving the undefined terms and relations of an axiomatic system which is not one of the axioms is potentially a theorem of the system. If it can be deduced from the axioms, it is in fact a theorem. If it can be proved false, then its denial or negation is a theorem. But what of a statement which can neither be proved true nor proved false? Clearly, if such a statement exists, it can only be because the axioms of the system are insufficient to permit the truth or falsity of every possible statement to be determined. And if such were the case, we could presumably add additional axioms until finally we had enough to enable us to prove or disprove every statement which could be made about the elements of the system.

On the other hand, if every statement can either be proved or disproved, then it is impossible to add any more axioms to the system, for any new axiom is simply some statement about the elements and relations of the system and therefore, by our supposition, either demonstrably true or demonstrably false. If it is true, it is simply a theorem, and including it as an axiom would add nothing to the set of axioms but would merely make the set dependent. If it is false, then including it as an axiom would introduce a contradiction into the system and make it inconsistent.

These ideas can be summarized as follows: A set of axioms is said to be **complete** if it is impossible to enlarge it by adding any other axiom which is consistent with, yet independent of, those already in the set. In any system with a complete set of axioms every statement can either be proved true or proved false.

It is often quite difficult to prove directly that a given set of axioms is complete, and the usual test for completeness is based upon the closely related idea of *categoricalness*, which we shall now investigate.

Suppose that we have two specific interpretations, or models, M_1 and M_2, of an axiomatic system, S, each containing the same number of elements. Obviously, the elements of M_1 and M_2 can be paired in many ways so that to each element of M_1 there corresponds a unique element of M_2 and vice versa. Now there are certain relations which exist among the elements of M_1 and certain relations which exist among the elements of M_2, and a particular 1:1 correspondence between the elements of M_1 and the elements of M_2 may or may not preserve these relations. For instance, if we consider the two models of the abstract system we introduced in Sec. 1.4, namely,

Model 1

	Ames	Berg	Cody	Dunn	Egan	Ford	Grey
Eagles	B	B	B				
Elks	B			B	B		
Masons	B					B	B
Moose		B		B		B	
Oddfellows		B			B		B
Shriners			B	B			B
Woodmen			B		B	B	

Model 2

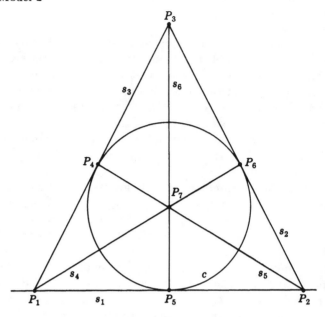

Fig. 1.1

we have, among many other possibilities, the following 1:1 correspondence:

M_1:	Ames	Berg	Cody	Dunn	Egan	Ford	Grey
M_2:	P_2	P_5	P_1	P_6	P_4	P_3	P_7

M_1:	Eagles	Elks	Masons	Moose	Oddfellows	Shriners	Woodmen
M_2:	s_2	s_5	s_1	C	s_3	s_4	s_6

Now in M_1 we have, among many other special properties, one described by the statement

"Cody, Dunn, and Grey all belong to the Shriners."

Under the given 1:1 correspondence this becomes the statement

"P_1, P_6, and P_7 all belong to s_4"

and in M_2 this is in fact true.   Hence at least this much of the structure of the two systems is preserved under the mapping effected by the given correspondence.   On the other hand, the statement

"Ames, Dunn, and Egan all belong to the Elks"

which is true in M_1, becomes, under the given correspondence,

"P_2, P_6, and P_4 all belong to s_5"

which is clearly false in M_2. In fact, in M_2 the points P_2, P_4, and P_6 belong to no one of the segments nor to the circle. Hence this aspect of the structure of the two systems is not preserved by the correspondence.

In general, then, an arbitrary $1:1$ correspondence between two models with the same number of elements will not preserve all, or even any, of the relations existing in the models. Nevertheless, in special cases there may be one or more $1:1$ correspondences which will preserve all details of the structure of two models. For example, it is not hard to verify that under the correspondence

M_1:	Ames	Berg	Cody	Dunn	Egan	Ford	Grey
M_2:	P_2	P_3	P_6	P_5	P_1	P_7	P_4

M_1:	Eagles	Elks	Masons	Moose	Oddfellows	Shriners	Woodmen
M_2:	s_2	s_1	s_5	s_6	s_3	C	s_4

every relation in M_1 goes over into a relation actually existing in M_2 and vice versa.

This leads us to the exceedingly important notion of an **isomorphism**: If there exists a $1:1$ correspondence between the elements of two systems which preserves all relations existing in either system, the correspondence is called an isomorphism and the two systems are said to be isomorphic. Since an isomorphism preserves all details of structure, we can say that two isomorphic systems are structurally identical and in fact differ only in the names given to the elements. Thus every detail of M_1 is correctly represented in M_2 if we merely rename the elements of M_2, using the second of the two correspondences above (see Fig. 1.5).

Unless two systems have the same number of elements, they obviously cannot be isomorphic, since no $1:1$ correspondence can exist between them. Moreover, even when the systems have the same number of elements, they usually are not isomorphic. The fact that many $1:1$ correspondences between the elements of the two systems are not isomorphisms, i.e., do not preserve relations, of course does not necessarily mean that the two systems are not isomorphic. Isomorphism requires only that there should exist *at least one* relation-preserving $1:1$ correspondence between the elements of the two systems.

We are now in a position to define what we mean by **categoricalness**: An axiomatic system is said to be categorical if and only if each of its models is isomorphic to every other model. To put it somewhat less technically, an axiomatic system is categorical if and only if all of its models are structurally identical or, in other words, if it has essentially just one concrete representation.

If an axiomatic system is categorical, it is necessarily complete.

To see this, let us investigate the contradictory possibility that an axiomatic system is categorical but incomplete. If the system is incomplete, there is at least one statement, σ, which can neither be proved nor disproved; that is, there is at least one statement such that both it and its negation are consistent with the given axioms.

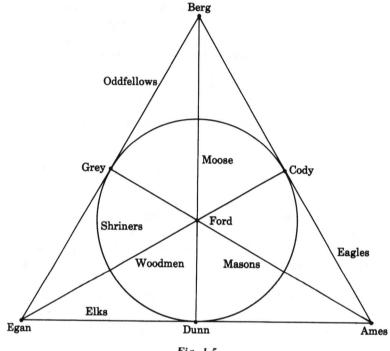

Fig. 1.5

Let us now consider two models of the given system, one in which σ is true and one in which σ is false. Since the system is categorical, by hypothesis these two models must be isomorphic. In other words, there exists a 1:1 correspondence between the elements of the two models such that corresponding statements in the two models are either both true or both false. But this is impossible, since by hypothesis the statement σ is true in one model and false in the other. Hence a categorical system cannot be incomplete, or in other words, *a categorical system is necessarily complete.*

To illustrate the ideas of completeness and categoricalness, we again use the axiomatic system introduced in Sec. 1.4. So far, we have given only two specific interpretations of it, and these we showed to be isomorphic. However, the system is not categorical, and therefore not complete, since it does admit of other representations which are not isomorphic to the ones we have so far given. Specifically, the following table

Model 3

P_1	P_3	P_2	P_3	P_2	P_1	P_1	P_2	P_3	P_1	P_4	P_7	P_{10}
P_6	P_5	P_4	P_4	P_6	P_5	P_4	P_5	P_6	P_2	P_5	P_8	P_{11}
P_7	P_8	P_9	P_7	P_8	P_9	P_8	P_7	P_9	P_3	P_6	P_9	P_{12}
P_{10}	P_{10}	P_{10}	P_{11}	P_{11}	P_{11}	P_{12}	P_{12}	P_{12}	P_{13}	P_{13}	P_{13}	P_{13}

in which the P's represent x's, the columns of P's represent y's, and an x and a y belong to each other if and only if the corresponding P occurs in the corresponding column, defines a system in which $A.1, \ldots, A.6$ are all satisfied. Since this model contains more elements than the ones we discussed above, it is clear that it cannot be isomorphic with them. Hence the set of axioms $A.1, \ldots, A.6$ is not categorical.

As an obvious statement, σ, which can neither be proved nor disproved in the given axiomatic system, we have the assertion σ: "If y_1 is a y, there are at most three x's which belong to y_1." As our models show, it is possible for this statement to be true in some interpretations (M_1 and M_2) and false in others (M_3). It is thus possible to enlarge the set of axioms, $A.1, A.2, A.3, A.4, A.5, A.6$, by adding the statement σ as a new axiom, $A.7$. For as M_1 and M_2 show, it is consistent with $A.1, \ldots, A.6$, and as M_3 shows, it is independent of $A.1, \ldots, A.6$. Moreover, it can be proved that the system consisting of $A.1, \ldots, A.6$ together with σ is categorical and hence complete.

Like independence, completeness and categoricalness are not essential characteristics of a set of axioms and, depending on the purpose for which the system is designed, may or may not be desirable. A categorical system admits of essentially just one model; a noncategorical system admits of a number of essentially different models. Thus when we investigate a noncategorical, or incomplete, system, we are in effect "killing many birds with one stone," inasmuch as any conclusions drawn from the axioms are automatically properties of every model of the system. For instance, if we work with the categorical system consisting of $A.1, \ldots, A.6$ together with the statement σ that every y belongs to at most three x's, we restrict ourselves to the almost trivial representation described equally well by M_1 or M_2. On the other hand, if we work with the incomplete system consisting of just $A.1, \ldots, A.6$, any conclusions we are able to draw imply properties not only of M_1 and M_2 but also of the essentially different model M_3 and of any other models which may exist, of which in fact there are infinitely many.

If, however, we are interested solely and exclusively in some one system, such as euclidean plane geometry, we would undoubtedly want to work with a categorical set of axioms having that particular system as its essentially unique model. Otherwise, we would be able to investigate only those properties of the system which were common to it and to all other representations of the set of axioms we were using. Briefly, we may say in summary that using a categorical set of axioms enables us to explore some one system completely, whereas using a noncategorical set of axioms allows us to investigate those properties which are common to a number of different systems.

EXERCISES

1. In how many different ways can a 1:1 correspondence be set up between two systems, each of which contains n elements?
2. Are two systems which are isomorphic to the same system isomorphic to each other?
3. Find another isomorphism between M_1 and M_2 different from the one given in the text.
4. Is there a 1:1 correspondence between M_1 and M_2 which does not preserve any of the relations existing in these models?
5. Show that the system defined in Exercise 1, Sec. 1.4, is isomorphic to M_1.
6. Show that the system described by the axioms of Exercise 9, Sec 1.4, with A.4 replaced by

    A.4'. S contains exactly three elements

    is categorical.
7. Show that the system described by the axioms of Exercise 9, Sec 1.4, with A.4 replaced by

    A.4''. S contains exactly four elements

    is not categorical.
8. Show that the system described in Exercise 7, Sec. 1.4, is categorical. [Hint: First establish the following theorems:

    (a) There is exactly one element, b, in S such that $b \, R \, x$ is true for one and only one x in S.

    (b) There is exactly one element, c, in S such that $c \, R \, x$ is true for two and only two elements in S.

    (c) There is exactly one element, d, in S such that $d \, R \, x$ is true for every element in S distinct from d.]

1.7 Finite Geometries. The set of axioms A.1, . . . , A.6, which we introduced in Sec. 1.4, has an intrinsic interest beyond its mere use in illustrating such notions as consistency, independence, completeness, and categoricalness. These axioms are, in fact, the basis for the study of the systems which are known collectively as **finite projective**

geometries, and although this study is somewhat aside from our main interest, we shall close this chapter with a brief introduction to it.

To acknowledge explicitly the motivation behind this study we shall restate the axioms of Sec. 1.4, using "point" and "line" as undefined terms in place of x and y and using "lies on," "passes through," and "contains" as equivalent terms for the undefined relation "belongs to":

*A.*1. If P_1 and P_2 are any two points, there is at least one line containing both P_1 and P_2.

*A.*2. If P_1 and P_2 are any two points, there is at most one line containing both P_1 and P_2.

*A.*3. If l_1 and l_2 are any two lines, there is at least one point which lies on both l_1 and l_2.

*A.*4. Every line contains at least three points.

*A.*5. If l is any line, there is at least one point which does not lie on l.

*A.*6. There exists at least one line.

From the work of the preceding sections we know that this set of axioms is consistent, independent, and incomplete. In passing we note that each of the axioms except *A*.3 expresses a familiar property of euclidean plane geometry. On the other hand, *A*.3 explicitly contradicts one of the best-known properties of euclidean plane geometry, namely, the existence of parallel lines, since it asserts that any two lines intersect. The fact that there exists a consistent set of axioms including one which so flatly contradicts intuition and "common sense" is a striking reminder of our earlier observation that axioms need not be "self-evident" or intuitively appealing.

We now turn our attention to some of the simpler theorems that follow from these axioms. However, in doing so we should bear in mind that although the theorems will involve the familiar terms "point" and "line" and although from time to time we shall use the conventional representations of points and lines to remind us graphically of the relations we are considering, nonetheless "point" and "line" are undefined terms without significance except as the axioms give them meaning. A point is not a small dot, nor even the limit of smaller and smaller dots made with sharper and sharper pencils. Nor is a line an indefinitely long, indefinitely thin mark made with an arbitrarily sharp pencil. Mathematically speaking, *a point is any object that has the properties the axioms say a point should have,* and *a line is any object that has the properties the axioms say a line should have.*

As preliminary results, we have the following almost obvious, yet nonetheless important, theorems.

Theorem 1. There exists at least one point.

Proof. By $A.6$ there exists at least one line, and by $A.4$ every line contains at least three points. Hence the assertion that at least one point exists is surely true.

Theorem 2. If l_1 and l_2 are any two lines, there is at most one point which lies on both l_1 and l_2.

Proof. To establish the theorem, let us assume the contrary and suppose it possible for two lines, say l_1 and l_2, to have two points, say P_1 and P_2, in common. This leads at once to a contradiction, however, since $A.2$ asserts that there is at most one line which contains each of two given points. Hence it follows that two lines can have at most one point in common.

Theorem 3. Two points determine exactly one line.

Proof. This follows immediately from $A.1$ and $A.2$.

Theorem 4. Two lines have exactly one point in common.

Proof. This is an immediate consequence of $A.3$ and Theorem 2.

Theorem 5. If P is any point, there is at least one line which does not pass through P.

Proof. By $A.6$ there exists at least one line, l. If this line does not pass through P, our proof is complete. Suppose, therefore, that l passes through P. By $A.4$, l contains at least one point besides P, say P'. By $A.5$ there is at least one point, say P'', which does not lie on l. By Theorem 3 there is a unique line, say l', which contains P' and P''. Moreover, l and l' are distinct, since l' contains P'' and l does not. Hence, by Theorem 4, l and l' have exactly one point in common, and this point is obviously P'. Therefore, P, which of course lies on l, cannot also lie on l'. In other words, l' is a line which does not contain P, which establishes the theorem.

Theorem 6. Every point lies on at least three lines.

Proof. Let P be an arbitrary point. By Theorem 5 there is at least one line, l, which does not pass through P, and by $A.4$ this line contains at least three points, say P_1, P_2, P_3. By Theorem 3 each of these points determines with P a unique line. Moreover, these lines are all distinct, for if two of them coincided, that line would have two points in common with l, which is impossible by Theorem 4. Hence there are at least three lines passing through an arbitrary point, P, as asserted.

It is interesting now to consider the pairs of axioms and theorems displayed in the following table:

Axiom 1 $\leftrightarrow$ Axiom 3
Axiom 2 $\leftrightarrow$ Theorem 2
Axiom 3 $\leftrightarrow$ Axiom 1
Axiom 4 $\leftrightarrow$ Theorem 6
Axiom 5 $\leftrightarrow$ Theorem 5
Axiom 6 $\leftrightarrow$ Theorem 1

Each statement in the first column has exactly the same structure as the corresponding statement in the second column and differs from it only in the interchange of the terms "point" and "line" and the use of a different one of the equivalent phrases "lies on," "passes through," and "contains." In other words, the statement obtained from any axiom by interchanging the terms "point" and "line" is either some other axiom or else a theorem which we have already proved and therefore, in either case, a true statement in our system.

This observation has far-reaching consequences in our particular axiomatic system, for consider any theorem and its proof: The theorem is, of course, some statement involving the terms "point," "line," and "lies on" (or some synonymous phrase). Likewise, the proof is a series of statements each containing these same terms and each justified, directly or indirectly, by one or more of the axioms. Suppose now that in each step of the proof we were to interchange the terms "point" and "line." Each of these new statements would have for its apparent justification one or more statements obtained from the axioms by similarly interchanging "point" and "line." But as we verified above, the statements obtained from the axioms by interchanging "point" and "line" throughout are in fact true in our system. Hence by the interchange of the terms "point" and "line," we have in a purely mechanical fashion constructed the proof of a new theorem, namely, the theorem obtained by replacing "point" by "line" and "line" by "point" throughout the statement of the original theorem!

Two statements in our system which differ only in the interchange of the words "point" and "line" are said to be **duals** of each other, and the fact that the dual of any theorem is also a theorem provable in a purely mechanical way simply by writing down the dual of each step in the original proof is known as the **principle of duality.** The importance of the principle of duality is that, once we have recognized it, each proof that we present actually establishes two theorems for us, namely, the one whose proof we have written out and also its dual.

The principle of duality is, of course, not characteristic of axiomatic systems in general, not even of those whose undefined terms are "point" and "line." And even in our system, if we were to add a new axiom, which we are able to do since our set of axioms is incomplete, we could not assert that the principle of duality held unless and until we either assumed the dual of the new axiom as an additional axiom or else proved it as a theorem.

We now turn our attention to certain enumerative theorems which give us information about the number of points and lines necessarily present in any finite projective geometry and about the way in which they are arranged.

> **Theorem 7.** If there exists one line which contains exactly n points, then every line contains exactly n points.

Proof. Let l be a line containing exactly n points, $P_1, P_2, \ldots . P_n$, and let l' be any other line. By Theorem 4, l and l' have exactly one point in common, and without loss of generality we may suppose this point to be P_1 (see Fig. 1.6). Now by $A.4$, l' contains at least one

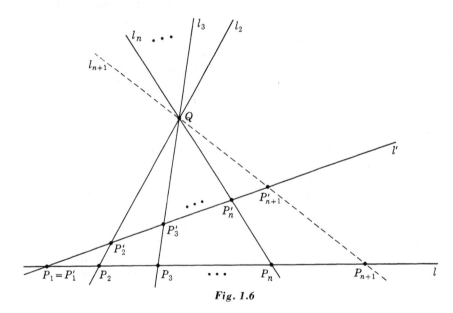

Fig. 1.6

point distinct from P_1, say P_2'. Moreover, by Theorem 4, P_2' is also distinct from $P_2, \ldots, P_n$. Now by Theorem 3, there is a unique line, say l_2, containing both P_2 and P_2', and by $A.4$, there is a third point, Q, distinct from P_2 and P_2', belonging to l_2, and also, by Theorem 4, distinct from $P_1, P_2, \ldots, P_n$. By Theorem 3, Q determines a unique line with each of the points $P_3, \ldots, P_n$, and by Theorem 4 these lines, say $l_3, \ldots, l_n$, are all distinct and of course distinct from l. Now by Theorem 4, the respective lines $l_3, \ldots, l_n$ intersect l' in unique points $P'_3, \ldots, P'_n$, necessarily distinct and also distinct from $P'_1 = P_1$ and P'_2. Hence l' contains at least n points.

To show that l' contains no more than n points, we assume the contrary and suppose that there is an additional point, P'_{n+1}, belonging to l'. By Theorem 3 there is a unique line, l_{n+1}, determined by Q and P'_{n+1}, and by Theorem 4 this line is distinct from $l_1, l_2, \ldots, l_n$, and of course from l. Therefore, by Theorem 4 there is a unique point common to l_{n+1} and l, and this point, say P_{n+1}, is distinct from $P_1, P_2, \ldots, P_n$. But this contradicts the hypothesis that l contained exactly n points. Hence l' cannot contain an additional point, P'_{n+1}, and the theorem is established.

Theorem 8.   If there exists one line which contains exactly n points, then exactly n lines pass through every point.

Proof. Let P be an arbitrary point. By Theorem 5 there is at least one line, l, which does not pass through P, and by Theorem 7 this line contains exactly n points, $P_1, P_2, \ldots, P_n$. By Theorem 3, P determines unique lines, $l_1, l_2, \ldots, l_n$, with the respective points $P_1, P_2, \ldots, P_n$ (see Fig. 1.7). Furthermore, by Theorem 4 these

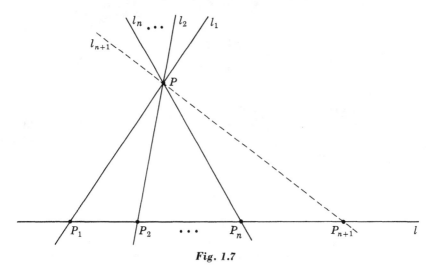

Fig. 1.7

lines are all distinct. Hence there are at least n lines passing through an arbitrary point, P. To show that there are exactly n lines belonging to P, we assume the contrary and suppose that there is at least one additional line, l_{n+1}, passing through P. By Theorem 4 this line must intersect l in a unique point, P_{n+1}, distinct from P_1, P_2, . . . , P_n. But this contradicts the hypothesis that l contains exactly n points. Hence l_1, l_2, . . . , l_n are the only lines which pass through P, and the theorem is established.

Theorem 9. If there exists one line which contains exactly n points, then the system contains exactly $n^2 - n + 1$ points.

Proof. By Theorem 1 there exists at least one point, P, and by Theorem 8 there are exactly n lines, l_1, l_2, . . . , l_n, passing through P. Moreover, by Theorem 3 every point in the system except P itself lies on exactly one line passing through P (Fig. 1.8). Furthermore, by

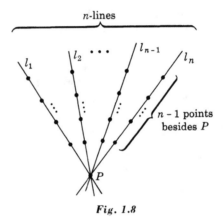

Fig. 1.8

Theorem 7 every line contains exactly n points. Hence each of the n lines belonging to P contains exactly $n - 1$ points besides P. Counting up all the points, then, we have $n - 1$ points on each of n lines plus the point P itself, or

$$n(n - 1) + 1 = n^2 - n + 1 \qquad \text{points}$$

as asserted.

Theorem 10. If there exists one line which contains exactly n points, then the system contains exactly $n^2 - n + 1$ lines.

Proof. By $A.6$ there exists at least one line, l, and by Theorem 7, l contains exactly n points, P_1, P_2, . . . , P_n. Moreover, by Theorem

4 every line in the system except l itself passes through exactly one of the points P_1, P_2, . . . , P_n (Fig. 1.9). By Theorem 8 exactly n lines pass through each of the points P_1, P_2, . . . , P_n or exactly $n - 1$ lines not counting l. Hence altogether there are $n - 1$ lines passing

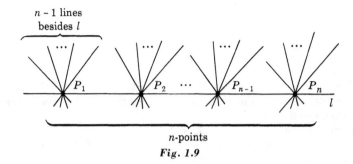

Fig. 1.9

through each of the n points on l plus l itself, or

$$n(n - 1) + 1 = n^2 - n + 1 \qquad \text{lines}$$

as asserted.

It is important to note that each of the last four theorems is of a conditional nature, that is, asserts merely that *if* there exists a geometry in which at least one line contains exactly n points, *then* certain other things are true. Nothing in the statements of these theorems asserts or implies the actual existence of geometries with these characteristics, and for particular values of n they may or may not exist.

In 1906 Veblen and Bussey showed* that for every value of n such that $n - 1$ is a power of a prime, a finite geometry satisfying $A.1$, . . . , $A.6$ actually exists. Thus there are geometries with

$$3, 4, 5, 6, 8, 9, 10, 12, \ldots \qquad \text{points per line}$$

Also, in 1949 it was shown by Bruck and Ryser† that if $n - 1$ is a number of the form $4k + 1$ or $4k + 2$ containing to an odd power a prime factor of the form $4m - 1$, then there is no geometry satisfying $A.1$, . . . , $A.6$ and containing exactly n points per line. Thus there are no finite projective geometries with

$$7, 15, 22, 23, \ldots \qquad \text{points per line}$$

* O. Veblen and W. H. Bussey, "Finite Projective Geometries," *Trans. Am. Math. Soc.*, vol. 7, pp. 241–259, 1906.
† R. H. Bruck and H. J. Ryser, "The Nonexistence of Certain Finite Projective Planes," *Can. J. Math.*, vol. 1, pp. 88–93, 1949.

If n is a number not covered by one or the other of these theorems, it is not known whether there exists a geometry satisfying $A.1, \ldots, A.6$ and containing exactly n points per line. In particular, it is not known whether there exists a finite projective geometry in which every line contains exactly 11 points.

EXERCISES

1. For what values of n less than 100 is it known that finite projective geometries exist? For what values of n less than 100 is it known that such geometries do not exist?
2. State the duals of Theorems 7, 8, 9, 10.
3. Which of the theorems of this section are provable if $A.4$ is replaced by the statement
    $A.4'$.   Every line contains at least two points.
4. Does the principle of duality hold in euclidean plane geometry?
5. Prove the following theorem: If $P_1, P_2, \ldots, P_n$ are distinct points on a line, l, and if P is a point that is not on l, then the lines determined by P and each of the points $P_1, P_2, \ldots, P_n$ are all distinct.
6. State and prove the dual of the theorem of Exercise 5.
7. A system consists of certain undefined objects called "points" and certain other objects called "lines" and a relation described variously as "belongs to," "lies on," or "passes through" which either does or does not exist between a given point and a given line. The properties of these objects are described by the following axioms:
    $A.1$.   Every line is a set of points and contains at least two points.
    $A.2$.   If P and Q are distinct points, there is one and only one line which contains them both.
    $A.3$.   If l is a line and P is a point that does not lie on l, there is one and only one line that contains P but contains no point of l, that is, is parallel to l.
    $A.4$.   Space contains at least three points that do not belong to the same line.
    Prove that this system is consistent and independent, but not categorical. Prove the following theorems about the system:
    If there exists one line which contains exactly n points, then
    (*a*) Every line contains exactly n points.
    (*b*) Every point lies on exactly $n + 1$ lines.
    (*c*) Space contains exactly n^2 points.
    (*d*) Space contains exactly $n(n + 1)$ lines.
8. Verify that if any line, that is, any y, together with the points, that is, the x's, which belong to the line are deleted from any of the models discussed in Sec. 1.6, the result is a model of the system described in Exercise 7. Does this suggest a meaningful interpretation that might be given to the statement, "Parallel lines meet at infinity"?

CREDITS FOR QUOTATIONS

1. Henri Poincaré, "The Foundations of Science," authorized translation by George Bruce Halstead, The Science Press, New York, 1913.
2. Mina Rees, "The Nature of Mathematics," *Science*, Oct. 5, 1962. Copyright 1962, American Association for the Advancement of Science.
3. George Polya, "Induction and Analogy in Mathematics," Princeton University Press, Princeton, N.J., 1954.
4. From "The Story of Philosophy," Copyright 1926, 1927, 1933 by Will Durant. Reprinted by permission.
5. Richard Brandon Kershner and Lee Roy Wilcox, "The Anatomy of Mathematics," The Ronald Press Company, New York, 1950.
6. Richard Courant and Herbert Robbins, "What Is Mathematics?" Oxford University Press, Fair Lawn, N.J., 1941.

2

EUCLIDEAN
GEOMETRY

2.1 Introduction. In the last chapter we investigated the general characteristics of the axiomatic method, using certain simple systems called finite geometries as illustrations. In this chapter we shall examine the axiomatic structure of euclidean geometry. Since we already possess an extensive knowledge of euclidean geometry, we shall not attempt a detailed development of the entire subject. Instead we shall restrict ourselves to the task of making clearer and more precise certain aspects of elementary geometry in which the treatment often falls short of acceptable modern standards. In doing this, we shall be concerned at least as much with the technical vocabulary and the axioms of euclidean geometry as with the proofs of theorems. In most cases, once careful definitions are given and a sufficiently powerful set of explicitly stated axioms is available, proofs will present no great problem, and we shall leave all but a few typical examples as exercises.

And now as we begin this portion of our work, a word of caution is perhaps in order. In undertaking a critical reexamination of anything as familiar as elementary geometry, there is always the danger that familiarity will breed, if not contempt, at least impatience and disinterest. What is the point, one may be inclined to ask, of struggling to make a careless definition more precise when everyone "knows" exactly what is meant anyway? Why should one be concerned with making a weak proof stronger when everybody "knows" that the theorem in question is true? Isn't preoccupation with the details of a field as well-established as geometry just quibbling and hair-splitting?

There are many answers to questions like these. In the first place, a critical reappraisal of elementary geometry does not imply that the facts of geometry are being called into question. Our concern here is not with *whether* certain theorems are true or false but rather with exactly *why* they are true or under just what assumptions or conditions they are true. Then there is the matter of pride, for it should be a source of pride to a person to be able to say exactly what he means, to be able to distinguish between careless and careful reasoning, and to be able to buttress his assertions with unassailable rather than merely plausible arguments. Finally, we must think of the students who will someday make their first exploration of geometry under our direction. For the most part they will come to us with only the most primitive notions of intuitive geometry. What we teach them will, in effect, define geometry for them and will be a significant factor in their ultimate appreciation of a logical argument. The better our understanding, the better able we shall be to lead them to a clear understanding of the facts of geometry and to provide them with sound models of precise expression and logical reasoning.

2.2 A Brief Critique of Euclid. There can be no question of the importance of Euclid's contributions to geometry and to human thought. The pattern of deductive organization which he established is the form in which almost all of present day mathematics is cast. Probably no book except the Bible has appeared in more editions or contributed more to the intellectual life of the world than has Euclid's "Elements." For over two thousand years, educated men in all fields, politicians, soldiers, and theologians no less than scholars and philosophers, have looked upon it as the epitome of rigor and its study as the best way to develop facility in logical reasoning. But it would be indeed surprising if the enormous increase in mathematical knowledge since the time of Euclid had not revealed flaws and weaknesses in his work, and such has been the case.

In the first place, instead of beginning with a few undefined concepts whose meaning was to come from the axioms, Euclid mistakenly attempted to define every term he used. This led him inevitably into some curious and quite unsatisfactory "definitions." For instance, the terms "point" and "line" he defined in this fashion:

A point is that which has no part.
A straight line is a line which lies evenly with the points on itself.

and for "angle" he gave this definition:

A plane angle is the inclination to one another of two straight lines in a plane which meet one another and do not lie in a straight line.

Although, admittedly, these do correspond in some vague way to our

intuitive ideas of "point," "line," and "angle," they actually beg the
question, inasmuch as their meaning depends upon such other technical terms as "part," "lies evenly," and "inclination."

Far more serious, however, is the fact that, careful as he was, Euclid
assumed and made essential use of a number of properties which he
did not include among his axioms and which cannot be derived from
his axioms. For example, Euclid's first proposition is the familiar
construction of an equilateral triangle on a given segment as base, and
he carried it out in the obvious way by striking arcs of the appropriate
radius from the ends of the segment as centers and then joining the
ends of the segment to the point of intersection of these arcs, as shown
in Fig. 2.1*a*. But he made no attempt to show, and indeed, from his

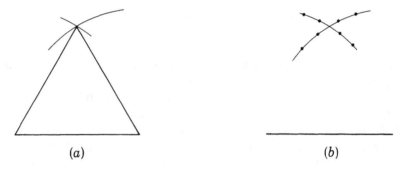

(*a*)   (*b*)

Fig. 2.1

axioms alone, could not have shown, that the two arcs intersect! The
parallel axiom guarantees that under certain conditions two lines will
intersect, but none of Euclid's axioms deals with conditions under
which two circles will or will not intersect. Conceivably, the points
on a circle may be spaced like beads on a string, and it may be possible
for one circle somehow to "slip through" another without the two
having a point in common (Fig. 2.1*b*). Of course, as Euclid himself
assumed, it *appears* that two circular arcs like those shown in Fig. 2.1*a*
do indeed intersect in a point, but only an axiom or an earlier theorem
can justify this conclusion. Questions of this sort come under the
general heading of *continuity considerations*, and this is one of the points
at which the traditional treatment of euclidean geometry must be
strengthened.

Euclid was also lax in his treatment of order relations. For instance,
"betweenness" seems to have been such a natural and familiar notion
to him that he apparently felt no need to spell out its properties
axiomatically. As a result, it was sometimes impossible for him to
establish with certainty the location of one point with respect to others

in a given discussion. This in turn opened the door to a number of striking paradoxes, of which the following is typical:

"Theorem."   Every triangle is isosceles.(!)

"*Proof.*"*   Let $\triangle ABC$ be an arbitrary triangle. Let the bisector of $\angle BAC$ and the perpendicular bisector of side BC be drawn, and let O be their point of intersection. Let A', B', C' be, respectively, the feet of the perpendiculars from O to the sides BC, CA, and AB, A' being also, by construction, the midpoint of the side BC (see Fig. 2.2)

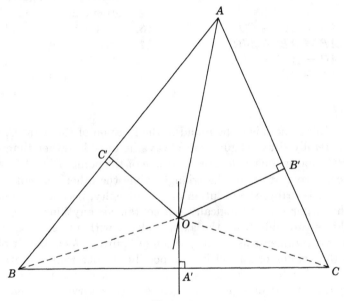

Fig. 2.2

Then

1. $A'O = A'O.$	1.	Identity.
2. $BA' = A'C.$	2.	Construction.
3. $\angle OA'B = \angle OA'C.$	3.	Both are right angles.
4. $\triangle OA'B \cong \triangle OA'C.$	4.	$S, A, S.$
5. $OB = OC.$	5.	Corresponding parts of congruent $\triangle$.
6. $AO = AO.$	6.	Identity.
7. $\angle C'AO = \angle B'AO.$	7.	AO bisects $\angle BAC.$

* We shall use conventional symbolism throughout this "proof" even though in later sections we shall modify it at certain points.

8. $\angle AC'O = \angle AB'O$.	8. Both are right angles.
9. $\triangle AC'O \cong \triangle AB'O$.	9. A, S, A.
10. $AC' = AB'$.	10. Corresponding parts of congruent $\triangle$.
11. $OC' = OB'$.	11. Corresponding parts of congruent $\triangle$.
12. $\angle OC'B = \angle OB'C$.	12. Both are right angles.
13. $OB = OC$.	13. Step 5.
14. $\triangle OC'B \cong \triangle OB'C$.	14. Right angle, hypotenuse, leg.
15. $C'B = B'C$.	15. Corresponding parts of congruent $\triangle$.
16. $AB = AC' + C'B$.	16. C' is between A and B.
17. $AB = AB' + B'C$.	17. Steps 10 and 15.
18. $AC = AB' + B'C$.	18. B' is between A and C.
19. $AB = AC$.	19. Steps 17 and 18.

Q.E.D.(!)

The fallacy here is to be found in the location of the point O, which in a carefully drawn figure will always lie outside rather than inside $\triangle ABC$† and will be so located that one of the points B' and C' will lie between two vertices of the triangle while the other will not. However, this is really no resolution of the difficulty, because, in a science which purports to be logically self-contained, anything that is true should be provable from the axioms alone, without recourse to such inductive arguments as the inspection of figures. As a matter of fact, from the axioms of Euclid it is impossible to determine whether the point O is inside or outside the triangle. Worse yet, it is impossible, using Euclid's axioms, even to give a satisfactory definition of the inside and outside of a triangle. Clearly, in a careful development of euclidean geometry, considerably more attention must be paid to order relations than is ordinarily the case.

Though the greatest weakness of the traditional development of euclidean geometry is probably its treatment (or lack of treatment!) of the concepts of continuity and order, there are other points, too, which need to be strengthened. In particular, the concepts of distance and angle are usually introduced without an adequate axiomatic foundation, and the important notion of congruence is ordinarily made to depend upon superposition which, in turn, involves the sophisticated ideas of motion and invariance. These, then, are some of the matters to which we shall give our attention in the rest of this chapter.

Through additional axioms, making explicit the intuitive ideas

† A proof of this is outlined in Exercise 10, Sec. 2.11.

Euclid did not formalize, and more precise definitions, we shall attempt to lay a foundation upon which more acceptable proofs of the familiar theorems of geometry can be constructed. In doing this, we shall follow the increasingly popular practice of blending plane and solid geometry at every opportunity. This makes possible certain economies in presentation and emphasizes the essential similarity of the two systems rather than their differences.

EXERCISES

1. Discuss the following quotation: "One of the pitfalls of working with a deductive system is too great familiarity with the subject matter of the system. It is this pitfall that accounts for most of the blemishes in Euclid's *Elements.*"

    Howard Eves, "An Introduction to the History of Mathematics," p. 126, Holt, Rinehart and Winston, Inc., New York, 1953.

2. Discuss the following definitions suggested by Heron (ca. 200 B.C.) as alternatives to Euclid's definition of a straight line:

    (*a*) A straight line is a line stretched to the utmost.

    (*b*) A straight line is a line which, when its ends remain fixed, itself remains fixed, when it is, as it were, turned around in the same plane.

    (*c*) A straight line is a line such that all its parts fit on all other parts alike.

    (*d*) A straight line is a line which with one other of the same species cannot complete a figure.

3. Discuss the following definitions used by Euclid in his "Elements":

    (*a*) A surface is that which has length and breadth only.

    (*b*) The extremities of a surface are lines.

    (*c*) A plane surface is a surface which lies evenly with the straight lines on itself.

    (*d*) A boundary is that which is the extremity of anything.

    (*e*) A figure is that which is contained by any boundary or boundaries.

4. Criticize the following argument:

    "Theorem." Every point in the interior of a circle lies on the circle.(!)

    "Proof." Let O be the center of an arbitrary circle of radius r, let P be any point in the interior of the circle, and let Q be the point on the line OP on the same side of O as P such that

    $$(OP)(OQ) = r^2$$

    Clearly, since OP is less than r, OQ must be greater than r, and hence P lies between O and Q. Let R be the midpoint of the segment PQ and let S

be one of the points in which the perpendicular bisector of PQ intersects the circle (Fig. 2.3).

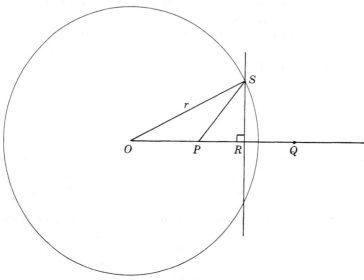

Fig. 2.3

1. $OP = OR - PR$.	1. P is between O and R.
2. $OQ = OR + RQ$.	2. R is between O and Q.
3. $OQ = OR + PR$.	3. R is the midpoint of PQ.
4. $(OP)(OQ) = (OR - PR)(OR + PR)$ $= (OR)^2 - (PR)^2$.	4. Substitution.
5. $(OR)^2 = (OS)^2 - (SR)^2$.	5. Pythagorean theorem.
6. $(PR)^2 = (PS)^2 - (SR)^2$.	6. Pythagorean theorem.
7. $(OP)(OQ) = [(OS)^2 - (SR)^2]$ $- [(PS)^2 - (SR)^2]$ $= (OS)^2 - (PS)^2$ $= r^2 - (PS)^2$.	7. Substitution from Steps 5 and 6 into Step 4.
8. $r^2 = r^2 - (PS)^2$.	8. Q was located so that $(OP)(OQ) = r^2$.
9. P lies on the circle.	9. From Step 8, the distance from P to S is zero; hence P and S are the same point.

Q.E.D.(!)

5. Criticize the following argument:

"Theorem."   Every obtuse angle is a right angle.(!)

"Proof."   Let $\angle DAE$ be an arbitrary obtuse angle, let B and C be points on the same side of the line AD as E such that $AB = AE$ and $ABCD$ is a rectangle, and let the perpendicular bisectors of AD and CE intersect in the point O (Fig. 2.4).

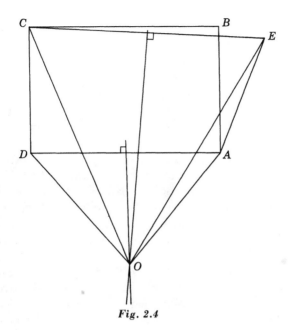

Fig. 2.4

1. $AE = AB = DC$.	1. B and C were located so that $ABCD$ is a rectangle in which $AB = AE$.
2. $EO = CO$.	2. O is on the perpendicular bisector of the segment CE.
3. $AO = DO$.	3. O is on the perpendicular bisector of the segment AD.
4. $\triangle AEO \cong \triangle DCO$.	4. S, S, S.
5. $\angle OAE = \angle ODC$.	5. Corresponding parts of congruent ⚠.
6. $\angle OAE = \angle OAD + \angle DAE$.	6. The whole is equal to the sum of its parts.
7. $\angle ODC = \angle ODA + \angle ADC$.	7. The whole is equal to the sum of its parts.
8. $\angle OAD + \angle DAE = \angle ODA + \angle ADC$.	8. Substitution from Steps 6 and 7 into Step 5.
9. $\angle OAD = \angle ODA$.	9. O is on the perpendicular bisector of the segment DA.
10. $\angle DAE = \angle ADC$.	10. Steps 8 and 9.
11. $\angle DAE$ is a right angle.	11. It is equal to $\angle ADC$ which is a right angle. Q.E.D.(!)

2.3 The Postulates of Connection. As we observed in Sec. 1.3, the formal study of geometry begins with a set of axioms, or postulates, as we shall henceforth call them, which assign properties to certain undefined objects. In our work we shall, naturally enough, take "point," "line," and "plane" as our undefined terms. Admittedly, we all have some understanding of these words. They suggest abstractions from familiar objects in the world around us, and our choice of postulates is motivated by what we believe are the properties of these objects. Nonetheless, as we use these terms in our work, we must remember that their meaning comes solely from the postulates and not from any reference to counterparts in the external world.

This does not mean, of course, that we shall not use figures in our discussions and proofs. We shall. But we shall use them only to *remind* us of conclusions already reached and to *suggest* further relations to be explored, never to justify assertions we have made. In this respect, the representations of points, lines, and planes which we shall draw from time to time are like the men in a game of chess. The shape, size, color, or material of the pieces is irrelevant, provided only that different pieces can be distinguished. Their meaning comes solely from the rules, i.e., postulates, of the game, and it is quite correct to say that each piece is simply a collection of properties. The only function of the physical pieces on a physical board is to remind the players of actions they have already taken and to suggest further moves it may be expedient to make. And just as some exceptional chess masters can play the game blindfolded, without benefit of board and men, so, ideally, a geometer should be able to dispense with the dots he calls "points" and the long thin marks he calls "lines" and carry out his analyses without the assistance of a drawing. That most of us are unable to do this is evidence only of our own shortcomings and does not alter the fact that the objects of geometric investigations, like chessmen, are simply collections of properties, independent of any particular representation or interpretation.

To accomplish our purposes, we shall need twenty-one postulates, but rather than listing them all at the outset, we shall introduce them in several groups. The first group consists of what are usually called the postulates of *incidence* or *connection*. These reflect our simplest intuitive ideas about how points, lines, and planes are related, as expressed by such words as "contains," "lies in," "intersects," and "determines."

In phrasing these, as well as certain other postulates and definitions to follow, it will be convenient to use the language of set theory, although in our work we shall actually need little more than an understanding of what is meant by the terms *subset, union,* and *intersection:*

Definition 1.   A set, S_1, is a subset of a set S if every member of S_1 is a member of S. A subset S_1 is called a proper subset of S if there is at least one member of S which is not a member of S_1.

Definition 2.   The union of the sets S_1 and S_2 is the set of all elements which are members of at least one of the sets S_1 and S_2.

Definition 3.   The intersection of the sets S_1 and S_2 is the set of all elements which are simultaneously members of S_1 and members of S_2.

In connection with the last definition, an interesting question arises: If the sets S_1 and S_2 have no elements in common, can we still speak of their intersection? Whether we can or not depends, of course, upon the conventions we agree to accept. For our purposes it is desirable to be able to speak, without exception, of the intersection of any two sets. Hence we shall introduce the so-called **null set** or **empty set,** that is, the set having no members, and we shall say that *the intersection of two sets which have no members in common is the empty set.*

Though the concept of a set containing no members may at first seem strange or artificial, it really is not and in fact is encountered in many different connections. For instance, a school might offer a course in advanced Latin conversation and find at the end of registration that no one had registered for it. The class, however, would continue to exist in the records of the school as a meaningful entity, even though it contained no students. It would, in other words, be the empty set. Or, more technically, if we are asked to solve the equation

$$\sqrt{x - 1} = 1 - \sqrt{x + 4}$$

we are really asked to determine the set of all values of x which convert the given equation into a true statement. Now by repeatedly squaring and collecting terms, we find that $x = 5$ is the only possibility for a solution. However, substitution shows that this value of x does not satisfy the given equation. Hence the solution set of the equation, though a perfectly meaningful concept, is the empty set.

Through the use of the empty set, we can now speak of the intersection of any sets whatsoever. The word *intersect*, however, we shall use only in speaking of sets whose intersection is not empty. Thus if we say that S_1 and S_2 intersect, we imply that there is at least one element common to both S_1 and S_2.

Our list of postulates begins with the following limited, though very natural, assertions about points and lines:

Postulate 1.   Every line is a set of points and contains at least two* points.

Postulate 2.   If P and Q are two points, there is one and only one line which contains them both.

Stated equivalently, Postulate 2 assures us that two points determine a unique line, and it is convenient to incorporate this fact into our notation.   Specifically, the line determined by the two points P and Q we shall denote equally well by

$$\overleftrightarrow{PQ} \quad \text{or} \quad \overleftrightarrow{QP}$$

Preparatory to stating the next two postulates, we need the following definition:

Definition 4.   The points of a set are said to be collinear if and only if there is a line which contains them all.

Three or more points which do not lie on the same line are said to be **noncollinear.**

The next four postulates introduce the last of our undefined terms, *plane:*

Postulate 3.   A plane is a set of points and contains at least three noncollinear points.

Postulate 4.   If P, Q, and R are three noncollinear points, there is one and only one plane which contains them all.

Postulate 5.   If two points of a line lie in a plane, then every point of the line lies in that plane.

Postulate 6.   If two planes have a point in common, their intersection is a line.

It is important to note that none of the first six postulates guarantees the existence of any points, lines, or planes.   They merely describe certain simple properties these objects have *if* they exist.   Hence, to

---

* As usual, we understand "two points" to mean "two distinct points."

be sure that our system is not empty, we need a postulate asserting that it actually contains something, and the last connection postulate does this. Before we can state Postulate 7, however, we must have the following definitions:

Definition 5.   The points of a set are said to be coplanar if and only if there is a plane which contains them all.

Definition 6.   The set of all points is called space.

Postulate 7.   Space contains at least four points which are non-collinear and noncoplanar.

Postulate 7 does not explicitly assert the existence of either lines or planes. However, Postulates 7 and 2 together imply that the system contains at least one line, and Postulates 7 and 4 together imply that the system contains at least one plane. It is interesting to note that if the word "noncollinear" is left out of Postulate 7, as is often the case, it is impossible to prove the existence of even one plane. In fact, it is easy to verify that a system consisting of $n(\geq 4)$ collinear points, the unique line which contains these points, and no planes satisfies not only Postulates 1 to 6 but also Postulate 7 with the word "noncollinear" omitted, since n collinear points are obviously noncoplanar if there exists no plane to contain them.

Knowing that our ultimate objective is to give a sound deductive structure to the body of familiar geometric knowledge, we have every reason to believe that lines, planes, and space all contain infinitely many points and that Postulates 1, 3, and 7 are remarkable understatements. However, there is nothing to be gained, and perhaps something to be lost, by ruling out finite geometries at this early stage, and our development proceeds more smoothly if the assertion that a line contains infinitely many points is introduced in its natural place among the postulates of distance in the next section.

The most important theorems which follow from Postulates 1 to 7 are all simple, and we shall merely state them, leaving their proofs as exercises:

Theorem 1.   If two lines intersect, their intersection is a point.

Theorem 2.   If a line intersects a plane which does not contain the line, the intersection is a point.

Theorem 3.   If two planes intersect, their intersection is a line.

Theorem 4.   A line and a point not on the line determine a unique plane.

Theorem 5.   Two intersecting lines determine a unique plane.

Theorem 6.   Space contains at least six lines.

Theorem 7.   Space contains at least four planes.

EXERCISES

1. Prove Theorem 1.   **2.** Prove Theorem 2.
3. Prove Theorem 3.   **4.** Prove Theorem 4.
5. Prove Theorem 5.   **6.** Prove Theorem 6.
7. Prove Theorem 7.

8. Verify that the system consisting of eight points,

$$P_1, P_2, P_3, P_4, P_5, P_6, P_7, P_8$$

twenty-eight lines, defined as the following sets of points,

$$
\begin{array}{llll}
l_1 = \{P_1,P_2\} & l_2 = \{P_1,P_3\} & l_3 = \{P_1,P_4\} & l_4 = \{P_1,P_5\} \\
l_5 = \{P_1,P_6\} & l_6 = \{P_1,P_7\} & l_7 = \{P_1,P_8\} & l_8 = \{P_2,P_3\} \\
l_9 = \{P_2,P_4\} & l_{10} = \{P_2,P_5\} & l_{11} = \{P_2,P_6\} & l_{12} = \{P_2,P_7\} \\
l_{13} = \{P_2,P_8\} & l_{14} = \{P_3,P_4\} & l_{15} = \{P_3,P_5\} & l_{16} = \{P_3,P_6\} \\
l_{17} = \{P_3,P_7\} & l_{18} = \{P_3,P_8\} & l_{19} = \{P_4,P_5\} & l_{20} = \{P_4,P_6\} \\
l_{21} = \{P_4,P_7\} & l_{22} = \{P_4,P_8\} & l_{23} = \{P_5,P_6\} & l_{24} = \{P_5,P_7\} \\
l_{25} = \{P_5,P_8\} & l_{26} = \{P_6,P_7\} & l_{27} = \{P_6,P_8\} & l_{28} = \{P_7,P_8\}
\end{array}
$$

and fourteen planes, defined as the following sets of points,

$$
\begin{array}{ll}
\pi_1 = \{P_1,P_2,P_3,P_4\} & \pi_2 = \{P_1,P_2,P_5,P_6\} \\
\pi_3 = \{P_1,P_2,P_7,P_8\} & \pi_4 = \{P_1,P_3,P_5,P_7\} \\
\pi_5 = \{P_1,P_3,P_6,P_8\} & \pi_6 = \{P_1,P_4,P_5,P_8\} \\
\pi_7 = \{P_1,P_4,P_6,P_7\} & \pi_8 = \{P_2,P_3,P_5,P_8\} \\
\pi_9 = \{P_2,P_3,P_6,P_7\} & \pi_{10} = \{P_2,P_4,P_5,P_7\} \\
\pi_{11} = \{P_2,P_4,P_6,P_8\} & \pi_{12} = \{P_3,P_4,P_5,P_6\} \\
\pi_{13} = \{P_3,P_4,P_7,P_8\} & \pi_{14} = \{P_5,P_6,P_7,P_8\}
\end{array}
$$

satisfies Postulates 1 to 7.

9. In Exercise 8
   (a) List the lines which lie in π_{10}.
   (b) List the planes which contain l_{10}.
   (c) List the planes which pass through P_5.
   (d) What lines, if any, contain P_8, are coplanar with l_6, and have the null set as their intersection with l_6?   What relation exists between these lines and l_6?

    (e) What is the intersection of π_{11} and π_{12}?

    (f) What is the union of l_{11} and l_{17}? Is the union of two lines always a plane?

    (g) What planes, if any, contain P_4 and have the null set as their intersection with π_9? What relation exists between these planes and π_9?

    (h) What is the union of π_5 and π_{10}? What is the union of π_5 and π_{11}? What appears to you to be the essential difference between these two unions?

10. Using Postulates 1 to 7, together with the following version of the parallel postulate,

    Through a given point not on a given line there passes one and only one line which is coplanar with but does not intersect the given line

prove the following theorem:

    If there exists one line which contains exactly n points, then

    (a) Every line contains exactly n points.

    (b) Every plane contains exactly n^2 points.

    (c) Space contains exactly n^3 points.

    (d) Space contains exactly $n^2(n^2 + n + 1)$ lines.

    (e) Exactly $n(n + 1)$ lines lie in each plane.

    (f) Exactly $n + 1$ lines pass through a given point and lie in a given plane containing the point.

    (g) Exactly $n^2 + n + 1$ lines pass through each point.

    (h) Exactly $n + 1$ planes pass through a given line.

    (i) Exactly $n^2 + n + 1$ planes pass through each point.

    (j) Space contains exactly $n(n^2 + n + 1)$ planes.

2.4 The Measurement of Distance. To guide us in our choice of postulates to establish the concept of distance, we should review the everyday notions of distance we intend to formalize. In the first place, all our experience points up the fundamental role of the unit of measurement. Until a unit is agreed upon, distances cannot be measured, but after one is chosen, it becomes possible, at least in theory, to assign a unique number to every pair of points as a measure of the distance between them. Moreover, this number is in all cases nonnegative and is zero if and only if the points of the pair are the same. For us, as individuals, our unit of measurement is provided by whatever ruler, yardstick, or tape we have at hand. Ultimately, however, it is defined by two points on a platinum bar kept somewhere in our country's national laboratory, the Bureau of Standards, in Washington, D.C. These ideas are all made precise in the first of our postulates of distance:

    Postulate 8. If A and A' are distinct points, there exists a correspondence which associates with each pair of

points in space a unique number, such that the
number assigned to a pair of points
(1) Is 0 if the points of the pair are the same.
(2) Is positive if the points of the pair are distinct.
(3) Is 1 if the points of the pair are the given points,
   A and A'.

The set $\alpha = \{A,A'\}$ consisting of the two points A and A' referred to
in Postulate 8 we call the **unit pair.** The number which corresponds
to a pair of points in accordance with Postulate 8 we call the **measure
of the distance between the points** relative to the unit pair
$\alpha = \{A,A'\}$. As a notation for the distance between the points P
and Q relative to the unit pair, α, we use either of the symbols

$$m_\alpha(P,Q) \qquad \text{or} \qquad m_\alpha(Q,P)$$

From our experience with the measurement of distance, it is clear
that many units are in common use. Distances can be measured in
inches, feet, yards, miles, or in numerous other, less common units.
But, as we well know, there is a remarkable similarity between measurements based on two different units. In fact, if a number of distances are measured first in terms of one unit, then in terms of another,
the two sets of measurements are always proportional, and the constant of proportionality is simply the length of one of the units in terms
of the other. For instance, measurements made in feet are always
three times as great as the corresponding measurements made in yards
and only one-twelfth as large as the corresponding measurements made
in inches. This familiar aspect of the measurement of distance we
also want to incorporate in our abstract formulation. Hence we
adopt the following postulate:

Postulate 9. If $\alpha = \{A,A'\}$ and $\beta = \{B,B'\}$ are two pairs of
distinct points, then for all pairs of points, P, Q,

$$m_\alpha(P,Q) = m_\alpha(B,B')m_\beta(P,Q)$$

From Postulate 9 it is obvious that if $m_\alpha(B,B') = 1$, then

$$m_\alpha(P,Q) = m_\beta(P,Q)$$

In other words, if we have a pair of points, B and B', such that
$m_\alpha(B,B') = 1$, then using $\{B,B'\}$ as a unit pair yields the same measurements as using the unit pair $\alpha = \{A,A'\}$. This, of course, is
equivalent to the important though obvious fact that copies of the
unit pair in the Bureau of Standards can be made and that rulers

based on these copies all yield identical measurements and can be used interchangeably.

Since it is possible to measure distances in terms of different units, it is important to know that certain simple combinations of distance measures which we frequently encounter do not depend upon the unit pair we employ. Specifically, in our discussion of order relations and "betweenness" in the next section, we shall need the following result:

Theorem 1.   If P, Q, and R are points such that for some unit pair, $\alpha = \{A,A'\}$,

$$m_\alpha(P,Q) + m_\alpha(Q,R) = m_\alpha(P,R)$$

then for any other unit pair, $\beta = \{B,B'\}$,

$$m_\beta(P,Q) + m_\beta(Q,R) = m_\beta(P,R)$$

Proof.   By Postulate 9 we have

$$m_\alpha(P,Q) = m_\alpha(B,B')m_\beta(P,Q)$$
$$m_\alpha(Q,R) = m_\alpha(B,B')m_\beta(Q,R)$$
$$m_\alpha(P,R) = m_\alpha(B,B')m_\beta(P,R)$$

Hence, substituting into the given relation,

$$m_\alpha(B,B')m_\beta(P,Q) + m_\alpha(B,B')m_\beta(Q,R) = m_\alpha(B,B')m_\beta(P,R)$$

and, dividing by the nonzero quantity $m_\alpha(B,B')$,

$$m_\beta(P,Q) + m_\beta(Q,R) = m_\beta(P,R)$$

as asserted.

In exactly the same way we can establish the following theorem:

Theorem 2.   If P, Q, R, and S are points such that for some unit pair, $\alpha = \{A,A'\}$,

$$\frac{m_\alpha(P,Q)}{m_\alpha(R,S)} = k$$

then for any other unit pair, $\beta = \{B,B'\}$,

$$\frac{m_\beta(P,Q)}{m_\beta(R,S)} = k$$

Neither Postulate 8 nor Postulate 9 tells us how to assign distance measures to pairs of points when a unit pair is chosen. They merely assert that such numbers exist and have certain properties. In life, these numbers are, of course, determined by our rulers or similar measuring devices, constructed, as we observed above, by direct or indirect reference to the unit pair in the Bureau of Standards. For our purpose we clearly need some theoretical equivalent of a physical ruler. As a first step in this direction we need to know that any given unit pair can, so to speak, be duplicated on any given line. More precisely, we need the property asserted by the next postulate:

Postulate 10. If P is an arbitrary point on an arbitrary line, l, and if $\alpha = \{A,A'\}$ is any unit pair, then there exists at least one point, Q, on l such that $m_\alpha(P,Q) = 1$.

In life, after a given unit pair has been duplicated on an initially unmarked ruler, the ruler is graduated with a number of equally spaced points of subdivision. Subsequently, the distance between two points is found by placing the ruler so that its edge passes through the given points, reading the numbers on its scale opposite the two points, and subtracting the smaller of these from the larger. This procedure is limited, of course, by the fact that physical rulers are necessarily of finite length and can bear only a finite number of scale marks. The "theoretical ruler" we have in mind, however, should be of unlimited extent and moreover should have every point "marked." Hence we are led to the following as the last of our postulates of distance:

Postulate 11 (*The Ruler Postulate*). Let P be an arbitrary point on an arbitrary line, l, and let $\alpha = \{A,A'\}$ be any unit pair. Then if Q is any point on l such that $m_\alpha(P,Q) = 1$, there is a unique one-to-one correspondence between the set of all real numbers and the set of all points on l such that
(1) The number 0 corresponds to the point P.
(2) The number 1 corresponds to the point Q.
(3) The measure of the distance between any points, R and S, on l relative to the unit pair $\alpha = \{A,A'\}$ is equal to the absolute value of the difference of the numbers which correspond to R and S.

Postulate 11 is arithmetic rather than geometric in character and has no counterpart among the axioms employed in the traditional presentation of euclidean geometry. Some mathematicians, inclining to the view that algebra and geometry should be kept apart until they are finally fused in analytic geometry, find Postulate 11 out of place and distasteful. It is an extremely powerful postulate, however, since it brings to bear on the problems of elementary geometry a large body of results from arithmetic. In particular, having available the well-known order and continuity properties of the real-number line makes possible a much simpler treatment of the troublesome questions of order and continuity which we noted in Sec. 2.2. Pedagogically speaking, although the beginning student in geometry will often not have seen careful proofs of the basic properties of the real numbers, his previous work in mathematics will certainly have made him familiar with these properties. This knowledge constitutes an additional tool for the study of geometry not available under alternative axiomatic treatments, and its use simultaneously provides the student with valuable opportunities for review and reinforcement.

The question of how many points there are on a general line is answered by Postulate 11. According to this postulate, the set of points on any line can be put in one-to-one correspondence with the set of real numbers, that is, there exists a correspondence in which a unique point corresponds to every real number and vice versa. Therefore, since there are infinitely many real numbers, it follows that on every line there are infinitely many points. Thus, from this point on, finite geometries are excluded from our discussion.

In what follows, we shall drop the loose term "theoretical ruler," which we used in the discussion motivating Postulate 11, and instead use the more accurate technical term **coordinate system:**

> **Definition 1.** A coordinate system on an arbitrary line, l, is a one-to-one correspondence between the set of all points on l and the set of all real numbers such that if P and Q are the points which correspond to 0 and 1, respectively, then the measure of the distance between any points, R and S, on l relative to the unit pair $\{P,Q\}$ is equal to the absolute value of the difference of the numbers which correspond to R and S.

In any coordinate system, the point which corresponds to the number 0 is called the **origin,** and the point which corresponds to the number 1 is called the **unit point.** The number which is assigned to any point by the coordinate system is called the **coordinate** of that point.

In most of our work we shall be using a single unit pair for measuring distances, and therefore it will be unnecessary to indicate the unit pair explicitly, as we do in the symbol $m_\alpha(P,Q)$. In such cases we shall use the simpler expressions

$$PQ \quad \text{or} \quad QP$$

instead of $m_\alpha(P,Q)$ or $M_\alpha(Q,P)$ to denote the measure of the distance between P and Q relative to the particular unit pair we have chosen.

EXERCISES

1. Prove Theorem 2.
2. If $m_{in.}(P,Q)$, $m_{ft}(P,Q)$, $m_{yd}(P,Q)$, . . . denote, respectively, the measures of the distance between the points P and Q in terms of unit pairs which define an inch, a foot, a yard, . . . , complete the following table in which each row refers to a different pair of points, P, Q.

$m_{in.}(P,Q)$	$m_{ft}(P,Q)$	$m_{yd}(P,Q)$	$m_{cm}(P,Q)$
	2		
8			
		1	
			100

3. Prove that if $[m_\alpha(P,Q)]^2 + [m_\alpha(Q,R)]^2 = [m_\alpha(P,R)]^2$, then

$$[m_\beta(P,Q)]^2 + [m_\beta(Q,R)]^2 = [m_\beta(P,R)]^2$$

4. If $m_\alpha(P,Q) + m_\alpha(Q,R) > m_\alpha(P,R)$, is it necessarily true that

$$m_\beta(P,Q) + m_\beta(Q,R) > m_\beta(P,R)$$

5. If $m_\alpha(P,Q)m_\alpha(P,R) = k$, what is $m_\beta(P,Q)m_\beta(P,R)$? Can you think of any application of this result?

2.5 Order Relations. With the ruler postulate, the problem of defining the concept of "betweenness" and related order relations is relatively simple:

Definition 1. A point B is said to be between the points A and C if and only if
(1) A, B, and C are distinct collinear points.
(2) $AB + BC = AC$.

From Theorem 1 of the last section, we know that if $AB + BC = AC$ for one coordinate system on the line l containing A, B, and C, then this relation is true in all coordinate systems on l. Hence, "betweenness" is independent of the coordinate system in terms of which the distances are measured.

As might be anticipated, the geometrical betweenness relation for points corresponds to the arithmetical betweenness relation for the coordinates of these points in any coordinate system. Specifically, we have the following important theorem:

> **Theorem 1.** Let A, B, and C be three points on a line, l, and let x, y, and z be, respectively, the coordinates of these points in a coordinate system on l. Then B is between A and C if and only if y is between x and z.

Proof. Let us suppose, first, that y is between x and z. Then either

$$x > y > z \qquad \text{or} \qquad x < y < z$$

In the first case we have

$$x - y > 0, \qquad y - z > 0, \qquad \text{and} \qquad x - z > 0$$

and therefore, from the definition of absolute values,

$$|x - y| = x - y$$
$$|y - z| = y - z$$
$$|x - z| = x - z$$

Also, by the ruler postulate,

$$|x - y| = AB$$
$$|y - z| = BC$$
$$|x - z| = AC$$

Hence, substituting,

$$AB + BC = |x - y| + |y - z|$$
$$= (x - y) + (y - z) = x - z = |x - z| = AC$$

Thus, by Definition 1, B is between A and C. An almost identical argument leads to the same conclusion if $x < y < z$. Hence we have proved that *if* y is between x and z, then B is between A and C.

To prove, conversely, that B is between A and C *only if* y is between x and z, that is, to prove that if B is between A and C, then necessarily y is between x and z, we begin with the definitive relation

$$(1) \qquad AB + BC = AC \qquad \text{or} \qquad |x - y| + |y - z| = |x - z|$$

From the arithmetic properties of real numbers, one and only one of the three numbers, x, y, z, is between the other two. Hence x, y, and z must satisfy one and only one of the six order relations

$$y > x > z \qquad x > y > z \qquad x > z > y$$
$$z > x > y \qquad z > y > x \qquad y > z > x$$

and we must determine which of these are consistent with (1). Suppose, first, that $y > x > z$.   Then $y - x > 0$, $x - z > 0$, $y - z > 0$, and therefore

$$|x - y| = y - x$$
$$|x - z| = x - z$$
$$|y - z| = y - z$$

Substituting these into (1), we find

$$(y - x) + (y - z) = x - z$$

or, simplifying,

$$2y = 2x$$

But this is impossible, since y and x are the coordinates of distinct points and hence cannot be equal.   Thus the coordinates of A, B, and C cannot satisfy the relation $y > x > z$.   In a similar fashion, it is easy to show that the relations

$$z > x > y, \qquad x > z > y, \qquad \text{and} \qquad y > z > x$$

are also inconsistent with (1), but that

$$x > y > z \qquad \text{and} \qquad z > y > x$$

are possible.   Thus the only order relations among the coordinates of A, B, and C which are possible if B is between A and C are those in which y is between x and z, as asserted.

> **Corollary 1.**   Of three collinear points, one and only one is between the other two.

With betweenness defined, we now can define what we mean by a **segment** and a **ray**:

> **Definition 2.**   If A and B are distinct points, the set consisting of A, B, and all points which are between A and B is called a segment.

The unique segment determined by two points A and B or, as we shall often say, the segment joining the points A and B, we shall denote by either of the symbols

$$\overline{AB} \qquad \text{or} \qquad \overline{BA}$$

The points A and B are called the **endpoints** of the segment $\overline{AB}$. The set consisting of all points of the segment except the endpoints A and B is called the **interior** of the segment.   The measure of the distance between A and B is called the **length** of the segment.   Segments having the same length are called **congruent** segments.   The fact that the segments $\overline{AB}$ and $\overline{CD}$ are congruent is indicated by writing

$$\overline{AB} \cong \overline{CD}$$

Obviously, every segment is congruent to itself, and segments congruent to the same segment are congruent to each other.

The following theorem is an immediate consequence of Theorem 1 and the definition of a segment:

Theorem 2. If A and B are two points on a line, l, and if, in any coordinate system on l, A and B have coordinates a and b such that $a < b$, then the segment $\overline{AB}$ is the same as the set of points whose coordinates, x, satisfy the relation $a \leqq x \leqq b$. If $b < a$, then the segment $\overline{AB}$ is the same as the set of points whose coordinates satisfy the relation

$$b \leqq x \leqq a$$

Definition 3. If A and B are two points, the set consisting of all points of the segment $\overline{AB}$ and all points, P, such that B is between A and P is called a ray.

The ray determined by the points A and B, in that order, or as we shall often say, the ray extending from A through B, we shall denote by the symbol

$$\overrightarrow{AB}$$

and we shall call the point A the **endpoint** of the ray $\overrightarrow{AB}$. In the symbol for a ray, the endpoint of the ray is always named first and the arrow always extends to the right. Rays with the same endpoint are said to be **concurrent.** Two concurrent rays which are collinear are called **opposite** rays, and each is said to be opposite to the other. The ray $\overrightarrow{BA}$ is of course not the ray which is opposite to $\overrightarrow{AB}$ since, though they are collinear, these rays do not have the same endpoint. The ray opposite to a given ray, $\overrightarrow{AB}$, is denoted by a symbol of the form $\overrightarrow{AX}$, where X is any point of the line $\overleftrightarrow{AB}$ which does not belong to the ray $\overrightarrow{AB}$. Figure 2.5 illustrates these observations.

At first glance, the definition of a ray, $\overrightarrow{AB}$, makes it appear that the point B is in some way exceptional. This is not the case, however, and, as Theorem 3 makes clear, the ray $\overrightarrow{AB}$ is the same as the ray $\overrightarrow{AC}$ or the ray $\overrightarrow{AD}$ or the ray $\overrightarrow{AE}$, . . . where $C, D, E, \ldots$ are any points of $\overrightarrow{AB}$ distinct from A (see Fig. 2.6).

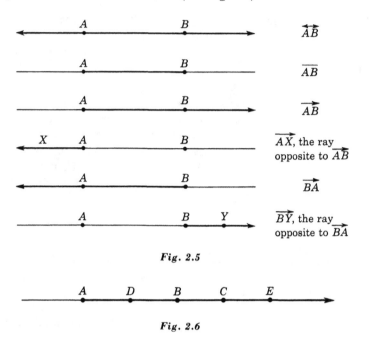

Fig. 2.5

Fig. 2.6

Like segments, rays too can be defined easily in terms of the coordinates of the points which they contain. Specifically, we have the following theorem:

Theorem 3. Let A and B be distinct points and let a and b be, respectively, the coordinates of these points in any coordinate system on $\overleftrightarrow{AB}$. Then if $a < b$, the ray $\overrightarrow{AB}$ is the same as the set of points whose coordinates, x, satisfy the condition $a \leqq x$. If $a > b$, the ray $\overrightarrow{AB}$ is the same as the set of points whose coordinates satisfy the condition $a \geqq x$.

Proof. Let a and b be, respectively, the coordinates of A and B in a coordinate system on the line $\overleftrightarrow{AB}$, and suppose first that $a < b$. If X is any point of the ray $\overrightarrow{AB}$, X is either a point of the segment $\overline{AB}$ or else is a point such that B is between A and X. If X is a point of

$\overline{AB}$, then by Theorem 2 the coordinate, x, of X must be such that

$$a \leqq x \leqq b$$

On the other hand, if B lies between A and X, then by Theorem 2 it follows that

$$a < b < x$$

Hence in either case it follows that

$$a \leqq x$$

Conversely, if $a \leqq x$, then either $a \leqq x \leqq b$ or $a < b < x$, and hence X either belongs to the segment $\overline{AB}$ or else is a point such that B lies between A and X. In either case, however, X belongs to $\overrightarrow{AB}$, by definition. Thus every point of the ray $\overrightarrow{AB}$ has a coordinate x, satisfying the condition $a \leqq x$, and conversely every point whose coordinate satisfies this condition is a point of the ray. An almost identical argument establishes the assertion of the theorem when $a > b$.

A closely related result is contained in the following theorem:

> **Theorem 4.** Let P be any point on an arbitrary line, l, and let p be the coordinate of P in any coordinate system on l. Then the set of points of l whose coordinates, x, satisfy the condition $x \leqq p$ and the set of points whose coordinates satisfy the condition $x \geqq p$ are opposite rays on l with common endpoint P.

One of the common operations in geometry is what is usually referred to as "laying off the distance d along the line l from the point A." In our development, the possibility of doing this is guaranteed by the following important theorem:

> **Theorem 5 (The Point-plotting Theorem).** If $\overrightarrow{AB}$ is a ray and d a positive number, there is exactly one point on $\overrightarrow{AB}$ and one point on the ray opposite to $\overrightarrow{AB}$ such that the distance from A to each of these points relative to a given unit pair is d.

Proof. By Postulate 10, there is a point U on the line $\overleftrightarrow{AB}$ such that $AU = 1$ relative to the given unit pair. By the ruler postulate, there is a coordinate system on $\overleftrightarrow{AB}$ in which A corresponds to the number 0 and U corresponds to the number 1. Furthermore, in this coordinate system there is a unique point, D, whose coordinate is the given positive number d and a unique point, D', whose coordinate is the negative number $-d$. Also, by the ruler postulate,

$$AD = |0 - d| = d \quad \text{and} \quad AD' = |0 - (-d)| = d$$

Hence the distance from A to each of the points D and D' is d, and clearly no other point on $\overleftrightarrow{AB}$ can be at this distance from A. Finally, by Theorem 4 the points D and D' are on opposite rays on the line $\overleftrightarrow{AB}$; hence one of them must be on the ray $\overrightarrow{AB}$ and one must be on the opposite ray, as asserted.

By taking the number d in Theorem 5 to be $k \cdot AB$, where k is positive, the following theorem can easily be established:

Theorem 6. If A and B are any two points and if k is an arbitrary positive number, there is a unique point, P, on $\overrightarrow{AB}$ such that $AP = k \cdot AB$. Moreover, if the coordinates of A and B in a coordinate system on $\overleftrightarrow{AB}$ are a and b, respectively, then the coordinate of P is the number

$$p = a + k(b - a)$$

Taking $k = \frac{1}{2}$ in Theorem 6, we have the following important special case:

Corollary 1. There is a unique point, P, between two given points, A and B, such that $AP = PB$. Moreover, if the coordinates of A and B are a and b, respectively, the coordinate of P is

$$p = \frac{a + b}{2}$$

The point P such that $AP = PB$ is called the **midpoint** of the segment $\overline{AB}$. The midpoint of a segment is said to **bisect** the segment, and more generally, any set of points whose intersection with a segment consists only of the midpoint of the segment is said to bisect the segment.

By considering two opposite rays with common endpoint A on a line, l (see Fig. 2.7), it is clear from Theorems 2 and 4 that a point such as A divides the rest of the points on any line through A into two sets, which might be called the "sides" of A, such that

(1) Any segment, such as $\overline{BC}$, whose endpoints are on the same side of A lies entirely on that side of A.
(2) Any segment, such as $\overline{CD}$, whose endpoints lie on opposite sides of A contains A in its interior.

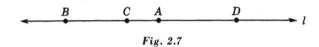

Fig. 2.7

In view of this obvious result, it is natural to ask if a line divides a plane into two parts and if a plane divides space into two parts the way a point divides a line.   Curiously enough, it is impossible to prove either of these statements from the postulates we have thus far adopted (or from the axioms and postulates of Euclid), and if we wish planes and space to have this property of separability we must introduce it via an additional postulate.   To aid us in phrasing the necessary new postulate concisely, we must first introduce the important concept of a **convex set**:

Definition 4.   A set of points is said to be convex if and only if any segment whose endpoints belong to the set lies entirely in the set.

Examples of sets which are convex as well as of sets which are not convex are shown in Fig. 2.8.

The most fundamental property of convex sets is contained in the following theorem:

Theorem 7.   The intersection of two convex sets is a convex set.

Proof.   Let S_1 and S_2 be two convex sets, and let $\overline{AB}$ be any segment whose endpoints lie in the intersection of S_1 and S_2.   From the definition of the intersection of two sets, it follows that both A and B lie in S_1 and also in S_2.   Therefore, since both S_1 and S_2 are convex, by hypothesis, the segment $\overline{AB}$ lies entirely in S_1 and also entirely in S_2. In other words, $\overline{AB}$ lies in the intersection of S_1 and S_2, as asserted.

Using the concept of a convex set, we can now state our earlier observations about the separability of lines as the following theorem:

Theorem 8.   Any point, A, divides the rest of the points on any line containing A into two classes such that
(1) Each set is convex.
(2) Any segment joining a point in one set to a point in the other contains A in its interior.

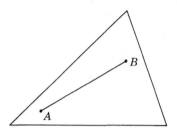

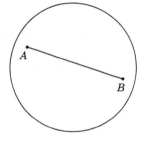
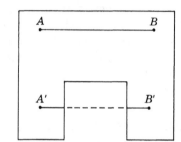

Fig. 2.8

As we implied above, the corresponding property for a plane must be postulated:

Postulate 12   (*The Plane-separation Postulate*).   For any plane and any line lying in the plane, the points of the plane which do not belong to the line form two sets such that
(1) Each set is convex.
(2) Any segment joining a point in one set to a point in the other intersects the given line.

Each of the convex sets determined by the given line in the given plane is called a **halfplane.**   The line itself, which by definition does not belong to either halfplane, is called the **edge** of each halfplane and is said to **separate** the plane into the two halfplanes.   Two or more points in the same halfplane are said to lie on the **same side** of the given line, and two points which lie in different halfplanes are said to lie on **opposite sides** of the given line.

With the plane-separation postulate available, it is possible to prove the corresponding separation property for space:

Theorem 9   (**The Space-separation Theorem**).   The points of space which do not lie in a given plane form two sets such that
(1) Each set is convex.
(2) Any segment joining a point in one set to a point in the other intersects the given plane.

Proof.   Let π be an arbitrary plane and let O be an arbitrary point which does not lie in π. Then, clearly, every point in space which is not a point of π must belong to one or the other of the following two non-empty sets:

(1) The set S_1 consisting of O and all points A_1 such that the segment $\overline{OA_1}$ does not intersect π.
(2) The set S_2 consisting of all points A_2, not in π, such that the segment $\overline{OA_2}$ intersects π.

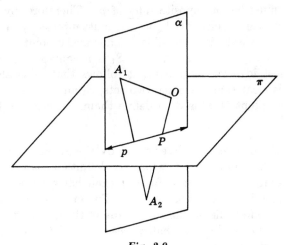

Fig. 2.9

To complete our proof, we must show that the sets S_1 and S_2 have the properties asserted by the theorem. To do this, suppose first that A_1 and A_2 are arbitrary points in S_1 and S_2, respectively. From the definition of S_2 it follows that $\overline{OA_2}$ has a point, P, in common with π. Hence if $A_1 = O$, it is obvious that $\overline{A_1 A_2}$ ($= \overline{OA_2}$) intersects π, as required. Let us continue, therefore, under the assumption that $A_1 \neq O$. Now since they have the point P in common, π and the plane, α, determined by A_1, A_2, and O (or any plane, α, containing

A_1, A_2, and O, if these points are collinear) must intersect in a line, p. Moreover, since $\overline{OA_1}$ has no point in common with π, it of course can have no point in common with p. Hence in the plane α, the points O and A_1 lie on the same side of p, while O and A_2 lie on opposite sides of p (see Fig. 2.9). Thus A_1 and A_2 are on opposite sides of p and therefore, by the plane-separation postulate, the segment $\overline{A_1A_2}$ must intersect p, and hence π, as required. To prove, further, that S_1 is convex, let A_1 and A_1' be any two points in S_1. If the segment $\overline{A_1A_1'}$ does not lie entirely in S_1, there must be at least one point, Q, between A_1 and A_1' which is either a point of S_2 or a point of π. If Q is a point of S_2, then by the first part of our proof, both the segment $\overline{A_1Q}$ and the segment $\overline{A_1'Q}$ must have a point in common with π. Moreover, since they lie on the opposite rays $\overrightarrow{QA_1}$ and $\overrightarrow{QA_1'}$, these points must be distinct. Hence the line $\overleftrightarrow{A_1A_1'}$ must have two points in common with π and therefore must lie entirely in π, which is impossible. On the other hand, if Q is a point of π, then the plane α determined by A_1, A_1', and O (or any plane α containing A_1, A_1', and O, if these points are collinear), having Q in common with π, must intersect π in a line, p; and in α, A_1 and A_1' must lie on opposite sides of p. Therefore, by the plane-separation postulate, either $\overline{OA_1}$ or $\overline{OA_1'}$ must intersect p, and hence π, which, by hypothesis, is impossible. Thus every point of $\overline{A_1A_1'}$ must be a point of S_1, or in other words, S_1 is a convex set, as asserted. By an almost identical argument, it follows that S_2 is also convex. Finally, it is clear that the same two sets, S_1 and S_2, are obtained no matter what point O is used to define them. Hence the theorem is established.

Each of the convex sets determined by the given plane is called a **halfspace.** The plane itself, which by definition does not belong to either halfspace, is called the **face** of each halfspace and is said to **separate** space into the two halfspaces. Two or more points in the same halfspace are said to lie on the **same side** of the given plane, and two points which lie in different halfspaces are said to lie on **opposite sides** of the given plane.

As an illustration of the use of the plane-separation postulate, we shall prove the following "obvious" result, which we will need in Sec. 2.7.

> **Theorem 10.**   If V is any point on the edge of a halfplane H, and if A, B, and X are three points in the union of H and its edge such that
>
> (1) No two of the points A, B, X are collinear with V

(2) A and B lie on opposite sides of $\overleftrightarrow{VX}$

then A and X lie on the same side of $\overleftrightarrow{VB}$, and B and

X lie on the same side of $\overleftrightarrow{VA}$.

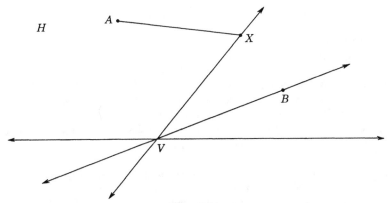

Fig. 2.10

Proof. By hypothesis, A, B, X and therefore the segment $\overline{AX}$ and the ray $\overrightarrow{VB}$ lie in or on the edge of the halfplane H (see Fig. 2.10). Therefore, if $\overline{AX}$ and $\overleftrightarrow{VB}$ intersect, their intersection must be a point on $\overrightarrow{VB}$ and not on the ray opposite to $\overrightarrow{VB}$. However, we are given that A and B lie on opposite sides of $\overleftrightarrow{VX}$. Hence, with the exception of the distinct points V and X, $\overline{AX}$ and $\overrightarrow{VB}$ lie on opposite sides of $\overleftrightarrow{VX}$ and therefore can have no point in common. Thus, since $\overline{AX}$ can intersect neither the ray $\overrightarrow{VB}$ nor the ray opposite to $\overrightarrow{VB}$, it follows that A and X are on the same side of $\overleftrightarrow{VB}$. Finally, an identical argument shows that B and X lie on the same side of $\overleftrightarrow{VA}$, as asserted.

EXERCISES

1. In a number of places, many geometry texts direct the reader to "Extend such and such a line until. . . ." Discuss the meaning, if any, which is attached to such a phrase.

2. Carry through the proof of the first part of Theorem 1 under the assumption that $x < y < z$.

3. In the proof of the second part of Theorem 1, show in detail why each of the relations

$$z > x > y, \quad x > z > y, \quad y > z > x$$

   is inconsistent with the hypothesis that

$$AB + BC = AC$$

4. Prove Corollary 1, Theorem 1.
5. Prove Theorem 2.
6. Prove Theorem 4.
7. Complete the proof of Theorem 6.
8. If P is between A and B, prove that the segments $\overline{AP}$ and $\overline{PB}$ have only the point P in common.
9. If A, B, and C are three collinear points such that $AC > AB$ and $AC > BC$, prove that B is between A and C.
10. What is the coordinate of the midpoint of the segment $\overline{AB}$ if the coordinates of A and B are, respectively, 2 and 3?   2 and -3? -2 and 3? -2 and -3?
11. If the coordinates of A and B are -1 and 5, respectively, what are the coordinates of the points which subdivide the segment $\overline{AB}$ into three segments of equal length? four segments of equal length? five segments of equal length?
12. In the proof of Theorem 10, why is it necessary to assume that A, B, and X all lie in the union of the halfplane H and its edge?

2.6   Angles and Angle Measurement.   Two quite different concepts of angle are encountered in elementary geometry.   One views an angle as the amount of rotation or difference of direction between two lines; the other considers an angle simply as a figure, that is, as a set of points.   Clearly, the ideas of rotation and direction are themselves quite complicated and difficult to make precise, and hence the first definition of an angle is much less satisfactory than the second, which is the one we shall adopt:

> **Definition 1.**   An angle is the union of two rays which have a common endpoint and do not lie in the same straight line.

Each of the rays which together form an angle is called a **side** of the angle, and the common endpoint of the two rays is called the **vertex** of the angle.   The angle formed by the rays $\overrightarrow{AB}$ and $\overrightarrow{AC}$ we shall denote by either of the symbols

$$\angle BAC \quad \text{or} \quad \angle CAB$$

or occasionally, when there is no possibility of confusion, simply by

$$\angle A$$

It should be noted that the figures which are often referred to as "zero angles" and "straight angles" are not angles according to our definition (see Fig. 2.11). Although these are of some importance in trigonometry, they serve no useful purpose in elementary geometry, and to include them would merely create exceptions to various fundamental theorems.

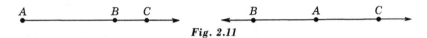

Fig. 2.11

The measurement of angles is based upon three postulates which bear a considerable resemblance to the postulates for the measurement of distance. The first asserts essentially just that angles can be measured:

Postulate 13. If R is any positive number, there exists a correspondence which associates with each angle in space a unique positive number between 0 and R.

The number R is called the **scale factor.** The number assigned to a given angle, $\angle BAC$, according to Postulate 13, is called the **measure of the angle relative to the scale factor R,** and is denoted by the

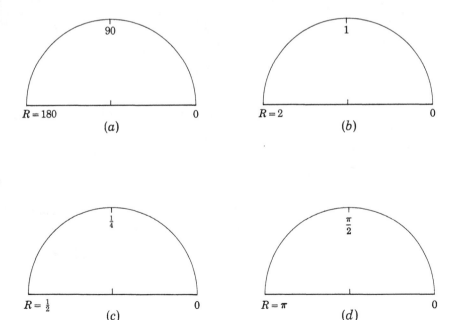

Fig. 2.12. (*a*) A degree protractor; (*b*) a right-angle protractor; (*c*) a revolution protractor; (*d*) a radian protractor.

symbol

$$m_R\angle BAC$$

The scale factor is the theoretical counterpart of the largest scale mark on the protractors of everyday experience. If $R = 180$, we are in effect measuring angles in degrees (see Fig. 2.12). Taking $R = 2$ amounts to measuring angles in terms of a right angle as a unit. Similarly, taking $R = \frac{1}{2}$ and $R = \pi$ is equivalent to measuring angles in terms of revolutions and radians, respectively.

The second postulate for angle measurement incorporates our intuitive conviction that angle measures based on different units, i.e., scale factors, are proportional:

Postulate 14.   If R and S are any positive numbers, then for every angle, $\angle BAC$,

$$m_R\angle BAC = \frac{R}{S}\, m_S\angle BAC$$

Finally, the third postulate for angle measurement provides us with a "theoretical protractor" by means of which the angle formed by two concurrent rays can be found by subtracting numbers associated with the respective rays (see Fig. 2.13):

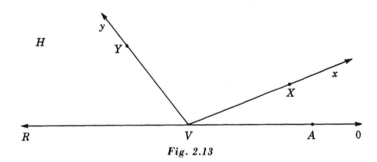

Fig. 2.13

Postulate 15   (*The Protractor Postulate*).   If H is any half-plane, $\overrightarrow{VA}$ any ray lying in the edge of H, and R any positive number, there is a one-to-one correspondence between the set of all numbers x for which $0 \leq x \leq R$ and the set of rays, $\overrightarrow{VX}$, which lie in the union of H and its edge, such that

(1) $\overrightarrow{VA}$ corresponds to the number 0.

(2) The ray opposite to $\overrightarrow{VA}$ corresponds to the number R.

(3) If X and Y are not collinear with V and if x and y are the numbers which correspond to $\overrightarrow{VX}$ and $\overrightarrow{VY}$, respectively, then $m_R \angle XVY = |x - y|$.

As we should expect, results analogous to Theorems 1 and 2, Sec. 2.4, are true for angles:

Theorem 1.   If $\overrightarrow{VA}$, $\overrightarrow{VB}$, and $\overrightarrow{VC}$ are rays such that for some scale factor, R,

$$m_R \angle AVB + m_R \angle BVC = m_R \angle AVC$$

then for any other scale factor, S,

$$m_S \angle AVB + m_S \angle BVC = m_S \angle AVC$$

Theorem 2.   If $\overrightarrow{VA}$, $\overrightarrow{VB}$, $\overrightarrow{VC}$, and $\overrightarrow{VD}$ are rays such that for some scale factor, R,

$$\frac{m_R \angle AVB}{m_R \angle CVD} = k$$

then for any other scale factor, S,

$$\frac{m_S \angle AVB}{m_S \angle CVD} = k$$

The point-plotting theorem, Theorem 5, Sec. 2.5, also has an important counterpart for angles:

Theorem 3   (**The Angle-construction Theorem**).   If H is a halfplane whose edge contains the ray $\overrightarrow{VA}$ and if r is any number between 0 and R, there is a unique ray, $\overrightarrow{VX}$, such that X is in H and $m_R \angle AVX = r$.

In general we shall restrict ourselves to some particular scale factor throughout any particular discussion. Hence the explicit indication of R will almost always be unnecessary, and we shall usually abbreviate the symbol $m_R \angle AVB$ to $m \angle AVB$.

In life, given three concurrent coplanar rays it is not always possible to say that one of them is between the other two. For instance, in the completely symmetrical case of three concurrent coplanar rays each pair of which forms an angle of measure 120, it is clear that no one can

properly be said to lie between the other two (see Fig. 2.14). There
are cases, however, in which there is a basis for asserting that an order

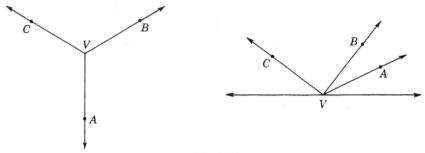

Fig. 2.14

relation exists between three concurrent coplanar rays, and the following definition allows us to identify these:

> **Definition 2.**   If three concurrent coplanar rays, $\overrightarrow{VA}$, $\overrightarrow{VB}$, $\overrightarrow{VC}$, are
> such that
> (1) No two of the points A, B, C are collinear with
> V
> (2) A, B, and C all lie in or on the edge of a half-
> plane whose edge passes through V
> (3) $m\angle AVB + m\angle BVC = m\angle AVC$
> then $\overrightarrow{VB}$ is said to lie between $\overrightarrow{VA}$ and $\overrightarrow{VC}$.

In view of Theorem 1, it is clear that if (3) holds for one scale factor, it
holds for every scale factor.   Hence the betweenness relation for rays
is independent of the scale factor used in measuring the angles.
 Just as we were able to restate the betweenness relation for points in
terms of the coordinates assigned to the points by the ruler postulate,
so we can restate the betweenness relation for rays in terms of the
coordinates assigned to the rays by the protractor postulate.   Specifically, we have the following result, whose proof follows closely the
proof of Theorem 1, Sec. 2.5:

> **Theorem 4.**   If $\overrightarrow{VA}$, $\overrightarrow{VB}$, and $\overrightarrow{VC}$ are three concurrent coplanar rays
> such that no two of the points A, B, C are collinear
> with V, then $\overrightarrow{VB}$ is between $\overrightarrow{VA}$ and $\overrightarrow{VC}$ if and only
> if under some correspondence described by the
> protractor postulate, the number assigned to $\overrightarrow{VB}$
> is between the numbers assigned to $\overrightarrow{VA}$ and $\overrightarrow{VC}$.

Closely associated with Theorems 3 and 4 is the following theorem and its corollary:

Theorem 5. If $\angle AVC$ is any angle and if k is any number between 0 and 1, there is a unique ray, $\overrightarrow{VB}$, between $\overrightarrow{VA}$ and $\overrightarrow{VC}$ such that $m\angle AVB = k \cdot m\angle AVC$. Moreover, if the numbers assigned to $\overrightarrow{VA}$ and $\overrightarrow{VC}$ in some correspondence described by the protractor postulate are a and c, respectively, the number assigned to $\overrightarrow{VB}$ is $b = a + k(c - a)$.

Corollary 1. If $\angle AVC$ is any angle, there is a unique ray, $\overrightarrow{VB}$, between $\overrightarrow{VA}$ and $\overrightarrow{VC}$ such that $m\angle AVB = m\angle BVC$. Moreover, if the numbers assigned to $\overrightarrow{VA}$ and $\overrightarrow{VC}$ by some correspondence described by the protractor postulate are a and c, respectively, then the number assigned to $\overrightarrow{VB}$ is $b = (a + c)/2$.

EXERCISES

1. Prove Theorem 1.   **2.** Prove Theorem 2.
3. Prove Theorem 3.
4. In applying Definition 2, it may be that there is more than one halfplane satisfying the requirements of the definition. Prove that if a betweenness relation exists for three rays, it is independent not only of the scale factor but also of the halfplane which is used to define it.
5. Prove Theorem 4.   **6.** Prove Theorem 5.
7. Complete the following table in which each row refers to a different angle:

$m_{180}\angle AVB$	$m_{\pi}\angle AVB$	$m_{2}\angle AVB$	$m_{\frac{1}{2}}\angle AVB$
120			
	$\pi/3$		
		$\frac{1}{2}$	
			$\frac{1}{4}$

8. If the coordinates of $\overrightarrow{VA}$ and $\overrightarrow{VB}$ are respectively 20 and 120, what are the coordinates of the rays $\overrightarrow{VX}$ such that (a) $m\angle AVX = \frac{1}{2}m\angle AVB$; (b) $m\angle AVX = \frac{1}{3}m\angle XVB$; (c) $m\angle XVB = \frac{1}{3}m\angle AVB$?

2.7   Further Properties of Angles.   For many purposes it is necessary to speak of the interior of an angle, and by analogy with the definition of a segment, we base our definition on the betweenness relation for rays:

Definition 1.   The interior of an angle, $\angle AVB$, is the set of all points, X, such that the ray $\overrightarrow{VX}$ lies between the ray $\overrightarrow{VA}$ and the ray $\overrightarrow{VB}$.

It appears from Fig. 2.15 that we could also define the interior of an

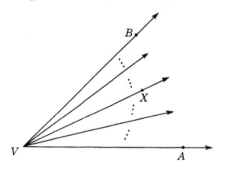

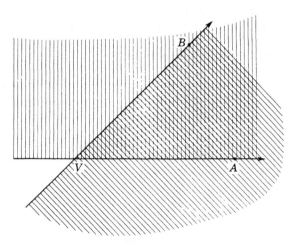

Fig. 2.15

angle as the intersection of two halfplanes, and in fact we have the following theorem:

Theorem 1. The interior of $\angle AVB$ is the intersection of the halfplane determined by $\overleftrightarrow{VA}$ and B and the halfplane determined by $\overleftrightarrow{VB}$ and A.

Proof. To prove this theorem, we must show that the set of points, S_2, defined by the theorem is exactly the same as the set of points, S_1, defined by Definition 1. To do this, suppose first that X is a point such that $\overrightarrow{VX}$ lies between $\overrightarrow{VA}$ and $\overrightarrow{VB}$. Then by definition there exists a halfplane, H, such that

(1) The edge of the halfplane H contains V.
(2) The points A, B, X lie in the union of H and its edge.
(3) No two of the points A, B, X are collinear with V.
(4) $m\angle AVX + m\angle XVB = m\angle AVB$.

At the outset, we observe that A and B lie on opposite sides of $\overleftrightarrow{VX}$. In fact, if A and B were on the same side of $\overleftrightarrow{VX}$, then by the protractor postulate a correspondence exists which assigns to $\overrightarrow{VX}$ the number 0 and assigns to $\overrightarrow{VA}$ and $\overrightarrow{VB}$, respectively, numbers a and b such that

$$m\angle AVX = |a - 0| = a$$
$$m\angle XVB = |b - 0| = b$$
$$m\angle AVB = |b - a|$$

But this is impossible, since for these values, condition (4) becomes

$$a + b = |b - a|$$

which holds for no positive numbers, a and b. Thus A and B lie on opposite sides of $\overleftrightarrow{VX}$, and we can invoke Theorem 10, Sec. 2.5, to show that X lies on the same side of $\overleftrightarrow{VB}$ as A and that X lies on the same side of $\overleftrightarrow{VA}$ as B. This proves that any point of S_1 is also a point of S_2.

Now suppose that X is a point of S_2, that is, suppose that X lies on the same side of $\overleftrightarrow{VA}$ as B and on the same side of $\overleftrightarrow{VB}$ as A. Then by the protractor postulate there exist two correspondences, in one of which $\overrightarrow{VA}$, $\overrightarrow{VB}$, and $\overrightarrow{VX}$ correspond, respectively, to the numbers 0,

b, and x (Fig. 2.16a) and

$$m\angle AVB = |b - 0| = b$$
$$m\angle AVX = |x - 0| = x$$
$$m\angle BVX = |b - x|$$

and in the other of which $\overrightarrow{VA}$, $\overrightarrow{VB}$, and $\overrightarrow{VX}$ correspond, respectively,

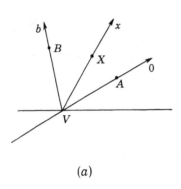

(a)   (b)

Fig. 2.16

to a', 0, x' (Fig. 2.16b) and

$$m\angle AVB = |a' - 0| = a'$$
$$m\angle AVX = |a' - x'|$$
$$m\angle BVX = |x' - 0| = x'$$

Now either $b < x$ or $b > x$. If $b < x$, then

$$m\angle BVX = |b - x| = x - b$$

and thus the measures of the angles are such that

$$m\angle AVB + m\angle BVX = b + (x - b) = x = m\angle AVX$$

But using the angle measures provided by the second correspondence, this becomes

$$a' + x' = |a' - x'|$$

which is impossible. Hence b cannot be less than x. On the other hand, $b > x$ implies that $m\angle BVX = |b - x| = b - x$, which leads to the conclusion that

$$m\angle AVX + m\angle XVB = x + (b - x) = b = m\angle AVB$$

This is consistent with the angle measures provided by the second correspondence; hence it follows that $\overrightarrow{VX}$ lies between $\overrightarrow{VA}$ and $\overrightarrow{VB}$. Thus any point which belongs to S_2 also belongs to S_1. This, coupled with our previous conclusion that any point which belongs to S_1 also belongs to S_2, proves that S_1 and S_2 are the same set, as asserted.

With Theorem 1 available, it is now an easy matter to establish the following properties of the interior of an angle:

Theorem 2.   The interior of an angle is a convex set.

Theorem 3.   If on each side of an angle a point other than the vertex is selected, every point between these points is in the interior of the angle.

The proof of the first of these theorems follows immediately from Theorem 7, Sec. 2.5, and the fact that a halfplane is a convex set. To prove the second, let V be the vertex of the given angle, let A and B be points other than V on the respective sides of the angle, and let P be any point between A and B (see Fig. 2.17). Since only one of three collinear points can be between the other two, it follows that B is not between A and P. Hence the segment $\overline{AP}$ does not intersect the line $\overleftrightarrow{VB}$, which means that P and A lie on the same side of $\overleftrightarrow{VB}$. Similarly,

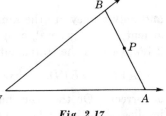

Fig. 2.17

P and B lie on the same side of $\overleftrightarrow{VA}$. Thus P lies in the interior of the angle, by Theorem 1. Incidentally, the last two results are examples of theorems which do not hold for "zero angles" and "straight angles."

Having now defined the interior of an angle, we find it natural to give a formal definition of the exterior of an angle:

Definition 2.   The exterior of an angle is the set of all points in the plane of the angle which do not belong to the angle and do not lie in the interior of the angle.

Most of our work with angles is associated with the important concept of the **congruence** of angles:

Definition 3.   Angles which have the same measure are said to be congruent.

The fact that angles such as $\angle ABC$ and $\angle DEF$ are congruent is indicated by writing

$$\angle ABC \cong \angle DEF$$

Clearly, every angle is congruent to itself, and angles which are congruent to the same angle are congruent to each other.

Since an angle is simply a set of points, it follows that angles are equal if and only if they consist of exactly the same points, that is, if

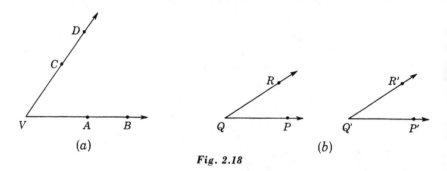

(a)   (b)

Fig. 2.18

and only if they are the same angle. Equal angles are therefore congruent, but in general, congruent angles are not equal. Thus in Fig. 2.18*a*, each of the statements

$$\angle AVC = \angle BVD, \qquad m\angle AVC = m\angle BVD, \qquad \angle AVC \cong \angle BVD$$

is correct. On the other hand, in Fig. 2.18*b*, assuming, as the figure implies, that the angles have the same measure, the assertions

$$m\angle PQR = m\angle P'Q'R' \qquad \text{and} \qquad \angle PQR \cong \angle P'Q'R'$$

are correct, but

$$\angle PQR = \angle P'Q'R'$$

is false.

Although we have excluded "straight angles" from our definition of angles, we nonetheless must often consider configurations like the one shown in Fig. 2.19 which suggest the notion of "straight angle."

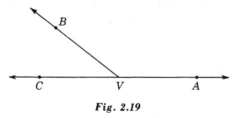

Fig. 2.19

To describe such figures in a way that will not be open to objection, we introduce the term **linear pair:**

Definition 4. Three concurrent rays, two of which are opposite rays, are said to form a linear pair of angles.

Clearly, a linear pair of angles is a special case of a pair of **adjacent angles:**

Definition 5. Two coplanar angles such that
    (1) They have a common vertex
    (2) They have one ray in common
    (3) The intersection of their interiors is the empty set
    are called adjacent angles.

With the understanding that henceforth we will use only the value $R = 180$ in the protractor postulate, that is, that we will always measure angles in degrees, we say that two angles whose measures add to 180 are **supplementary** and each is a **supplement** of the other.   An angle whose measure is 90 is called a **right angle.**   Clearly, all right angles are congruent.   Two angles whose measures add to 90 we call **complementary,** and we say that each is a **complement** of the other.   An angle whose measure is less than 90 is called **acute.** An angle whose measure is more than 90 is called **obtuse.**   Two angles whose sides form two pairs of opposite rays are called **vertical angles.**

Using the preceding definitions, the proofs of the following theorems are immediate.

Theorem 4. If two angles form a linear pair, they are supplementary.

Theorem 5. If the two angles of a linear pair are congruent, each is a right angle.

Theorem 6. Supplementary adjacent angles form a linear pair.

Theorem 7. Supplements of congruent angles are congruent.

Theorem 8. Complements of congruent angles are congruent.

Theorem 9. Vertical angles are congruent.

The concept of a right angle is of course closely related to the concept of **perpendicularity:**

Definition 6. The lines determined by two rays which form a right angle are called perpendicular lines.

Since segments as well as rays determine unique lines, it is natural to extend the last definition as follows:

Definition 7. Two sets, each of which is a segment, a ray, or a line and which determine two perpendicular lines, are said to be perpendicular.

We indicate that two sets, S_1 and S_2, are perpendicular by the notation

$$S_1 \perp S_2$$

In drawings we indicate that two sets are perpendicular by the symbolism shown in Fig. 2.20.

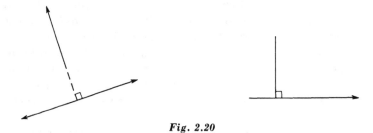

Fig. 2.20

Clearly, two perpendicular sets can intersect in at most one point and need not have any point in common. If two perpendicular sets have a point, F, in common and if P is a point in one of the sets but not in the other, we say that F is the **foot** of the perpendicular from P to the set which does not contain P.

A first fundamental theorem on perpendicular lines is the following:

Theorem 10. At each point of a given line there is one and only one line which is perpendicular to the given line and lies in a given plane containing the line.

Proof. Let P be any point on an arbitrary line, l, in an arbitrary plane, π, and let A be any point on l distinct from P. Then in one of the halfplanes, H, determined in π by l there is, by the angle-construction theorem, a unique ray, $\overrightarrow{PQ}$, such that $\angle APQ$ is a right angle (see Fig. 2.21). Therefore, by definition, $\overleftrightarrow{PQ} \perp \overleftrightarrow{PA}$, and we have shown the existence of at least one line in π which is perpendicular to l at P.

Similarly, there is a ray $\overrightarrow{PQ'}$ extending into the other halfplane determined in π by l such that $\angle APQ'$ is a right angle. Hence $\overleftrightarrow{PQ'} \perp \overleftrightarrow{PA}$, and to complete the proof of the theorem we must show that $\overleftrightarrow{PQ}$ and $\overleftrightarrow{PQ'}$ are the same line, i.e., we must show that Q, P, and Q' are collinear. But this follows at once from Theorem 6, since $\angle QPA$ and

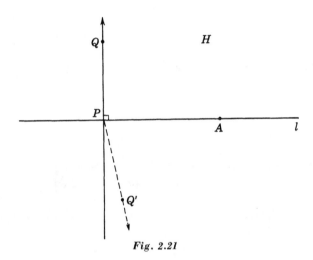

Fig. 2.21

$\angle APQ'$ are supplementary adjacent angles. Therefore, in π there is a unique line perpendicular to l at P, as asserted. The question of the existence and uniqueness of a line perpendicular to a given line through a point not on that line must be postponed to a later section (Theorem 5, Sec. 2.9).

EXERCISES

1. If $\overrightarrow{PA}$ and $\overrightarrow{PB}$ are two concurrent noncollinear rays and if $\overrightarrow{PA'}$ and $\overrightarrow{PB'}$ are the rays opposite to $\overrightarrow{PA}$ and $\overrightarrow{PB}$, respectively, show that if $\overrightarrow{PC}$ is between $\overrightarrow{PA}$ and $\overrightarrow{PB}$, then the ray $\overrightarrow{PC'}$ which is opposite to $\overrightarrow{PC}$ is between $\overrightarrow{PA'}$ and $\overrightarrow{PB'}$.

2. If $\overrightarrow{PA}$ and $\overrightarrow{PA'}$ are opposite rays and if B, C, and D are three points on the same side of $\overleftrightarrow{AA'}$ such that $\overrightarrow{PB}$ is between $\overrightarrow{PA}$ and $\overrightarrow{PC}$ and $\overrightarrow{PD}$ is between $\overrightarrow{PC}$ and $\overrightarrow{PA'}$, show that $\overrightarrow{PC}$ is between $\overrightarrow{PB}$ and $\overrightarrow{PD}$.

3. Prove that if three coplanar lines, l_1, l_2, and l_3, are concurrent in a point,

P, there exist points L_1, L_2, and L_3 on l_1, l_2, and l_3, respectively, such that $\overrightarrow{PL_2}$ is between $\overrightarrow{PL_1}$ and $\overrightarrow{PL_3}$.

4. If V is the vertex of an angle, if A_1 and B_1 are points on one side of the angle such that A_1 is between V and B_1, and if A_2 and B_2 are points on the other side of the angle such that A_2 is between V and B_2, show that $\overleftrightarrow{A_1B_2}$ and $\overleftrightarrow{A_2B_1}$ intersect in a point in the interior of the angle.

5. Give a definition of the exterior of an angle which does not depend on the definition of the interior of an angle.

6. Prove Theorem 4.   7. Prove Theorem 5.
8. Prove Theorem 6.   9. Prove Theorem 7.
10. Prove Theorem 8.   11. Prove Theorem 9.
12. State and prove the converse of Theorem 9.

2.8 Triangles and Polygons. The figure most commonly encountered in elementary geometry is probably the **triangle:**

> *Definition 1.* The union of the three segments determined by three noncollinear points is called a triangle.

The triangle determined by the three noncollinear points A, B, and C is denoted by the symbol

$$\triangle ABC$$

The points A, B, C are called the **vertices** of $\triangle ABC$. The segments $\overline{AB}$, $\overline{BC}$, and $\overline{AC}$ are called the **sides** of $\triangle ABC$. The angles $\angle ABC$, $\angle BCA$, and $\angle CAB$ are called the **angles** of $\triangle ABC$. A triangle one of whose angles is a right angle is called a **right triangle.** A triangle two of whose sides are congruent is called an **isosceles triangle.** If the three sides of a triangle are congruent in pairs, the triangle is said to be **equilateral.** Any side of a triangle and the angle whose vertex is not a point of that side are said to be **opposite** each other. In a right triangle, the side opposite the vertex of the right angle is called the **hypotenuse** and the other two sides are called the **legs.** For any triangle, the perpendicular segment from any vertex to the line determined by the opposite side is called an **altitude.***

Since a triangle is a set of points and since an angle is also a set of points, it is proper to ask if a triangle contains its angles. The answer, of course, is "No," because obviously, among the points belonging to each angle of a triangle there are infinitely many, such as B', B'', . . . , C', C'', . . . , in Fig. 2.22, which are not points of the triangle. Thus, although it is proper to speak of "the angles of a triangle" or "the

---

* Recalling the remark we made at the end of the proof of Theorem 10, Sec. 2.7, we note that the existence of altitudes has not yet been established.

angles determined by a triangle," it is incorrect to speak of "the angles contained in a triangle."

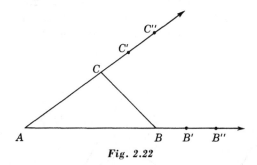

Fig. 2.22

After a triangle has been defined, it is natural to define the **interior** and **exterior** of a triangle:

Definition 2.   The interior of a triangle is the intersection of the interiors of the three angles of the triangle.

Definition 3.   The exterior of a triangle is the set of all points in the plane of the triangle which do not belong either to the triangle or to its interior.

The following theorem is an immediate consequence of Definition 2:

Theorem 1.   The interior of a triangle is a convex set.

The idea of a triangle leads naturally to the concepts of **quadrilaterals** and **polygons** in general:

Definition 4.   If A, B, C, and D are four coplanar points such that
(1) No three of the points are collinear
(2) None of the segments $\overline{AB}$, $\overline{BC}$, $\overline{CD}$, $\overline{DA}$ intersects any other at a point which is not one of its endpoints
then the union of the segments $\overline{AB}$, $\overline{BC}$, $\overline{CD}$, and $\overline{DA}$ is called a quadrilateral.

The four points A, B, C, D are called the **vertices** of the quadrilateral. The four segments $\overline{AB}$, $\overline{BC}$, $\overline{CD}$, $\overline{DA}$ are called the **sides** of the quadrilateral. Two vertices of a quadrilateral which are endpoints of the same side are said to be **consecutive;** two vertices which are not consecutive are said to be **opposite.** Two sides of a quadrilateral which have a common endpoint are said to be **consecutive;** two sides which are not consecutive are said to be **opposite.** The segments joining opposite vertices of a quadrilateral are called **diagonals.**

As Fig. 2.23 suggests, four coplanar points determine more than one quadrilateral, since a reordering of the same four vertices will, in

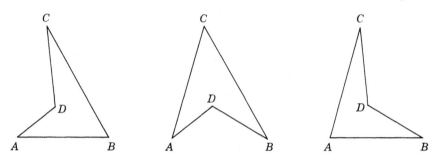

Fig. 2.23. Different quadrilaterals with the same vertices.

general, change the segments which collectively define the quadrilateral. Thus a quadrilateral cannot be identified by its vertices alone and is determined only when a list of its vertices and a list of its sides are given. We thus agree that we will always list the vertices of a quadrilateral so that

(1) Adjacent vertices in the list are endpoints of the same side.
(2) The first and last vertices in the list are endpoints of the same side.

With this convention it is now possible to designate the quadrilateral whose vertices are the points A, B, C, and D and whose sides are the segments $\overline{AB}$, $\overline{BC}$, $\overline{CD}$, and $\overline{DA}$ simply as

Quadrilateral $ABCD$

Thus it is correct to say, for instance, that

Quadrilateral $ABCD$ = quadrilateral $BCDA$
or   Quadrilateral $ABCD$ = quadrilateral $CBAD$

but incorrect to say that

Quadrilateral $ABCD$ = quadrilateral $ACBD$

Although the interior of a triangle is always a convex set, this is not true for quadrilaterals:

Definition 5. A quadrilateral is said to be convex if and only if each of its sides lies in the edge of a halfplane which contains the rest of the quadrilateral.

Definition 6. The interior of a convex quadrilateral is the intersection of the halfplanes, each of which has a side of the quadrilateral in its edge and contains the rest of the quadrilateral.

With these definitions, the following result is obvious:

Theorem 2. The interior of a convex quadrilateral is convex.

If a quadrilateral is convex, the angles determined by pairs of consecutive sides are called the **angles** of the quadrilateral. Two angles of a convex quadrilateral are said to be **consecutive** if they have a side of the quadrilateral in common. Two angles of a convex quadrilateral which are not consecutive are said to be **opposite**. A convex quadrilateral each of whose angles is a right angle is called a **rectangle.**
   The definition of a **polygon** follows closely the definition of a quadrilateral and in fact includes both triangles and quadrilaterals as special cases:

Definition 7. If $P_1, P_2, P_3, \ldots, P_n$ are n (≥ 3) coplanar points and if the n segments $\overline{P_1P_2}, \overline{P_2P_3}, \ldots, \overline{P_{n-1}P_n}, \overline{P_nP_1}$ are such that
   (1) No two segments with a common endpoint are collinear
   (2) No two segments intersect except possibly at their endpoints
   then the union of the segments $\overline{P_1P_2}, \overline{P_2P_3}, \ldots,$ $\overline{P_{n-1}P_n}, \overline{P_nP_1}$ is called a polygon.

The n points $P_1, P_2, P_3, \ldots, P_n$ are called the **vertices** of the polygon. The n segments $\overline{P_1P_2}, \overline{P_2P_3}, \ldots, \overline{P_{n-1}P_n}, \overline{P_nP_1}$ are called the **sides** of the polygon. For polygons, the concepts of consecutive and nonconsecutive vertices, and consecutive and nonconsecutive sides, are defined precisely as they are for quadrilaterals. However, for polygons with more than four sides, we do not speak of opposite vertices and opposite sides. Like quadrilaterals, a general polygon is not determined by the list of its vertices unless the sides are also identified. This we do by listing the vertices so that
   (1) Vertices which are adjacent in the list are endpoints of the same side.
   (2) The first and last vertices are endpoints of the same side.
A polygon is said to be **convex** if and only if each of its sides lies in the edge of a halfplane which contains all the rest of the polygon.

The **interior** of a *convex* polygon is the intersection of all the half-planes, each of which has a side of the polygon in its edge and contains the rest of the polygon.   The **angles** of a *convex* polygon are the angles determined by pairs of consecutive sides.   A convex polygon whose angles are all congruent and whose sides are all congruent is called a **regular polygon**.

EXERCISES

1.   Give a definition of the exterior of a triangle which does not depend upon the concept of the interior of a triangle.
2.   Is a convex quadrilateral a convex set?
3.   Prove that opposite vertices of a convex quadrilateral lie on opposite sides of the diagonal of the quadrilateral which does not contain the two vertices.
4.   Can you give a definition of your intuitive notion of the interior of a nonconvex quadrilateral?
5.   A side of a polygon is said to be a supporting side if it lies in the edge of a halfplane which contains the rest of the polygon.   What is the maximum and minimum number of supporting sides which each of the following figures can have?   (*a*) A nonconvex quadrilateral, (*b*) a nonconvex pentagon, (*c*) a nonconvex hexagon, (*d*) a nonconvex *n*-gon.
6.   Prove that the interior of any convex quadrilateral lies in the interior of each of its angles.   Is this true for polygons of any number of sides?

2.9   The Congruence Postulate.   We have already defined what is meant by congruent segments and by congruent angles.   Using these concepts, we can now define the important notion of congruent triangles:

Definition 1.   A one-to-one correspondence between the vertices of the same or different triangles such that corresponding sides and corresponding angles are congruent is called a congruence between the triangles.

Definition 2.   Triangles are congruent if and only if there exists a congruence between them.

Clearly, any triangle, $\triangle ABC$, is congruent to itself, since the identity correspondence

$$A \leftrightarrow A, \qquad B \leftrightarrow B, \qquad C \leftrightarrow C$$

is obviously one in which corresponding angles and corresponding sides are congruent. It may also happen, of course, that a triangle is congruent to itself via a nonidentical correspondence. In general, among the six possible one-to-one correspondences between the vertices of two triangles, none will be a congruence. The triangles are congruent, however, if *at least one* of the six one-to-one correspondences is a congruence. To emphasize that the existence of a congruence between triangles is the basis for asserting that the triangles are congruent, we shall use the conventional symbol for congruent triangles, namely,

$$\triangle ABC \cong \triangle DEF$$

always with the understanding that the indicated correspondence

$$A \leftrightarrow D, \qquad B \leftrightarrow E, \qquad C \leftrightarrow F$$

is a congruence. Thus, even when $\triangle ABC \cong \triangle DEF$ is a true statement, the statement $\triangle ABC \cong \triangle DFE$ is false unless the correspondence

$$A \leftrightarrow D, \qquad B \leftrightarrow F, \qquad C \leftrightarrow E$$

is also a congruence.* The advantage of this convention is that with it, corresponding parts of congruent triangles can be read directly from the notation itself, without reference to a figure.

According to our definitions, before we can assert that triangles are congruent, we must know that six simpler congruence relations hold, namely, three congruences between corresponding angles and three congruences between corresponding sides. Experience suggests, however, that less than this will suffice and that if certain sets of three of the six possible congruence relations between corresponding parts of the triangles hold, then the others hold also. This cannot be proved from the postulates we have thus far adopted, but it provides the motivation for our next postulate.

> **Postulate 16** (*The Congruence Postulate*). If there exists a one-to-one correspondence between two triangles or between a triangle and itself in which two sides and the angle determined by these sides in one triangle are congruent to the corresponding parts of the second triangle, then the correspondence is a congruence and the triangles are congruent.

Using this postulate, we can show that certain other subsets of the six fundamental congruence relations between corresponding parts of

---

* Under this convention we thus have the interesting situation that although $\triangle DEF = \triangle DFE$, there are occasions when we cannot substitute one of these names for the other.

triangles are sufficient to guarantee that the triangles themselves are congruent. Specifically, we have the following important theorems:*

Theorem 1. If there exists a one-to-one correspondence between two triangles or between a triangle and itself in which two angles and the side common to the two angles in one triangle are congruent to the corresponding parts of the other triangle, then the correspondence is a congruence and the triangles are congruent.

Proof. Let $A \leftrightarrow A'$, $B \leftrightarrow B'$, $C \leftrightarrow C'$ be a correspondence between $\triangle ABC$ and $\triangle A'B'C'$ in which

$$\angle BAC \cong \angle B'A'C', \qquad \overline{AB} \cong \overline{A'B'}, \qquad \angle ABC \cong \angle A'B'C'$$

If $\overline{BC} \cong \overline{B'C'}$, the triangles are congruent by Postulate 16. Suppose,

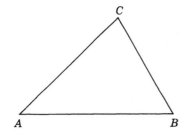

Fig. 2.24

therefore, that $\overline{BC}$ is not congruent to $\overline{B'C'}$ and, for definiteness, let $BC < B'C'$ (see Fig. 2.24).

1. There exists a point, C'', between B' and C' such that $\overline{BC} \cong \overline{B'C''}$.	1. Point-plotting theorem.
2. $\triangle ABC \cong \triangle A'B'C''$.	2. Postulate 16.
3. $\angle B'A'C'' \cong \angle BAC$.	3. Definition of congruent triangles.
4. $\angle BAC \cong \angle B'A'C'$.	4. Hypothesis.
5. $\angle B'A'C'' \cong \angle B'A'C'$.	5. Steps 3 and 4.
6. C' and C'' are on the same side of $\overleftrightarrow{A'B'}$.	6. B' is not between C' and C''.

---

* To expedite the development, Theorems 1 and 4 are often accepted without proof as additional congruence postulates. In particular, this is done by the School Mathematics Study Group.

7. $\overrightarrow{A'C'} = \overrightarrow{A'C''}$.	**7.** Angle-construction theorem.
8. $C' = C''$.	**8.** Theorem 1, Sec. 2.3.

But this is impossible, since C'' is between B' and C'.   Therefore, the assumption that $\overline{BC}$ and $\overline{B'C'}$ are not congruent must be abandoned, and hence $\triangle ABC \cong \triangle A'B'C'$, by Postulate 16.   Q.E.D.

Theorem 2.   In $\triangle ABC$, if $\overline{AB} \cong \overline{AC}$, then $\triangle ABC \cong \triangle ACB$ and $\angle ABC \cong \angle ACB$.

Proof.   Under the correspondence $A \leftrightarrow A$, $B \leftrightarrow C$, $C \leftrightarrow B$ we have $\overline{AB} \cong \overline{AC}$, $\angle BAC \cong \angle CAB$, and $\overline{AC} \cong \overline{AB}$.   Therefore, by Postulate 16, $\triangle ABC \cong \triangle ACB$, and from the definition of congruent triangles, $\angle ABC \cong \angle ACB$, as asserted.

Theorem 3.   In $\triangle ABC$, if $\overline{AB} \cong \overline{AC}$, then the segment determined by A and the midpoint of $\overline{BC}$ is perpendicular to $\overline{BC}$

Proof.   In $\triangle ABC$, let D be the midpoint of $\overline{BC}$, so that $\overline{BD} \cong \overline{CD}$. By hypothesis, $\overline{AB} \cong \overline{AC}$, and by Theorem 2, $\angle ABD \cong \angle ACD$. Therefore, by Postulate 16, $\triangle ABD \cong \triangle ACD$, and hence

$$\angle BDA \cong \angle CDA$$

Since these are congruent angles forming a linear pair, each is a right angle.   Therefore, $\overline{AD} \perp \overline{BC}$, as asserted.

Corollary 1.   The perpendicular bisector of the base of an isosceles triangle passes through the vertex opposite that side.

Theorem 4.   If there exists a one-to-one correspondence between two triangles, or between a triangle and itself, in which three sides of one triangle are congruent to the corresponding sides of the other triangle, the correspondence is a congruence and the triangles are congruent.

Proof.   Let $A \leftrightarrow A'$, $B \leftrightarrow B'$, $C \leftrightarrow C'$ be a correspondence between $\triangle ABC$ and $\triangle A'B'C'$, in which $\overline{AB} \cong \overline{A'B'}$, $\overline{BC} \cong \overline{B'C'}$, $\overline{AC} \cong \overline{A'C'}$. If $\angle BAC \cong \angle B'A'C'$, the triangles are congruent by Postulate 16. Suppose, therefore, that $\angle BAC$ is not congruent to $\angle B'A'C'$, and for definiteness, let $m\angle BAC < m\angle B'A'C'$ (see Fig. 2.25).

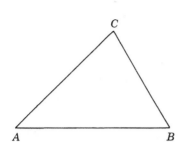

Fig. 2.25

1. There is a ray, $\overrightarrow{A'D'}$, between $\overrightarrow{A'B'}$ and $\overrightarrow{A'C'}$ such
   that $\angle B'A'D' \cong \angle BAC$.

2. On $\overrightarrow{A'D'}$ there is a point,
   C'', such that $\overline{A'C''} = \overline{AC}$.

3. C' and C'' are distinct points
   on the same side of $\overleftrightarrow{A'B'}$.

4. $\triangle BAC \cong \triangle B'A'C''$.

5. $\overline{B'C''} \cong \overline{BC} \cong \overline{B'C'}$.

6. $\triangle C'A'C''$ is isosceles.

7. $B', C',$ and C'' are not collinear.

8. $\triangle C'B'C''$ is isosceles.

9. $\therefore$ the perpendicular bisector of $\overline{C'C''}$ must pass
   through both A' and B'.

10. $\therefore$ $\overleftrightarrow{A'B'}$ is the perpendicular
    bisector of $\overline{C'C''}$.

1. Theorem 5, Sec. 2.6.

2. Point-plotting theorem.

3. $\overrightarrow{A'D'}$ is on same side of $\overleftrightarrow{A'B'}$
   as C' and $\overrightarrow{A'D'} \neq \overrightarrow{A'C'}$.

4. Postulate 16.

5. Definition of congruent triangles.

6. $\overline{C'A'} \cong \overline{C''A'}$, by definition
   of C''.

7. Steps 3 and 5 and the point-
   plotting theorem.

8. Step 5.

9. Corollary 1, Theorem 3.

10. Postulate 2.

But this is impossible because, by step 3, C' and C'' are on the same side of $\overleftrightarrow{A'B'}$. Thus the assumption that $\angle BAC$ and $\angle B'A'C'$ are not congruent must be abandoned, and therefore, by Postulate 16,

$$\triangle ABC \cong \triangle A'B'C'$$

Q.E.D.

With Postulate 16 and Theorems 1 and 4, the standard applications of congruence present no problem and, with the exception of the following important result, we shall pursue the matter no further.

Theorem 5. Through a given point not on a given line there is one
and only one line perpendicular to the given line.

Proof. Let $\overleftrightarrow{AC}$ be any line and P any point not on $\overleftrightarrow{AC}$. To prove
the theorem we must first show the existence of at least one line which
contains P and is perpendicular to $\overleftrightarrow{AC}$; then we must show that there
is only one such line. To show that there is at least one perpendicular
from P to $\overleftrightarrow{AC}$, let B be any point between A and C. If $\overleftrightarrow{PB} \perp \overleftrightarrow{AC}$, the
first part of our proof is complete. If $\overleftrightarrow{PB}$ is not perpendicular to $\overleftrightarrow{AC}$,
then one or the other of the supplementary angles, $\angle ABP$ and $\angle PBC$,
say $\angle ABP$, must be acute (see Fig. 2.26). Now by the angle-con-
struction theorem, there is a point, Q, on the opposite side of $\overleftrightarrow{AC}$ from P

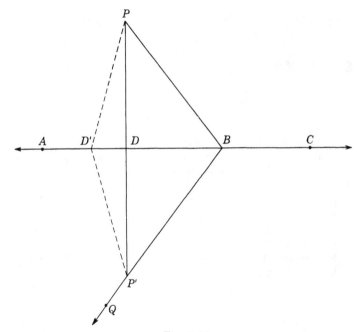

Fig. 2.26

such that $m\angle ABQ = m\angle ABP$. Also, by the point-plotting theorem,
there is a point, P', on $\overrightarrow{BQ}$ such that $BP' = BP$. By the plane-separation postulate, the segment $\overline{PP'}$ must intersect $\overleftrightarrow{AC}$ in a point, D, and
by Postulate 16, $\triangle PBD \cong \triangle P'BD$. Hence $\angle PDB \cong \angle P'DB$ and
therefore, since these congruent angles form a linear pair, each must be

a right angle. Thus $\overleftrightarrow{PP'} \perp \overleftrightarrow{AC}$, and there exists at least one perpendicular to $\overleftrightarrow{AC}$ from P, as asserted.

To show further that there is only one line which contains P and is perpendicular to $\overleftrightarrow{AC}$, assume the contrary and suppose that a second line $\overleftrightarrow{PD'}$ is also perpendicular to $\overleftrightarrow{AC}$ from P. Now if P' is the point on the ray opposite to $\overrightarrow{DP}$ such that $P'D = PD$, it follows from Postulate 16 that $\triangle PDD' \cong \triangle P'DD'$. Therefore, being corresponding parts of congruent triangles, $\angle PD'D \cong \angle P'D'D$, and hence $\angle P'D'D$ is also a right angle. Being supplementary adjacent angles, $\angle PD'D$ and $\angle P'D'D$ thus form a linear pair, or in other words, P, D', and P' are collinear. But this is impossible, since the distinct lines $\overleftrightarrow{PD}$ and $\overleftrightarrow{PD'}$ cannot have two distinct points, P and P', in common. Hence the perpendicular to $\overleftrightarrow{AC}$ from P is unique, as asserted.

EXERCISES

1. How many 1:1 correspondences can exist between a triangle and itself? In particular cases, how many of these can be congruences?
2. Prove Corollary 1, Theorem 3.
3. Prove that if there exists a 1:1 correspondence between two right triangles in which the hypotenuse and an acute angle of one triangle are congruent to the corresponding parts of the other triangle, the correspondence is a congruence and the triangles are congruent.
4. Give definitions of congruent convex quadrilaterals and congruent convex n-gons.
5. State and prove a theorem giving necessary and sufficient conditions for two convex quadrilaterals to be congruent.
6. At the present stage of our work, can we prove the existence of a rectangle with consecutive sides of prescribed length?

2.10 The Parallel Postulate. Parallel lines are sometimes described as lines which have the same direction or lines which are everywhere equidistant, but neither of these notions is really satisfactory. The first is inadequate because it makes parallelism depend upon the more sophisticated idea of direction, which actually requires the concept of parallel lines for its explanation. The second suffers from somewhat the same defect, since it is based on the not entirely simple idea of equidistance. Moreover, it is not as general as we might wish, for in hyperbolic geometry, as we shall see in Chap. 4, the locus of points equidistant from a given line is a curve and *not* a line. The best definition of parallel lines is one based on the simple idea of nonintersection:

> **Definition 1.**   Two coplanar lines which do not intersect are called parallel lines and each is said to be parallel to the other.

If p and q are parallel lines, we denote this fact by writing

$$p\|q$$

Since there are two requirements which parallel lines must meet, namely, that they lie in the same plane and have no point in common, it is clear that lines may fail to be parallel for either of two reasons:

1.   They may lie in the same plane but intersect.
2.   They may have no point in common but not lie in the same plane.

Two lines related as in (2) are called **skew lines** and are said to be **skew** to each other.

Of great importance in the study of parallel lines is the notion of a **transversal** of two lines:

> **Definition 2.**   A transversal of two coplanar lines is a line which intersects the union of the two lines in two points.

Clearly, since a transversal of two lines must meet the union of the two lines in *two* points, a line which passes through the intersection of two lines cannot be a transversal of the lines.

In the figure consisting of two coplanar lines, p and q, and a transversal, t, there are various pairs of angles to which we will need to refer. If P and Q are the points in which t intersects p and q, respectively (Fig. 2.27), if A and B are points of p on opposite sides of P, and if C and D are points of q on the same side of $\overleftrightarrow{PQ}$ as A and B, respectively, then

$$\angle APQ, \angle DQP \quad \text{and} \quad \angle BPQ, \angle CQP$$

are said to be pairs of **alternate interior angles,** and

$$\angle APQ, \angle CQP \quad \text{and} \quad \angle BPQ, \angle DQP$$

are said to be pairs of **consecutive interior angles.** If α and β are two alternate interior angles, then either angle and the angle vertical to the other are said to form a pair of **corresponding angles.** From

Theorem 7, Sec. 2.7, it is obvious that if the angles of one pair of alternate interior angles are congruent, then the same thing is true of the other pair of alternate interior angles and of all pairs of corresponding angles. Moreover, in this case consecutive interior angles are supplementary.

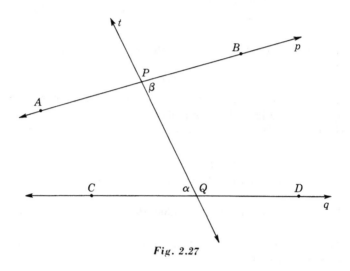

Fig. 2.27

A fundamental question which arises at the outset is "Do parallel lines exist?" That they do is a consequence of the following important theorem:

Theorem 1. If a pair of alternate interior angles formed by two coplanar lines and a transversal are congruent, the lines are parallel.

Proof. Let p and q be two coplanar lines, let t be a transversal of p and q meeting p and q in P and Q, respectively, and let a pair of alternate interior angles formed by these lines be congruent. Now if p and q are parallel, the assertion of the theorem is true, and there is nothing more to prove. Suppose, therefore, that p and q intersect, say in the point R (see Fig. 2.28). From the definition of a transversal, it follows that R is not a point of t. Now let S be a point on p on the opposite side of P from R. By the point-plotting theorem, there is a unique point, T, on the ray opposite to $\overrightarrow{QR}$ such that $\overline{TQ} \cong \overline{RP}$. Moreover, by hypothesis, $\angle TQP \cong \angle RPQ$. Therefore, since

$$\overline{QP} \cong \overline{PQ}$$

it follows by Postulate 16 that

$$\triangle QPT \cong \triangle PQR$$

Hence $\angle QPT \cong \angle PQR$, and therefore, since $\angle PQR \cong \angle QPS$, it follows that $\angle QPT \cong \angle QPS$. Since T and S are, from the definition of T, on the same side of $\overleftrightarrow{PQ}$, it follows from the angle-construction theorem

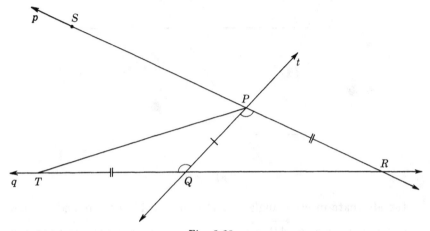

Fig. 2.28

that $\overrightarrow{PT} = \overrightarrow{PS}$. Hence T, P, and S are collinear, that is, T is a point of p. However, T and R are obviously distinct points since they lie on opposite sides of $\overleftrightarrow{PQ}$. Therefore, the distinct lines p and q have the distinct points T and R in common, which is impossible. Hence the assumption that p and q intersect must be abandoned. In other words, $p\|q$, as asserted.

An important special case of the last theorem is contained in the following corollary:

Corollary 1.   Two coplanar lines which are perpendicular to the same line are parallel.

The existence of parallel lines is now evident; more specifically, we have the following result:

Theorem 2.   In the plane determined by a given point and a given line not containing the point, there exists a line which passes through the given point and is parallel to the given line.

Proof.   Let $\overleftrightarrow{AB}$ be any line and let P be any point which is not on $\overleftrightarrow{AB}$.   Now by the angle-construction theorem, there exists a point, Q, on the opposite side of $\overleftrightarrow{AP}$ from B such that $m\angle APQ = m\angle PAB$ (see Fig. 2.29).   The lines $\overleftrightarrow{QP}$ and $\overleftrightarrow{AB}$ therefore have the property that

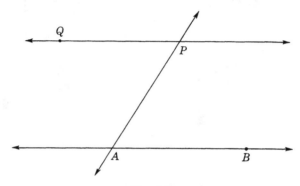

Fig. 2.29

the alternate interior angles, $\angle APQ$ and $\angle PAB$, determined by them and the transversal $\overleftrightarrow{AP}$, are congruent.   Hence, by the last theorem, $\overleftrightarrow{PQ} \parallel \overleftrightarrow{AB}$.   In other words, there exists at least one parallel to $\overleftrightarrow{AB}$ through P, as asserted.

It is now natural to ask, "How many lines are there which pass through P and are parallel to $\overleftrightarrow{AB}$?"   We know that there is at least one, but perhaps there are more.   If we could prove the converse of Theorem 2, namely, that if two lines are parallel, then the alternate interior angles determined by these lines and any transversal are necessarily congruent, the uniqueness of the parallel to $\overleftrightarrow{AB}$ through P would follow at once.   However, with the postulates we have at our disposal it is impossible to prove this; hence if we wish the parallel to a given line through a given point to be unique, we must postulate it:

Postulate 17   **(*The Parallel Postulate*).**   There is at most one line parallel to a given line through a given point not on that line.

Since Theorem 2 assures us that there is at least one parallel to a given line through a given point not on that line and since Postulate 17 tells us that there is at most one, it follows that the parallel is, in fact,

unique. We must remember, however, that instead of accepting
Postulate 17, we could have postulated the existence of, say, *two*
parallels to a given line through a given point not on that line. In
Chap. 4 we shall do just this, opening up for our study the interesting
noneuclidean geometry known as hyperbolic geometry.

With the parallel postulate available, it is now possible to prove the
converse of Theorem 2:

Theorem 3.   If two lines are parallel, the alternate interior angles
determined by these lines and any transversal are
congruent.

Proof.   Let p and q be parallel lines cut by a transversal, t, in points
P and Q, respectively. Let Q' be any point of q distinct from Q, and
let P' be any point of p on the opposite side of $\overleftrightarrow{PQ}$ from Q' (see Fig.
2.30). Now if $\angle Q'QP \cong \angle P'PQ$, the theorem is true and there is nothing more to prove. Hence suppose that $m\angle Q'QP \neq m\angle P'PQ$. Then

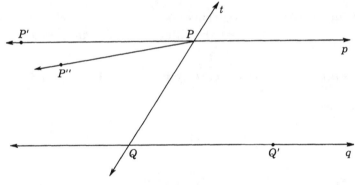

Fig. 2.30

by the angle-construction theorem, there is a unique ray, $\overrightarrow{PP''}$, extending into the halfplane determined by $\overleftrightarrow{PQ}$ and P' such that

$$m\angle P''PQ = m\angle Q'QP$$

Thus $\angle P''PQ \cong \angle Q'QP$ and by Theorem 1, $\overleftrightarrow{PP''} \| \overleftrightarrow{QQ'}$. But since
$m\angle P''PQ \neq m\angle P'PQ$, it follows that $\overleftrightarrow{PP''} \neq \overleftrightarrow{PP'}$. Hence $\overleftrightarrow{PP''}$ and
$\overleftrightarrow{PP'}$ are two distinct lines each parallel to $\overleftrightarrow{QQ'}$ through P. This contradicts the parallel postulate, however. The assumption that
$\angle Q'QP$ and $\angle P'PQ$ are not congruent must therefore be abandoned,
and the theorem is established.

Various other theorems now follow easily.   For instance,

Theorem 4.   A transversal which is perpendicular to one of two parallel lines is perpendicular to the other also.

Theorem 5.   A line which lies in the plane of two parallel lines and intersects one of the lines in a single point intersects the other in a single point also.

Theorem 6.   If each of two coplanar lines is parallel to a third line, they are parallel to each other.

The last theorem can be proved without the assumption that the two lines are coplanar,* but to do so requires a knowledge of theorems concerning perpendicular lines and planes which we have not yet discussed.

It is often convenient to extend the notion of parallelism to include segments and rays, as well as lines:

Definition 3.   Two segments are said to be parallel if and only if the lines which they determine are parallel.

Definition 4.   Two rays, $\overrightarrow{AB}$ and $\overrightarrow{CD}$, are said to be parallel if and only if

(1) The lines $\overleftrightarrow{AB}$ and $\overleftrightarrow{CD}$ are parallel.

(2) B and D lie on the same side of $\overleftrightarrow{AC}$.

Definition 5.   Two rays, $\overrightarrow{AB}$ and $\overrightarrow{CD}$, are said to be antiparallel if and only if

(1) The lines $\overleftrightarrow{AB}$ and $\overleftrightarrow{CD}$ are parallel.

(2) B and D lie on opposite sides of $\overleftrightarrow{AC}$.

With parallel segments defined, it is now possible to define the important type of quadrilateral known as a **parallelogram:**

Definition 6.   A quadrilateral each of whose sides is parallel to the side opposite it is called a parallelogram.

The following important theorems are now easy to prove:

* See Theorem 11, Sec. 2.15.

Theorem 7.   Opposite sides of any parallelogram are congruent.

Theorem 8.   If two sides of a quadrilateral are congruent and parallel, the quadrilateral is a parallelogram.

Theorem 9.   The diagonals of a parallelogram bisect each other.

Theorem 10.   If three or more parallel lines determine congruent segments on one transversal, they determine congruent segments on every transversal.

EXERCISES

1. Prove the existence of skew lines.
2. Prove that a parallelogram is a convex quadrilateral.
3. Prove that opposite angles of a parallelogram are congruent.
4. Prove the existence of a parallelogram having consecutive sides of prescribed length.
5. Prove Theorem 4.   6. Prove Theorem 5.
7. Prove Theorem 6.   8. Prove Theorem 7.
9. Prove Theorem 8.   10. Prove Theorem 9.
11. Prove Theorem 10.
12. Prove that two parallel lines are everywhere equidistant.
13. Prove that on either of two intersecting lines there are points whose perpendicular distance from the other line is arbitrarily large.
14. Prove that if two coplanar angles have their sides respectively parallel or respectively antiparallel, the angles are congruent. What relation, if any, exists between two coplanar angles which have two sides parallel and two sides antiparallel?

2.11 Further Properties of Triangles. In this section we shall investigate a number of important results involving triangles, including Pasch's theorem, the exterior-angle theorem, the triangle inequality, and the angle-sum theorem.

One of the simplest yet most important properties of triangles is contained in Theorem 1. Obvious as it is, it cannot be proved from the axioms of Euclid. The German mathematician Moritz Pasch (1843–1931) was the first to recognize this explicitly and to accept this result as a postulate. For this reason, it is commonly referred to as **Pasch's axiom.** Because of our choice of postulates, in particular because of the plane-separation postulate, we are able to prove Pasch's axiom; hence we shall refer to it hereafter as **Pasch's theorem.**

Theorem 1.   A line which lies in the plane of a triangle and contains an interior point of one side of the triangle but
does not pass through any vertex of the triangle contains an interior point of exactly one other side of the
triangle.

Proof.   Let $\triangle ABC$ be any triangle and let l be a line in the plane of
$\triangle ABC$ which intersects one side, say $\overline{AB}$, in an interior point, P, but
does not pass through A, B, or C.   By the plane-separation postulate,
l determines two halfplanes, H_1 and H_2, in the plane of $\triangle ABC$ (see
Fig. 2.31).   Moreover, since $\overline{AB}$ contains a point of l, namely P, in

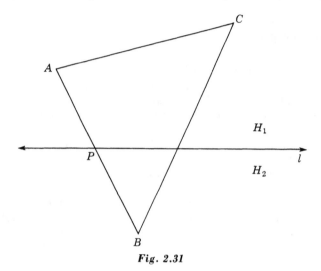

Fig. 2.31

its interior, A and B lie in opposite halfplanes, say A in H_1 and B in
H_2.   Now by hypothesis, C is not a point of l; hence it must lie either
in H_1 or in H_2.   If it lies in H_1, it lies on the same side of l as A and on
the opposite side of l from B.   Hence $\overline{AC}$ cannot intersect l but $\overline{BC}$
must.   On the other hand, if C lies in H_2, it lies on the same side of l as
B and on the opposite side of l from A.   Hence $\overline{BC}$ cannot intersect l
but $\overline{AC}$ must.   In either case, l intersects one and only one of the
remaining sides of $\triangle ABC$, as asserted.

The following result, which is an easy consequence of Pasch's
theorem, is also of considerable utility:

Corollary 1.   A line which passes through one of the vertices of a
triangle and contains an interior point of the triangle
contains an interior point of the side opposite the
vertex through which the line passes.

The angles of any triangle are often referred to explicitly as the **interior angles** of the triangle. At any vertex of a triangle there are two angles each of which forms with the interior angle at that vertex a linear pair. These are called the **exterior angles** at that vertex. The interior angle at any vertex is said to be **adjacent** to each of the exterior angles at that vertex. The interior angles which are not adjacent to an exterior angle are said to be **nonadjacent** or **opposite** to it. Thus in Fig. 2.32, $\angle ABC$ is the interior angle adjacent to each

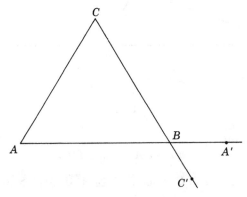

Fig. 2.32

of the exterior angles $\angle ABC'$ and $\angle CBA'$, and $\angle CAB$ and $\angle ACB$ are the interior angles nonadjacent to these exterior angles.

With the foregoing definitions, we can now state and prove the important result known as the **weak form of the exterior-angle theorem:**

Theorem 2. The measure of any exterior angle of a triangle is greater than the measure of either nonadjacent interior angle.

Proof. Given $\triangle ABC$ and any exterior angle, say $\angle CBD$, where D is a point on the ray opposite to $\overrightarrow{BA}$. Let M be the midpoint of $\overline{BC}$. Then by the point-plotting theorem, there is a point, A', on the ray opposite to $\overrightarrow{MA}$ such that $\overline{A'M} \cong \overline{MA}$. Thus A' lies on the same side of $\overleftrightarrow{BC}$ as D (see Fig. 2.33). Moreover, not only M but also A' lies on the same side of $\overleftrightarrow{AD}$ as C; for otherwise $\overline{MA'}$ would contain a point of $\overleftrightarrow{AD}$ distinct from A, and hence $\overleftrightarrow{AA'}$ and $\overleftrightarrow{AD}$ would have two distinct points in common. Thus A' lies not only on the D side of

$\overleftrightarrow{BC}$ but also on the C side of $\overleftrightarrow{AD}$, and therefore, by Theorem 1, Sec. 2.7, it lies in the interior of $\angle CBD$. Hence $\overrightarrow{BA'}$ is between $\overrightarrow{BD}$ and

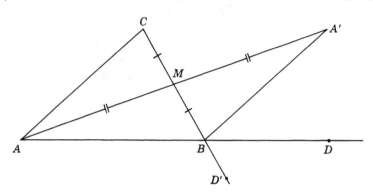

Fig. 2.33

$\overrightarrow{BC}$, and therefore

$$m\angle MBA' < m\angle MBD$$

Now $\triangle AMC \cong \triangle A'MB$, since $\overline{AM} \cong \overline{A'M}, \overline{MC} \cong \overline{MB}$, and

$$\angle AMC \cong \angle A'MB$$

Hence $$m\angle MCA = m\angle MBA'$$

and therefore, substituting into the last inequality,

$$m\angle MCA < m\angle MBD = m\angle CBD$$

To prove that $m\angle CAB < m\angle CBD$, we first repeat the preceding argument to show that $m\angle CAB < m\angle ABD'$ where D' is a general point on the ray opposite to $\overrightarrow{BC}$. Then, since $\angle ABD'$ and $\angle CBD$ are vertical angles, it follows that

$$m\angle ABD' = m\angle CBD$$

and finally, by substitution,

$$m\angle CAB < m\angle CBD$$

as asserted.

> *Corollary 1.* In any right triangle, the angles which are not right angles are acute.

The next four theorems are concerned with inequalities which exist among the measures of the parts of general triangles.

Theorem 3.   If the measures of two sides of a triangle are unequal, then the measures of the angles opposite these sides are unequal in the same order.

Proof.   Given $\triangle ABC$ with $AB > AC$, we are to prove that $m\angle ACB > m\angle ABC$. To do this, we observe that by the point-plotting theorem, there is a unique point between A and B such that $AD = AC$ (see Fig. 2.34). More-over, by Theorem 3, Sec. 2.7, D lies in the interior of $\angle ACB$ and hence $\overrightarrow{CD}$ is between $\overrightarrow{CA}$ and $\overrightarrow{CB}$. Therefore

$$m\angle ACB = m\angle ACD + m\angle DCB$$

and since $m\angle DCB > 0$,

$$m\angle ACB > m\angle ACD$$

Also, since $\triangle ACD$ is isosceles, it follows that $m\angle ACD = m\angle ADC$ and therefore, from the last inequality,

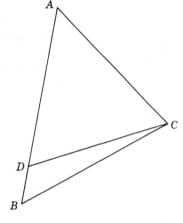

Fig. 2.34

(1) $m\angle ACB > m\angle ADC$

Moreover, from the last theorem, applied to $\triangle ACD$, we have

$$m\angle ADC > m\angle DBC = m\angle ABC$$

Hence, combining this inequality with (1),

$$m\angle ACB > m\angle ABC$$

as asserted.

Theorem 4.   If the measures of two angles of a triangle are unequal, the measures of the sides opposite these angles are unequal in the same order.

Proof.   In $\triangle ABC$ let $m\angle ACB > m\angle ABC$. Then there are three possibilities to consider:

$$AB < AC$$
$$AB = AC$$
$$AB > AC$$

The first of these is clearly impossible, for if $AB < AC$, then by the last theorem, $m\angle ACB < m\angle ABC$, contrary to hypothesis. Similarly, if $AB = AC$, then $\triangle ABC$ is isosceles, and by Theorem 2, Sec. 2.9,

$m\angle ACB = m\angle ABC$, which is also contrary to hypothesis. This leaves $AB > AC$ as the only possibility, as asserted.

Corollary 1.   The shortest segment joining a point to a line which does not contain the point is the segment which is perpendicular to the line.

Theorem 5.   If one side of a triangle is at least as long as either of the other two sides, then the foot of the perpendicular from the opposite vertex to the line determined by that side is an interior point of that side.

Proof.   In $\triangle ABC$ let $AB \geqq AC$ and $AB \geqq BC$ and let D be the foot of the perpendicular from C to $\overleftrightarrow{AB}$ (see Fig. 2.35).   Now by the

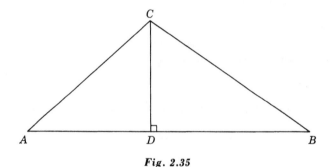

Fig. 2.35

corollary of the last theorem, the shortest segment from A to $\overleftrightarrow{CD}$ is the perpendicular segment $\overline{AD}$; hence

$$(2) \qquad\qquad\qquad AD < AC$$

Similarly, the shortest segment from B to $\overleftrightarrow{CD}$ is the perpendicular segment $\overline{BD}$; hence

(3)	$BD < BC$
But by hypothesis	$AC \leqq AB$
Hence, from (2)	$AD < AB$
Also, by hypothesis,	$BC \leqq AB$
Hence, from (3)	$BD < AB$

Thus for the three collinear points A, D, B, $\overline{AB}$ is longer than either $\overline{AD}$ or $\overline{BD}$.   Hence D must be between A and B, as asserted.

Theorem 6 (*The Triangle Inequality*). The sum of the lengths of any two sides of a triangle is greater than the length of the third side.

Proof. In $\triangle ABC$ let $\overline{AB}$ be a side which is at least as long as either of the other two sides. Then, by hypothesis,

$$AB \geqq AC \quad \text{and} \quad AB \geqq BC$$

Hence, adding the appropriate quantity to the respective left members, it follows immediately that

$$AB + BC > AC \quad \text{and} \quad AB + AC > BC$$

To establish the third inequality, namely $AC + CB > AB$, let D be the foot of the perpendicular from C to $\overleftrightarrow{AB}$. By the last theorem, D lies between A and B and hence

$$(4) \qquad\qquad AD + DB = AB$$

Also, by the corollary of Theorem 4,

$$AC > AD$$
and
$$CB > DB$$
Hence, adding, $\qquad AC + CB > AD + DB$
and finally, using (4),
$$AC + CB > AB$$

as asserted.

The triangle inequality is often taken as one of the postulates defining the concept of distance. However, with the postulates we have chosen, it can be proved as a theorem and need not be assumed.

It is important to note that the proofs of the last five theorems involved only Postulates 1 to 16 and theorems derived from these postulates and did not involve the parallel postulate, Postulate 17. Since Postulates 1 to 16 are all included among the postulates of hyperbolic geometry, it is clear that Theorems 1 to 5 hold true in hyperbolic geometry as well as in euclidean geometry. In particular, we shall make frequent use of the exterior-angle theorem in our work in hyperbolic geometry in Chap. 4.

The next theorem is one of the most famous in euclidean geometry. In fact, it can almost be said to characterize euclidean geometry, for if it is assumed, the parallel postulate can be proved.

Theorem 7 (*The Angle-sum Theorem*). The sum of the measures of the three angles of any triangle is 180.

Proof. Let $\triangle ABC$ be any triangle, let p be the line parallel to $\overleftrightarrow{AC}$ through the vertex B, and let D be a point on p such that D and C are on the same side of $\overleftrightarrow{AB}$ (see Fig. 2.36). Since $\overleftrightarrow{AC}\|p$, the segment $\overline{AC}$

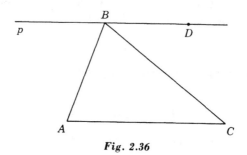

Fig. 2.36

cannot intersect p. Hence A and C are on the same side of p or $\overleftrightarrow{BD}$, and therefore, by Theorem 1, Sec. 2.7, $\overrightarrow{BC}$ is between $\overrightarrow{BA}$ and $\overrightarrow{BD}$. Thus, from the definition of betweenness for rays,

$$m\angle DBA = m\angle DBC + m\angle CBA$$

Furthermore, since $\overleftrightarrow{AC}\|p$, it follows from Theorem 3, Sec. 2.10, that

$$m\angle DBC = m\angle ACB$$

and therefore, substituting,

$$m\angle DBA = m\angle ACB + m\angle CBA$$

Also, since $\angle CAB$ and $\angle DBA$ are consecutive interior angles,

$$m\angle CAB + m\angle DBA = 180$$

Finally, eliminating $m\angle DBA$ between the last two equations,

$$m\angle CAB + m\angle ACB + m\angle CBA = 180$$

as asserted.

Using the last result, it is possible to prove the following stronger form of the exterior-angle theorem, which is valid only in euclidean geometry, however, since it depends on the parallel postulate:

Theorem 8. The measure of any exterior angle of a triangle is equal to the sum of the measures of the two non-adjacent interior angles.

EXERCISES

1. Prove that if $\angle C$ in $\triangle ABC$ is obtuse, then the foot of the perpendicular from A to $\overleftrightarrow{BC}$ is not a point of the segment $\overline{BC}$.

2. If three numbers, $a = BC$, $b = AC$, and $c = AB$ satisfy two of the three inequalities appearing in Theorem 6, is there necessarily a triangle having sides of length a, b, and c?

3. Assuming Pasch's theorem as a postulate, prove the plane-separation postulate as a theorem.

4. Prove Pasch's theorem from the following weaker assumption: "A line which lies in the plane of a triangle and contains an interior point of one side of the triangle but does not pass through any vertex of the triangle contains an interior point of *at least* one other side of the triangle."

5. Prove the following stronger form of Corollary 1, Theorem 1: "If X is a point in the interior of $\angle BAC$ then $\overrightarrow{AX}$ contains an interior point of $\overline{BC}$."

6. Without using the angle-sum theorem, prove the following theorem: "If $\triangle ABC$ and $\triangle A'B'C'$ are such that

$$\overline{AB} \cong \overline{A'B'}, \qquad \angle ABC \cong \angle A'B'C', \qquad \angle BCA \cong \angle B'C'A'$$

then $\qquad\qquad\qquad \triangle ABC \cong \triangle A'B'C'$

7. Discuss the following proof of the angle-sum theorem:

Let the sum of the measures of the angles of an arbitrary triangle be x.

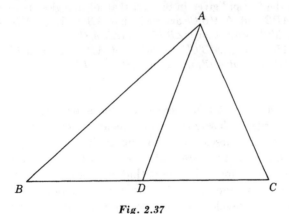

Fig. 2.37

Then if $\triangle ABC$ is any triangle and if D is an arbitrary point on one of its sides, say $\overline{BC}$ (Fig. 2.37), we have, by hypothesis,

$$m\angle ABD + m\angle BDA + m\angle DAB = x$$
$$m\angle ACD + \angle CDA + m\angle DAC = x$$

and $\qquad\quad m\angle ABC + m\angle BCA + m\angle CAB = x$

Moreover $\qquad\qquad\quad m\angle DAB + m\angle DAC = m\angle BAC$

and $\qquad\qquad\qquad\quad m\angle BDA + m\angle ADC = 180$

Hence, adding the first two equations and using the last three equations to simplify the result,

$$2x = (m\angle ABD + m\angle BDA + m\angle DAB)$$
$$+ (m\angle ACD + m\angle CDA + m\angle DAC)$$
$$= m\angle ABD + (m\angle BDA + m\angle CDA)$$
$$+ (m\angle DAB + m\angle DAC) + m\angle ACD$$
$$= m\angle ABD + 180 + m\angle BAC + m\angle ACD$$
$$= x + 180$$

Therefore $x = 180$, as asserted.

Does this proof depend on the parallel postulate? Does this proof depend on an assumption equivalent to the parallel postulate? If so, what is the assumption?

8. If $\triangle ABC$ is an arbitrary triangle, prove that the bisectors of $\angle A$ and $\angle B$ intersect in a point in the interior of the triangle. Then prove that the bisectors of the angles of a triangle are concurrent.

9. If $\triangle ABC$ is an arbitrary triangle and if A' and A'' are, respectively, points distinct from B and C on the rays opposite to $\overrightarrow{BA}$ and $\overrightarrow{CA}$, prove that the bisectors of $\angle CBA'$ and $\angle BCA''$ intersect in the interior of $\angle A$. Hence prove that the bisectors of any two exterior angles of a triangle and the bisector of the interior angle at the remaining vertex are concurrent.

10. If the bisectors of an angle of a triangle and the perpendicular bisector of the opposite side do not coincide, prove that their intersection is always a point in the exterior of the triangle. (This fact is required to overthrow the "proof" given in Sec. 2.2 that all triangles are isosceles.)

11. If $\triangle ABC$ and $\triangle A'B'C'$ are such that $AB = AC = A'B' = A'C'$ and $BC < B'C'$, prove that $m\angle BAC < m\angle B'A'C'$.

12. If $\triangle ABC$ and $\triangle A'B'C'$ are such that $AB = AC = A'B' = A'C'$ and $m\angle BAC < m\angle B'A'C'$, prove that $BC < B'C'$.

2.12 Similarity. As the next definition shows, the geometric concept of similarity is closely related to the arithmetic concept of proportionality. To the Greeks, who had no arithmetic theory of irrational numbers and who therefore had to discuss proportionality in essentially geometric terms, this meant that similarity was one of the more difficult topics in geometry. For us, who have borrowed freely from arithmetic in our development of geometry, this is not the case, though of course it would be if we were to include in our work the proofs of all the arithmetic facts we use.

Definition 1.   A one-to-one correspondence between the vertices of two convex polygons or between the vertices of a single convex polygon and themselves such that corresponding angles are congruent and the lengths of corresponding sides are proportional is called a similarity, and the polygons are said to be similar.

The fact that a polygon with vertices $A_1A_2, \ldots, A_n$ is similar to a polygon with vertices $B_1, B_2, \ldots, B_n$ is indicated by writing

Polygon $A_1A_2 \cdots A_n \sim$ polygon $B_1B_2 \cdots B_n$

Clearly, a congruence is simply a similarity for which the constant of proportionality is 1. In particular, every polygon is similar to itself under the identity correspondence with proportionality constant 1.

The following theorems follow at once from the arithmetic properties of proportion:

Theorem 1. If one polygon is similar to a second under a similarity with proportionality constant k, then the second polygon is similar to the first with proportionality constant $1/k$.

Theorem 2. If one polygon is similar to a second with proportionality constant k_1 and if the second is similar to a third with proportionality constant k_2, then the first is similar to the third with proportionality constant k_1k_2.

Corollary 1. Polygons which are similar to the same polygon are similar to each other.

Our development of the concept of similarity is based upon the following theorem:

Theorem 3. If a line parallel to one side of a triangle intersects the other two sides in points which are interior points of those sides, then the lengths of the segments into which one side is divided are proportional to the lengths of the corresponding segments of the other side.

Proof. Let $\triangle ABC$ be an arbitrary triangle, let P be an arbitrary interior point of one side, say $\overline{AB}$, and let l be the line which is parallel to $\overleftrightarrow{BC}$ and contains P. By Pasch's theorem, since l cannot intersect $\overline{BC}$, it must intersect $\overline{AC}$, say at the point Q. To prove the theorem, we must show that the ratio of the lengths of the segments into which $\overline{AB}$ is divided by P is equal to the ratio of the lengths of the segments into which $\overline{AC}$ is divided by Q, that is, we must show that

$$\frac{AP}{PB} = \frac{AQ}{QC}$$

To do this, we observe first that by the point-plotting theorem (see Fig. 2.38), we can determine $n - 1$ points, $P_1, P_2, \ldots, P_{n-1}$, between A and P such that

$$AP_1 = P_1P_2 = \cdots = P_{n-1}P = \frac{AP}{n} = \varepsilon$$

Moreover, if m is the quotient and r is the remainder when PB is

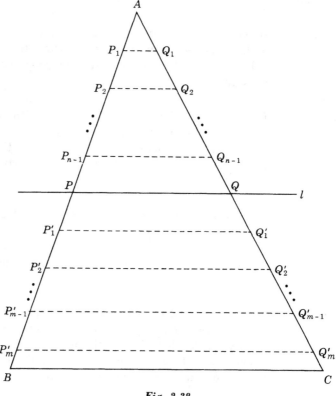

Fig. 2.38

divided by ε, we can determine m points, $P_1', P_2', \ldots, P_m'$, between P and B such that

$$PP_1' = P_1'P_2' = \cdots = P_{m-1}'P_m' = \varepsilon, \qquad P_m'B = r \leqq \varepsilon$$

Now the parallels to $\overleftrightarrow{BC}$ through $P_1, P_2, \ldots, P_{n-1}$ each intersect $\overline{AQ}$, say in the points $Q_1, Q_2, \ldots, Q_{n-1}$, and the parallels to $\overleftrightarrow{BC}$ through $P_1', P_2', \ldots, P_m'$ each intersect $\overline{QC}$, say in $Q_1', Q_2', \ldots, Q_m'$. Moreover, according to Theorem 10, Sec. 2.10,

$$AQ_1 = Q_1Q_2 = \cdots = Q_{n-1}Q = \frac{AQ}{n} = \varepsilon'$$

and $\quad QQ_1' = Q_1'Q_2' = \cdots = Q_{m-1}'Q_m' = \varepsilon', \qquad Q_m'C = r' \leqq \varepsilon'$

Thus we have

$$AP = n\varepsilon, \qquad PB = m\varepsilon + r$$

and $\qquad\qquad AQ = n\varepsilon', \qquad QC = m\varepsilon' + r'$

Hence $\qquad\qquad \dfrac{PB}{AP} = \dfrac{m\varepsilon + r}{n\varepsilon} = \dfrac{m}{n} + \dfrac{r}{n\varepsilon} = \dfrac{m}{n} + \dfrac{r}{AP}$

and $\qquad\qquad \dfrac{QC}{AQ} = \dfrac{m\varepsilon' + r'}{n\varepsilon'} = \dfrac{m}{n} + \dfrac{r'}{n\varepsilon'} = \dfrac{m}{n} + \dfrac{r'}{AQ}$

Now the absolute value of the difference between the two ratios PB/AP and QC/AQ is

$$\left|\frac{PB}{AP} - \frac{QC}{AQ}\right| = \left|\left(\frac{m}{n} + \frac{r}{AP}\right) - \left(\frac{m}{n} + \frac{r'}{AQ}\right)\right| = \left|\frac{r}{AP} - \frac{r'}{AQ}\right|$$

Furthermore, $\quad r \leqq \varepsilon = \dfrac{AP}{n} \qquad$ and $\qquad r' \leqq \varepsilon' = \dfrac{AQ}{n}$

that is, $\qquad\qquad \dfrac{r}{AP} \leq \dfrac{1}{n} \qquad$ and $\qquad \dfrac{r'}{AQ} \leq \dfrac{1}{n}$

Thus, overestimating, we have for all positive integers, n,

$$\left|\frac{PB}{AP} - \frac{QC}{AQ}\right| = \left|\frac{r}{AP} - \frac{r'}{AQ}\right| < \frac{r}{AP} + \frac{r'}{AQ} \leqq \frac{2}{n}$$

However, the difference $|PB/AP - (QC/AQ)|$ is obviously independent of n. Hence the only way it can be less than $2/n$ for all positive integers, n, is for it to be zero. Thus

$$\frac{PB}{AP} = \frac{QC}{AQ}$$

as asserted.

The proof of the following theorem is an easy application of Theorem 3.

Theorem 4. The lengths of corresponding segments determined on two transversals by three parallel lines are proportional.

It is now appropriate to ask if similar, noncongruent triangles actually exist. The question, of course, is not trivial for, as we shall see in Chap. 4, similar triangles do *not* exist in hyperbolic geometry, except in the special case when the constant of proportionality is 1 and similarity is, in fact, just congruence. That there are triangles in euclidean geometry which are similar but not congruent is suggested by our past experience in geometry and guaranteed by the following theorem:

Theorem 5.   If a triangle and a positive number, k, are given, there exists a triangle which is similar to the given triangle with proportionality constant k.

Proof.   Let $\triangle ABC$ be the given triangle. If $k = 1$, $\triangle ABC$ is similar to itself, and the assertion of the theorem is verified. Suppose next that $k < 1$. Then by the point-plotting theorem, there is a point, D, between A and B such that $AD = k \cdot AB$ (see Fig. 2.39a).

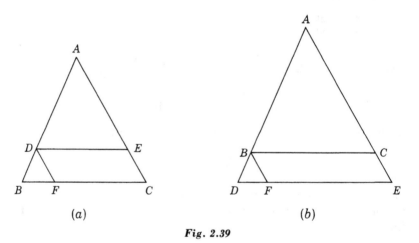

(a)   (b)

Fig. 2.39

Now the line through D parallel to $\overleftrightarrow{BC}$ can neither intersect $\overline{BC}$ nor pass through A. Hence, by Pasch's theorem, it must intersect $\overline{AC}$ at a point, E, between A and C. Since $\overleftrightarrow{DE} \parallel \overleftrightarrow{BC}$, it follows that

$$\angle ADE \cong \angle ABC \qquad \text{and} \qquad \angle AED \cong \angle ACB$$

and, of course, $\angle DAE \cong \angle BAC$. Also, since $AD = k \cdot AB$, it follows, by Theorem 3, that $AE = k \cdot AC$. To prove that $\triangle ADE \sim \triangle ABC$, it therefore remains only to prove that

$$DE = k \cdot BC$$

Now the line through D parallel to $\overleftrightarrow{AC}$ must intersect $\overline{BC}$ in some point,

F.   Hence, by Theorem 3, since $AD = k \cdot AB$, it follows that

$$FC = k \cdot BC$$

Moreover, since $DECF$ is a parallelogram, Theorem 7, Sec. 2.10 assures us that $FC = DE$.   Therefore, substituting,

$$DE = k \cdot BC$$

which completes the proof of the theorem when $K < 1$.

If $k > 1$, there is a point, D, on $\overleftrightarrow{AB}$ such that B is between A and D and $AD = k \cdot AB$.   The parallel to $\overleftrightarrow{BC}$ through D must now intersect $\overrightarrow{AC}$ at a point, E, such that C is between A and E.   As before, $\angle BAC \cong \angle DAE$, $\angle ABC \cong \angle ADE$, $\angle ACB \cong \angle AED$, and by Theorem 3, $AE = k \cdot AC$, since $AD = k \cdot AB$.   Therefore, to prove that $\triangle ADE \sim \triangle ABC$, it remains to prove that

$$DE = k \cdot BC$$

Now the line through B parallel to $\overleftrightarrow{AC}$ must intersect $\overline{DE}$ in some point F.   Hence, by Theorem 3, since $AD = k \cdot AB$, it follows that

$$DE = k \cdot FE$$

Finally, since $BCEF$ is a parallelogram, it follows that

$$FE = BC$$

Therefore

$$DE = k \cdot BC$$

which completes the proof that when $k > 1$, $\triangle ADE \sim \triangle ABC$ with proportionality constant k.   Thus we have shown that for every positive number, k, there exists a triangle similar to any given triangle with proportionality constant k.

In order for two triangles to be similar, the definition of similarity requires that corresponding angles be congruent and corresponding sides have proportional lengths.   However, certain subsets of these six conditions are sufficient to establish similarity, as the following theorems show:

Theorem 6.   A one-to-one correspondence between two triangles such that the lengths of corresponding sides are proportional is a similarity.

Proof.   Let $A \leftrightarrow A'$, $B \leftrightarrow B'$, $C \leftrightarrow C'$ be a correspondence between $\triangle ABC$ and $\triangle A'B'C'$ such that

(1) $AB = k \cdot A'B'$, $BC = k \cdot B'C'$, $AC = k \cdot A'C'$

By the last theorem, there exists a triangle, $\triangle A''B''C''$, similar to

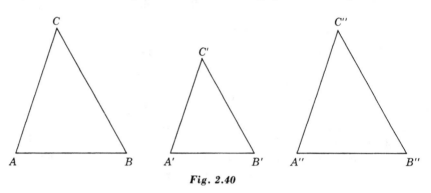

Fig. 2.40

$\triangle A'B'C'$ with proportionality constant k (see Fig. 2.40), which means that

(2) $A''B'' = k \cdot A'B'$, $B''C'' = k \cdot B'C'$, $A''C'' = k \cdot A'C'$

and

(3) $\angle A''B''C'' \cong \angle A'B'C'$, $\angle B''C''A'' \cong \angle B'C'A'$
$\angle C''A''B'' \cong \angle C'A'B'$

Therefore, from (1) and (2),

$AB = A''B''$, $BC = B''C''$, $AC = A''C''$

and hence, by Theorem 4, Sec. 2.9, $\triangle ABC \cong \triangle A''B''C''$.   Thus

$\angle ABC \cong \angle A''B''C''$, $\angle BCA \cong \angle B''C''A''$, $\angle CAB \cong \angle C''A''B''$

and, using (3),

$\angle ABC \cong \angle A'B'C'$, $\angle BCA \cong \angle B'C'A'$, $\angle CAB \cong \angle C'A'B'$

These, with (1), complete the proof that $\triangle ABC \sim \triangle A'B'C'$, as asserted.

Theorem 7.   A one-to-one correspondence between two triangles such that the lengths of two sides of one triangle are proportional to the lengths of the corresponding sides of the other and such that the angles determined by these pairs of sides are congruent is a similarity.

Theorem 8. A one-to-one correspondence between two triangles such that two angles of one triangle are congruent to the corresponding angles of the other triangle is a similarity.

Of the many theorems involving the notion of similarity, the following are among the most important:

Theorem 9. In any right triangle, the altitude to the hypotenuse forms with the respective legs and the corresponding segments of the hypotenuse two triangles which are similar to each other and to the original triangle.

Proof. Let $\triangle ABC$ be a right triangle with right angle at C and let D be the foot of the perpendicular to $\overleftrightarrow{AB}$ from the vertex of the right angle, C (see Fig. 2.41). By Theorem 5, Sec. 2.11, D lies between A and B. To prove the theorem, we must prove, in part, that the correspondence

$$A \leftrightarrow A, \qquad C \leftrightarrow B, \qquad D \leftrightarrow C$$

is a similarity between $\triangle ACD$ and $\triangle ABC$ and that the correspondence

$$C \leftrightarrow A, \qquad B \leftrightarrow B, \qquad D \leftrightarrow C$$

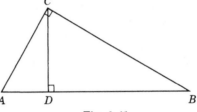

Fig. 2.41

is a similarity between $\triangle CBD$ and $\triangle ABC$. To prove the first of these, we observe that $\angle CAD \cong \angle BAC$, since they are the same angle, and that $\angle ADC \cong \angle ACB$, since each is a right angle. Hence, by Theorem 8,

$$\triangle ACD \sim \triangle ABC$$

To prove the second assertion, we observe likewise that

$$\angle CBD \cong \angle ABC \qquad \text{and} \qquad \angle CDB \cong \angle ACB$$

Hence, again by Theorem 8,

$$\triangle CBD \sim \triangle ABC$$

Finally, since $\triangle ACD$ and $\triangle CBD$ are both similar to $\triangle ABC$, they are, by Corollary 1, Theorem 2, similar to each other, as asserted.

Theorem 10. If one triangle is similar to another triangle with proportionality constant k, then the altitudes of the

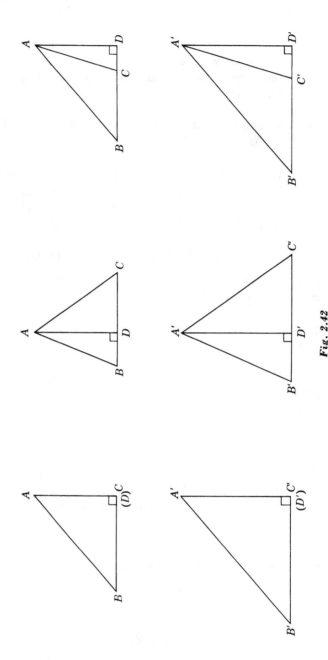

Fig. 2.42

116

first triangle are proportional to the corresponding altitudes of the second triangle with proportionality constant k.

Proof. Let $A \leftrightarrow A'$, $B \leftrightarrow B'$, $C \leftrightarrow C'$ be a similarity between $\triangle ABC$ and $\triangle A'B'C'$ with proportionality constant k. Then, by definition,

$$AB = k \cdot A'B', \qquad BC = k \cdot B'C', \qquad AC = k \cdot A'C'$$

Let $\overline{AD}$ and $\overline{A'D'}$ be the respective altitudes from A and A' to the lines determined by the opposite sides, $\overline{BC}$ and $\overline{B'C'}$ (see Fig. 2.42). If $\angle ACB$ and $\angle A'C'B'$ are right angles, then D coincides with C, D' coincides with C',·

$$AD = AC, \qquad A'D' = A'C'$$

and, since $AC = k \cdot A'C'$, we have

$$AD = k \cdot A'C' = k \cdot A'D'$$

as asserted. On the other hand, if $\angle ACB$ and $\angle A'C'B'$ are not right angles, then D does not coincide with C, D' does not coincide with C', and we can therefore consider $\triangle ACD$ and $\triangle A'C'D'$. In these triangles $\angle ADC \cong \angle A'D'C'$, since each is a right angle, and

$$\angle ACD \cong \angle A'C'D'$$

since these are either corresponding angles of the original similar triangles (if $\angle ACB$ and $\angle A'C'B'$ are acute) or else supplements of these corresponding angles (if $\angle ACB$ and $\angle A'C'B'$ are obtuse). Hence, by Theorem 8,

$$\triangle ACD \sim \triangle A'C'D'$$

Moreover, since $AC = k \cdot A'C'$ by hypothesis, k is the proportionality constant of this similarity, and hence

$$AD = k \cdot A'D'$$

as asserted.

EXERCISES

1. If B is a point on the side $\overrightarrow{VA}$ of the angle $\angle AVC$ such that A is between V and B, prove that the line which contains B and is parallel to $\overleftrightarrow{AC}$ intersects $\overrightarrow{VC}$ at a point, D, such that C is between V and D.
2. Prove that the perimeters of similar triangles are proportional to the lengths of any pair of corresponding sides. Is this result true for similar polygons in general?
3. Prove that the perimeters of similar triangles are proportional to the lengths of any pair of corresponding altitudes.

4. Prove Theorem 1.
5. Prove Theorem 2.
6. Prove Theorem 4.
7. Prove Theorem 7.
8. Prove Theorem 8.
9. What additional conditions, if any, must be fulfilled in order that a similarity may exist between two convex quadrilaterals whose corresponding angles are congruent? between two convex n-gons whose corresponding angles are congruent?
10. What additional conditions, if any, must be fulfilled in order that a similarity may exist between two convex quadrilaterals whose corresponding sides are proportional? between two convex n-gons whose corresponding sides are proportional?

2.13   The Pythagorean Theorem.   In this section we shall consider briefly what is probably the most famous, and perhaps the most important, theorem in mathematics, the **Theorem of Pythagoras** (580?–500? B.C.), for which literally scores of different proofs have been given.

> *Theorem 1.*   In any right triangle, the square of the length of the hypotenuse is equal to the sum of the squares of the lengths of the other two sides.

Proof.   Let $\triangle ABC$ be a right triangle with right angle at C, let the lengths of $\overline{BC}$, $\overline{CA}$, and $\overline{AB}$ be a, b, and c, respectively, and let D be the foot of the altitude from C to the hypotenuse (see Fig. 2.43).

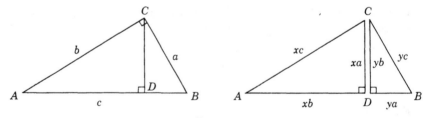

Fig. 2.43

Then by Theorem 9, Sec. 2.12,

$$\triangle ACD \sim \triangle ABC \qquad \text{and} \qquad \triangle CBD \sim \triangle ABC$$

Hence there exist proportionality constants, x and y, such that

$$
\begin{aligned}
b = AC &= x \cdot AB = x \cdot c & a = CB &= y \cdot AB = y \cdot c \\
CD &= x \cdot BC = x \cdot a & BD &= y \cdot BC = y \cdot a \\
AD &= x \cdot AC = x \cdot b & CD &= y \cdot AC = y \cdot b
\end{aligned}
$$

Now from the first equation in each of these sets we have, respectively,

$$x = \frac{b}{c} \qquad y = \frac{a}{c}$$

Moreover, since D is between A and B, we have

$$AD + DB = AB$$
or
$$x \cdot b + y \cdot a = c$$

Hence, substituting for x and y,

$$\frac{b}{c} b + \frac{a}{c} a = c$$
or finally
$$b^2 + a^2 = c^2$$

as asserted.

The converse of the theorem of Pythagoras is also an important result:

Theorem 2.   If the square of the length of one side of a triangle is equal to the sum of the squares of the lengths of the other two sides, the triangle is a right triangle with the right angle opposite the first, or longest, side.

Proof.   In $\triangle ABC$ let the lengths of $\overline{BC}$, $\overline{AC}$, and $\overline{AB}$ be a, b, and c, respectively, with $a^2 + b^2 = c^2$. Now consider two perpendicular rays, $\overrightarrow{DG}$ and $\overrightarrow{DH}$. On $\overrightarrow{DG}$ there is a point, E, such that

$$DE = CA = b$$

and on $\overrightarrow{DH}$ there is a point, F, such that $DF = CB = a$ (see Fig. 2.44).

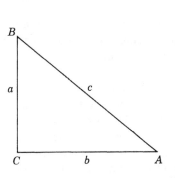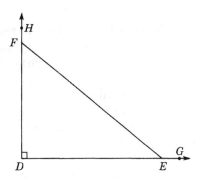

Fig. 2.44

Then since $\triangle DEF$ is a right triangle, it follows from the theorem of Pythagoras that

$$a^2 + b^2 = (EF)^2$$

But we are given that $a^2 + b^2 = c^2$. Therefore $c^2 = (EF)^2$, or since c and EF are both positive,

$$c = EF$$

Hence, by Theorem 4, Sec. 2.9,

$$\triangle ABC \cong \triangle EFD$$

Thus
$$\angle BCA \cong \angle FDE$$

In other words, $\angle BCA$ is a right angle and $\triangle ABC$ is a right triangle, as asserted.

EXERCISE

1. Look up and discuss critically another proof of the Pythagorean theorem and its converse.

2.14   The Measurement of Area. With every triangle there is associated a set of points, called a **triangular region,** consisting of the points of the triangle together with the points which comprise its interior. The triangle itself is said to be the **boundary** of the region. The interior of the triangle is called the **interior** of the region. Extending this obvious definition, it would seem natural to say that a polygonal region is the union of any polygon and its interior. However, we have not defined the interior of a general polygon, only the interior of a convex polygon. Hence we adopt the following definition:

> **Definition 1.** A polygonal region is the union of a finite number of coplanar triangular regions, no two of which have any interior points in common.

The union of all segments which lie in the boundary of one of the component triangular regions and have at most a finite number of points in common with any other triangular region is called the **boundary** of the polygonal region. The set of all points of a polygonal region which are not boundary points is called the **interior** of the region. As Fig. 2.45a shows, the set of triangular regions forming a particular polygonal region is not unique. Moreover, as Fig. 2.45b shows, although the boundary of a polygonal region is, by definition, the union of certain segments, the boundary may not be a polygon. When the boundary of a polygonal region is a polygon, the word

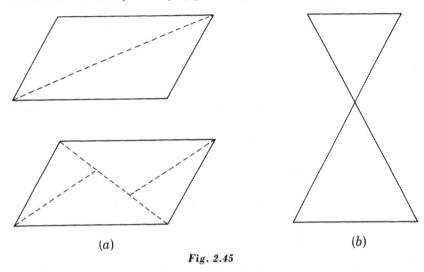

(a) (b)

Fig. 2.45

"polygon" is often used to denote not only the boundary but also the region itself. This makes for a certain economy of expression, but it tends to obscure the distinction between the two concepts and should be avoided if there is any possibility of confusion.

Our fundamental concern with polygonal regions is their measurement, that is, the assignment to them of numbers which will have the ordinary properties of area. To enable us to do this, we adopt four more postulates. Postulate 18 merely asserts that in some way, a unique measure can be assigned to each polygonal region. Postulate 19 goes further and asserts that if two polygonal regions do not have a polygonal region in common, then the measure of their union, i.e., their sum, is the sum of their individual measures. Postulate 20 guarantees that regions whose boundaries are congruent triangles all have the same measure. Finally, Postulate 21 describes how to assign measures by asserting the truth of the familiar formula for the area of a rectangular region.

> **Postulate 18.** If S is any square region, there exists a correspondence which associates with every polygonal region a unique positive number such that the number assigned to the given square, S, is 1.

The square S to which the number 1 is assigned by the correspondence described in Postulate 18 is called the **unit square**. The number assigned to any polygonal region by the correspondence is called the **area** of the region relative to the unit square S.

Postulate 19.   If a polygonal region, R, is the union of two polygonal regions, R_1 and R_2, whose intersection consists at most of a finite number of segments and isolated points, then relative to any unit square the area of R is the sum of the areas of R_1 and R_2.

Postulate 20.   If two triangles are congruent, then relative to any unit square the corresponding triangular regions have the same area.

Postulate 21.   If S is a square region whose sides are of length 1 in terms of some unit pair, α, then the area of any rectangular region, R, relative to S is the product of the lengths of two consecutive sides of R measured in terms of α.

It is now an easy matter to derive the familiar formula for the area of a triangular region. Preparatory to this, however, it is convenient to obtain the formula for the area of the region bounded by a right triangle:

Theorem 1.   The area of the region bounded by a right triangle is one half the product of the lengths of its legs.

Proof. Let $\triangle QRS$ be a right triangle with right angle at R. Let q and s be, respectively, the lengths of $\overline{RS}$ and $\overline{QR}$, and let A be the area of $\triangle QRS$. Now, by Exercise 4, Sec. 2.10, there exists a rectangle,

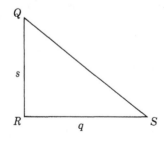
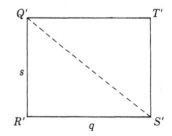

Fig. 2.46

$Q'R'S'T'$, such that $Q'R' = s$ and $R'S' = q$ (see Fig. 2.46). Moreover, the diagonal $\overline{Q'S'}$ forms with the sides of the rectangle two triangles, $\triangle Q'R'S'$ and $\triangle S'T'Q'$, each of which is congruent to $\triangle QRS$

and hence, by Postulate 20, of area A. Furthermore, R' and T' lie on opposite sides of $\overleftrightarrow{Q'S'}$. Hence the triangular regions determined by $\triangle Q'R'S'$ and $\triangle S'T'Q'$, respectively, have only the segment $\overline{Q'S'}$ in common. Therefore, by Postulate 19, the area of the rectangular region $Q'R'S'T'$ is equal to the sum of the areas of the triangular regions $Q'R'S'$ and $S'T'Q'$. Hence, using Postulate 21,

$$qs = A + A$$
or
$$A = \tfrac{1}{2}qs$$

as asserted.

Theorem 2. The area of the region bounded by any triangle is the product of the lengths of any side of the triangle and the length of the altitude to that side.

Proof. Let $\triangle BCD$ be any triangle; let h be the length of the altitude, $\overline{BF}$, from any vertex, B, to the line determined by the opposite side, $\overline{CD}$; let b be the length of the side $\overline{CD}$; and let A be the area of the triangle (see Fig. 2.47). If F coincides with C or D, then $\triangle BCD$

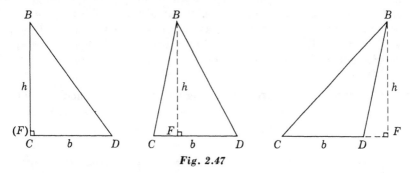

Fig. 2.47

is a right triangle, the altitude, $\overline{BF}$, is the vertical leg, and by Theorem 1,

$$A = \tfrac{1}{2}hb$$

If F falls between C and D, the altitude, $\overline{BF}$, is the common side of two right triangles, $\triangle BCF$ and $\triangle BFD$, which define regions whose areas, by Theorem 1, are respectively

$$\tfrac{1}{2}h(CF) \quad \text{and} \quad \tfrac{1}{2}h(FD)$$

Since these triangular regions have only the segment $\overline{BF}$ in common, the sum of their areas is, by Postulate 19, the area of the triangular region BCD; that is,

$$A = \tfrac{1}{2}h(CF) + \tfrac{1}{2}h(FD) = \tfrac{1}{2}h(CF + FD) = \tfrac{1}{2}h(CD) = \tfrac{1}{2}hb$$

Finally, if F is not a point of the segment $\overline{CD}$, say F is a point such that D is between C and F, then the triangular regions BCD and BDF have only $\overline{BD}$ in common. Hence, by Postulate 19,

Area of region BCD + area of region BDF = area of region BCF

or, using Theorem 1,

$$\begin{aligned}
A + \tfrac{1}{2}h(DF) &= \tfrac{1}{2}h(CF) \\
A &= \tfrac{1}{2}h(CF - DF) \\
&= \tfrac{1}{2}h(CD) \\
&= \tfrac{1}{2}hb
\end{aligned}$$

Thus, in all cases the area of $\triangle BCD$ is given by the formula

$$A = \tfrac{1}{2}hb$$

as asserted.

Once Theorem 2 is established, it is a simple matter to obtain the usual formulas for the areas of regions bounded by such familiar polygons as parallelograms, trapezoids, or any of the regular polygons, and we shall leave such derivations as exercises. As a final result in this section we shall prove the following important theorem on the areas of triangular regions whose boundaries are similar triangles:

Theorem 3. The areas of triangular regions bounded by similar triangles are proportional to the squares of the lengths of any pair of corresponding sides.

Proof. Let $B \leftrightarrow B'$, $C \leftrightarrow C'$, $D \leftrightarrow D'$ be a similarity between $\triangle BCD$ and $\triangle B'C'D'$ with proportionality constant k. Let the lengths of the sides of the two triangles be

$$\begin{aligned}
BC &= d & B'C' &= d' \\
CD &= b & C'D' &= b' \\
BD &= c & B'D' &= c'
\end{aligned}$$

Let the lengths of the altitudes from B and B' in the respective triangles be h and h', and let A and A' be, respectively, the areas of the triangles. Then, from the definition of a similarity,

$$b = kb', \qquad c = kc', \qquad d = kd'$$

or

$$k = \frac{b}{b'} = \frac{c}{c'} = \frac{d}{d'}$$

and, by Theorem 10, Sec. 2.12,

$$h = kh'$$

Also, by Theorem 2,

$$A = \tfrac{1}{2}hb \quad \text{and} \quad A' = \tfrac{1}{2}h'b'$$

Therefore

$$A = \tfrac{1}{2}(kh')(kb') = k^2(\tfrac{1}{2}h'b') = k^2 A' \quad \text{or} \quad \frac{A}{A'} = k^2$$

Hence, substituting for k^2,

$$\frac{A}{A'} = \frac{b^2}{b'^2} = \frac{c^2}{c'^2} = \frac{d^2}{d'^2}$$

as asserted.

EXERCISES

1. Prove that every convex polygon is the boundary of a polygonal region.
2. Obtain a formula for the area of a trapezoid.
3. Prove that two congruent convex quadrilaterals have the same area.
4. Since the set of triangular regions forming a given polygonal region is not unique, it is conceivable that considering the polygonal region as the union of different triangular regions might lead to different values for its area. Outline a proof of the fact that the area of a given polygonal region is independent of the particular set of triangular regions into which it is subdivided.

2.15 Lines and Planes in Space. The familiar properties of lines and planes in space can all be established by essentially traditional proofs based on the postulates we have already assumed. However, since a knowledge of these results and of the arguments which support them will be essential in our study of the geometry of four dimensions in the next chapter, we shall review them in some detail. Our discussion begins with the following definition:

> **Definition 1.** A line and a plane not containing the line are said to be perpendicular to each other if they intersect and if every line which passes through their intersection and lies in the given plane is perpendicular to the given line.

At the outset it is not clear that there exists even one line and one plane which are perpendicular. It is true, of course, that at any point, P, on a line, l, there are many different lines each perpendicular to l. In fact, in each plane containing l there is a unique line which passes through P and is perpendicular to l (see Fig. 2.48). However,

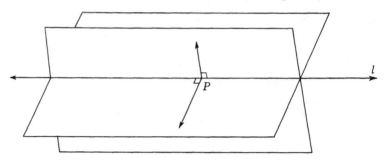

Fig. 2.48

until we prove it, we do not know that these perpendiculars all lie in the same plane nor, even if they do, are we certain that every line through P in this plane is perpendicular to l. Hence the following theorems are of fundamental importance:

> **Theorem 1.** If a line is perpendicular to each of two intersecting lines at their point of intersection, it is perpendicular to every line which passes through their intersection and lies in the plane they determine.

Proof. Let l_1, l_2, and λ be three lines such that l_1 and l_2 are each perpendicular to λ at the point P, and let π be the plane determined by l_1 and l_2. To prove the theorem, we must show that if l_3 is any line which lies in π and passes through P, then l_3 is perpendicular to λ. This is true by hypothesis if l_3 is either l_1 or l_2. Hence we may suppose that l_3 is distinct from either l_1 or l_2. Then on l_1, l_2, and l_3 there are rays, r_1, r_2, and r_3, respectively, with the common endpoint P, such that r_3 lies between r_1 and r_2. Let B_1 and B_2 be points on r_1 and r_2, respectively, such that

$$PB_1 = PB_2$$

and let B_3 be the intersection of $\overline{B_1B_2}$ and r_3 (see Fig. 2.49). Finally, let A and A' be points on λ on opposite sides of P such that

$$PA = PA'$$

Then

1. $\angle APB_1 \cong \angle APB_2$ 　　　　 1. All are right angles.
　 $\cong \angle A'PB_1 \cong \angle A'PB_2$.
2. $\triangle APB_1 \cong \triangle APB_2$ 　　　　 2. $s, \angle, s$.
　 $\cong \triangle A'PB_1 \cong \triangle A'PB_2$.

3. $AB_1 = AB_2$
     $= A'B_1 = A'B_2.$

        3.   Corresponding parts of congruent triangles.

4. $\triangle AB_1B_2 \cong \triangle A'B_1B_2.$

        4.   s, s, s.

5. $\angle AB_1B_2 \cong \angle A'B_1B_2.$

        5.   Reason 3.

6. $\triangle AB_1B_3 \cong \triangle A'B_1B_3.$

        6.   s, $\angle$, s.

7. $AB_3 = A'B_3.$

        7.   Reason 3.

8. $\triangle APB_3 \cong \triangle A'PB_3.$

        8.   s, s, s.

9. $\angle APB_3 \cong \angle A'PB_3.$

        9.   Reason 3.

10. $\angle APB_3$ and $\angle A'PB_3$ are right angles.

     10.   Congruent angles forming a linear pair.

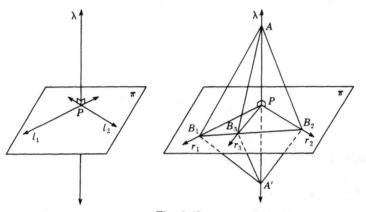

Fig. 2.49

Hence l_3 is perpendicular to λ, as asserted.

Theorem 2.   All lines which are perpendicular to a given line at a given point lie in the same plane.

Proof.   Let λ be an arbitrary line, let l_1 and l_2 be two lines each perpendicular to λ at the point P, and let π be the plane determined by l_1 and l_2 (see Fig. 2.50). To prove the theorem, let us assume the contrary and suppose that l_3 is a line which is perpendicular to λ but does not lie in the plane, π, determined by l_1 and l_2. Now the plane, σ, determined by l_3 and λ has the point P in common with the plane π. Hence these two planes must have a line, say l_3', in common; and by Theorem 1, l_3' must be perpendicular to λ at P. Clearly, l_3' and l_3 are distinct, since l_3' lies in π, whereas l_3 does not. Hence, in the plane σ there are two lines, l_3' and l_3, each perpendicular to λ at the point P. However, this contradicts Theorem 10, Sec. 2.7. Therefore, the assumption that there is a line perpendicular to λ at P which does not lie in π must be abandoned, and the theorem is established.

Putting Theorems 1 and 2 together, we are able to answer the questions we raised earlier about the existence and uniqueness of a plane

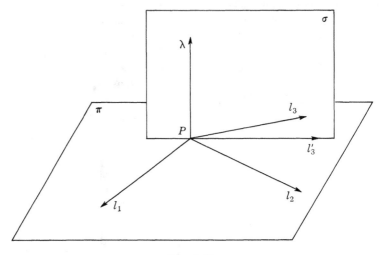

Fig. 2.50

perpendicular to a given line at a given point of the line:

Theorem 3.   At each point of a given line there is a unique plane which contains the given point and is perpendicular to the given line.

It is easy now to prove the following generalization of Theorem 3:

Theorem 4.*   There is a unique plane which passes through a given point and is perpendicular to a given line.

Proof.   If the given point, P, lies on the given line, λ, the assertion of the theorem follows immediately from Theorem 3.   Suppose, therefore, that P does not lie on λ.   Then, by Theorem 5, Sec. 2.9, there is a unique line, l, which contains P and is perpendicular to λ.   Let F be the intersection of λ and l, and let π be the unique plane which is perpendicular to λ at F.   Since π contains all lines which are perpendicular to λ at F, it must contain l; hence the existence of at least one plane which contains P and is perpendicular to λ has been established.   To show that there is no other plane with these properties, assume the contrary and suppose that there is a second plane, π', which passes

---

* To expedite the development, this theorem is taken as a postulate by the School Mathematics Study Group in its text, "Geometry with Coordinates."

through P and is perpendicular to λ. By definition, π' must intersect λ at a point, F'. If F' and F are the same point, we have two planes perpendicular to λ at the same point, which is impossible by Theorem 3. On the other hand, if F' and F are distinct, then $\overleftrightarrow{PF}$ and $\overleftrightarrow{PF'}$ are two distinct lines, each containing P and each perpendicular to λ, which contradicts Theorem 5, Sec. 2.9. Hence the supposition that there is a second plane which contains P and is perpendicular to λ must be abandoned, and the theorem is established.

Not only is there a unique plane which is perpendicular to a given line at a given point of the line, but there is also a unique line which is perpendicular to a given plane at a given point of the plane:

> **Theorem 5.** At each point of a given plane there is a unique line which contains the given point and is perpendicular to the given plane.

Proof. Let P be a given point in a given plane, π, let l_1 be any line which lies in π and passes through P, and let σ be the unique plane which is perpendicular to l_1 at P (see Fig. 2.51). Clearly, σ cannot

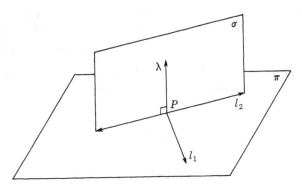

Fig. 2.51

contain l_1; hence, since it has P in common with π, it must intersect π in a line, l_2, distinct from l_1. Now in σ there is a unique line, λ, which is perpendicular to l_2 at P. Moreover, since l_1 is perpendicular to σ, λ is also perpendicular to l_1. Hence, by Theorem 1, λ is perpendicular to π at the point P. To prove that there is only one such line, assume the contrary and suppose that λ' is a second line which is perpendicular to π at P. Now λ and λ' determine a plane, σ', which, perforce, intersects π in a line, l_3, containing P, and, by hypothesis, in σ' both λ and λ' are perpendicular to l_3 at P. But this contradicts Theorem 10,

Sec. 2.7.   Hence there cannot be a second line perpendicular to π at P, and the theorem is established.   If P is not a point of π, there is also a unique line which contains P and is perpendicular to π, but this fact is more conveniently established after we have discussed certain parallel relations between lines and planes in space.

Definition 2.   A line and a plane are parallel if and only if they do not intersect.

Definition 3.   Two planes are parallel if and only if they do not intersect.

The existence of parallel planes is guaranteed by the following theorem:

Theorem 6.   Two planes which are perpendicular to the same line are parallel.

Proof.   Let π_1 and π_2 be two planes perpendicular to a line, l, at the respective points F_1 and F_2.   Obviously, since there is only one plane perpendicular to a given line at a given point, F_1 and F_2 must be distinct.   To prove that π_1 and π_2 are parallel, let us assume the contrary and suppose that they have a point, P, in common.   Then $\overleftrightarrow{PF_1}$ and $\overleftrightarrow{PF_2}$ are both perpendicular to l.   But this contradicts Theorem 5, Sec. 2.9.   Hence the supposition that π_1 and π_2 have a point, P, in common must be abandoned, and therefore these planes are parallel, as asserted.

Analogous to Theorem 6 is the following important result:

Theorem 7.*   Two lines which are perpendicular to the same plane are parallel.

Proof.   Let l and m be two lines perpendicular to a plane, π, at the points L and M, respectively.   To prove them parallel, let n be the line which lies in π and is perpendicular to $\overleftrightarrow{LM}$ at M, let A be any point on l distinct from L, and let B and C be points on n on opposite sides of M (see Fig. 2.52) such that

$$BM = CM$$

---

* This theorem is also assumed as an axiom by the School Mathematics Study Group.

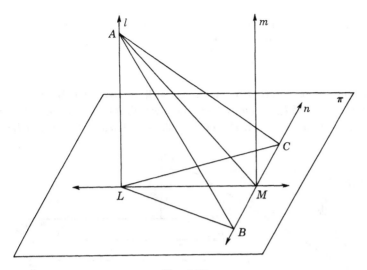

Fig. 2.52

Then

1.	$\angle LMB \cong \angle LMC.$	**1.**	Each is a right angle, since $n \perp \overleftrightarrow{LM}.$
2.	$\triangle LMB \cong \triangle LMC.$	**2.**	$s, \angle, s.$
3.	$LB = LC.$	**3.**	Corresponding parts of congruent triangles.
4.	$\angle ALB \cong \angle ALC.$	**4.**	Each is a right angle, since $l \perp \pi.$
5.	$\triangle ALB \cong \triangle ALC.$	**5.**	$s, \angle, s.$
6.	$AB = AC.$	**6.**	Reason 3.
7.	$\triangle AMB \cong \triangle AMC.$	**7.**	$s, s, s.$
8.	$\angle AMB \cong \angle AMC.$	**8.**	Reason 3.
9.	$\angle AMB$ and $\angle AMC$ are right angles.	**9.**	Congruent angles forming a linear pair.
10.	$n \perp \overline{AM}.$	**10.**	Step 9.
11.	$\overleftrightarrow{AM}$ lies in the plane of $\overleftrightarrow{LM}$ and $m.$	**11.**	Theorem 2.
12.	l lies in the plane of $\overleftrightarrow{LM}$ and $m.$	**12.**	Postulate 6.
13.	$l \Vert m.$	**13.**	Corollary 1, Theorem 1, Sec. 2.10.

Q.E.D.

The next two theorems are sometimes taken for granted and used without proof or even explicit mention.

> **Theorem 8.** If a plane intersects one of two parallel lines in a single point, it intersects the other in a single point also.

Proof. Let l_1 and l_2 be two parallel lines, let σ be the plane determined by l_1 and l_2, and let π be a plane which intersects one of the lines, say l_1, in a single point, P (see Fig. 2.53). Clearly, l_2 cannot lie in π,

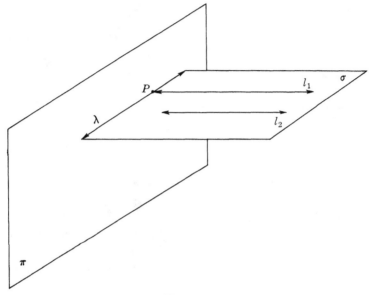

Fig. 2.53

for then π, having l_2 and P in common with σ, would coincide with σ, contrary to the hypothesis that l_1 has a single point in common with π. To prove the theorem, it is thus sufficient to show that l_2 is not parallel to π. Now σ and π, having P in common, must intersect in a line, say λ, containing P but necessarily distinct from l_1. Assuming that l_2 is parallel to π, it is clear that it cannot intersect λ, since λ is contained in π. Hence, in σ the distinct lines l_1 and λ both pass through P and are parallel to l_2. But this contradicts the parallel postulate; therefore the assumption that l_2 does not intersect π must be rejected. Thus l_2, being neither contained in nor parallel to π, must intersect π in a single point, as asserted.

> **Theorem 9.** If a line intersects one of two parallel planes in a single point, it intersects the other in a single point also.

Proof. Let π_1 and π_2 be two parallel planes and let l be a line intersecting one of the planes, say π_1, in a single point, P_1 (see Fig. 2.54).

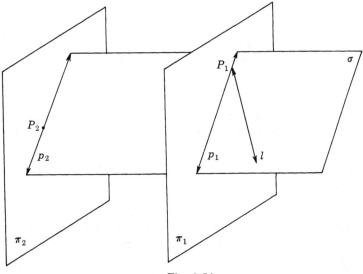

Fig. 2.54

Clearly l cannot lie in π_2, for then π_1 and π_2 would have the point P_1 in
common, contrary to the hypothesis that they are parallel.   Hence to
prove the theorem we need only show that l cannot be parallel to π_2.
Suppose, then, that l is parallel to π_2, and let P_2 be any point in π_2.
The point P_2 and the line l determine a plane, σ, which perforce
intersects π_2 in a line, p_2, and intersects π_1 in a line, p_1, containing
P_1.   Clearly, p_1 is parallel to p_2, for if these lines intersected, their
intersection would be a point common to π_1 and π_2.   Moreover, l is
parallel to p_2, since, by hypothesis, l is parallel to π_2.   Thus in the
plane σ both l and p_1 contain P_1 and are parallel to p_2.   This contradicts the parallel postulate, however; hence we must reject the assumption that l is parallel to π_2.   Thus, since l is not contained in π_2 and is
not parallel to π_2, it must intersect π_2 in a single point, as asserted.

It is now an easy matter to prove the following theorem:

Theorem 10.   If a plane is perpendicular to one of two parallel
lines, it is perpendicular to the other line also.

Proof.   Let l and m be two parallel lines and let π be a plane which
is perpendicular to one of the lines, say l, at the point L.   Then by
Theorem 8, π must intersect m in a single point, say M.   If m is perpendicular to π there is nothing more to prove.   Hence let us suppose

that m is not perpendicular to π, and let m' be the unique line which is perpendicular to π at M. Then by Theorem 7, m' is parallel to l, and therefore we have two distinct lines, m and m', each containing M and each parallel to l. This contradicts the parallel postulate, however; hence the supposition that m is not perpendicular to π must be abandoned, and the theorem is established.

Using Theorem 10 together with Theorem 7, it is now easy to prove the general form of Theorem 6, Sec. 2.10:

Theorem 11. Two lines which are parallel to a third line are parallel to each other.

Using Theorem 10, we can now prove the following extension of Theorem 5:

Theorem 12. There is a unique line which contains a given point and is perpendicular to a given plane.

Proof. Let P be a given point and π a given plane. If P is a point of π, the assertion of the theorem follows at once from Theorem 5. Suppose, therefore, that P is not a point of π, and let Q be an arbitrary point of π. By Theorem 5, there is a unique line, q, which is perpendicular to π at Q. If q contains P, the existence of a line containing P and perpendicular to π has been established. If q does not pass through P, then by the parallel postulate, there is a unique line, p, which contains P and is parallel to q, and by Theorem 10, p is perpendicular to π. Thus in every case the existence of a line containing P has been established. To prove that there is only one such line, assume the contrary and suppose that there are two lines, each containing P and each perpendicular to π at the points F_1 and F_2, respectively. Then in the plane determined by F_1, F_2, and P, there are two lines which pass through P and are perpendicular to $\overleftrightarrow{F_1F_2}$. But this contradicts Theorem 5, Sec. 2.9. Hence the assumption of a second perpendicular to π from P must be rejected, and the theorem is established.

The next theorem makes an assertion about planes analogous to the assertion of the parallel postulate.

Theorem 13. There is a unique plane which is parallel to a given plane and contains a given point not in the given plane.

Proof. Let π be an arbitrary plane and let P be an arbitrary point which does not lie in π. Then by Theorem 12 there is a unique line, l, which contains P and is perpendicular to π. Moreover, there is a unique plane which contains P and is perpendicular to l, and by Theorem 6 this plane, σ, is parallel to π (see Fig. 2.55). To prove

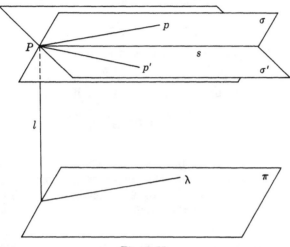

Fig. 2.55

that σ is the only plane which contains P and is parallel to π, assume that there is a second plane, say σ'. Then by Postulate 6, σ and σ' intersect in a line, s. Now let λ be any line in π which is not coplanar with s. Then the plane, ρ, determined by P and λ intersects σ and σ' in two lines, p and p', respectively, which have P in common. Moreover, since σ and σ' are each parallel to π, neither p nor p' can intersect λ. Hence, in ρ there are two lines, p and p', each passing through P and each parallel to λ. Since this contradicts the parallel postulate, the assumption of a second plane containing P and parallel to π must be abandoned, and the theorem is established.

Using Theorem 13, it is now easy to prove the following results:

Theorem 14. Two planes which are parallel to a third plane are parallel to each other.

Theorem 15. If a plane intersects one of two parallel planes in a line, it intersects the other plane in a line also, and the two lines of intersection are parallel.

Theorem 16. A line which is perpendicular to one of two parallel planes is perpendicular to the other plane also.

As a final topic in this section, we shall touch briefly on **dihedral angles:**

> *Definition 4.* A dihedral angle is the union of a line and two halfplanes which have the line as edge and do not lie in the same plane.

The common edge of the two halfplanes is called the **edge** of the dihedral angle, and the two halfplanes are called the **faces** of the dihedral angle. If $\overleftrightarrow{PQ}$ is the edge of a dihedral angle and if A and B are points in the respective faces, then the dihedral angle is denoted by the symbol

$$\angle A\text{-}PQ\text{-}B$$

> *Definition 5.* A plane angle of a dihedral angle is the angle formed by the two rays in which any plane perpendicular to the edge of the dihedral angle intersects the faces of the angle.

Since a dihedral angle has many plane angles, it is important to know whether or not they are all congruent, for if they are, then their common measure is a natural measure of the dihedral angle itself. The next theorem answers this question.

> *Theorem 17.* Any two plane angles of a dihedral angle are congruent.

Proof. Let $\angle A\text{-}PQ\text{-}B$ be an arbitrary dihedral angle, let $\angle U_1 V_1 W_1$ be any one of its plane angles, let V_2 be the vertex of any other of its plane angles, and on the respective sides of $\angle V_2$ let U_2 and W_2 be points such that (see Fig. 2.56)

$$U_2 V_2 = U_1 V_1 \quad \text{and} \quad W_2 V_2 = W_1 V_1$$

Then

1. $\overleftrightarrow{V_1 U_1} \| \overleftrightarrow{V_2 U_2}$.	1. Corollary 1, Theorem 1, Sec. 2.10.
2. $U_1 U_2 V_2 V_1$ is a parallelogram.	2. Theorem 8, Sec. 2.10.
3. $U_1 U_2 = V_1 V_2$.	3. Theorem 7, Sec. 2.10.
4. $\overleftrightarrow{V_1 W_1} \| \overleftrightarrow{V_2 W_2}$.	4. Reason 1.

5. $W_1W_2V_2V_1$ is a parallelogram.	5. Reason 2.
6. $W_1W_2 = V_1V_2$.	6. Reason 3.
7. $\overleftrightarrow{U_1U_2} \| \overleftrightarrow{W_1W_2}$.	7. Theorem 11.
8. $U_1U_2W_2W_1$ is a parallelogram.	8. Reason 2.
9. $U_1W_1 = U_2W_2$.	9. Reason 3.
10. $\triangle U_1V_1W_1 \cong \triangle U_2V_2W_2$.	10. $s,\ s,\ s$.
11. $\angle U_1V_1W_1 \cong \angle U_2V_2W_2$.	11. Corresponding parts of congruent triangles.

Q.E.D.

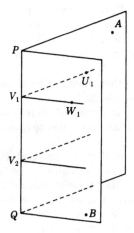 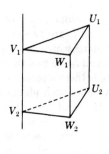

Fig. 2.56

The last theorem now justifies the following definitions:

Definition 6. The measure of a dihedral angle is the measure of any of its plane angles.

Definition 7. A dihedral angle whose measure is 90 is called a right dihedral angle.

Definition 8. The planes determined by the faces of a right dihedral angle are said to be perpendicular.

With these definitions it is not difficult to establish the following theorems:

Theorem 18. If a line is perpendicular to a plane, then any plane containing the line is perpendicular to the given plane.

Theorem 19. If two intersecting planes are each perpendicular to a third plane, their line of intersection is perpendicular to that plane.

EXERCISES

1. Prove Theorem 11.
2. Prove Theorem 14.
3. Prove Theorem 15.
4. Prove Theorem 16.
5. Prove Theorem 18.
6. Prove Theorem 19.
7. Prove that if two planes are perpendicular, then any line in either of the planes which is perpendicular to the line of intersection of the planes is perpendicular to the other.
8. If l_1 and l_2 are two skew lines, prove that there is always one and only one plane which contains l_1 and is parallel to l_2.
9. If l_1 and l_2 are two skew lines and if P is a point which is not on either l_1 or l_2, is there a plane which contains P and is parallel to l_1 and l_2? If so, how many such planes are there?
10. (a) If l_1 and l_2 are two skew lines and if P is a point which is not on either l_1 or l_2, is there always a line which contains P and intersects l_1 and l_2? May there be such a line? May there be more than one such line? Under what conditions, if any, will such a line fail to exist?
    (b) If l_1, l_2, and l_3 are three lines which are mutually skew, prove that there are infinitely many lines each of which intersects l_1, l_2, and l_3.
11. If l_1 and l_2 are two intersecting lines and if l_1' and l_2' are two intersecting lines such that $l_1' \| l_1$ and $l_2' \| l_2$, prove that the plane determined by l_1 and l_2 is parallel to the plane determined by l_1' and l_2'.

2.16   Circles. In this section we shall conclude our reexamination of euclidean geometry with an investigation of some of the more important properties of the sets of points which are called **circles:**

Definition 1. The set of all points which lie in a given plane, π, and whose distance from a given point, P, in π is a given positive number, r, is called a circle.

The point P is called the **center** of the circle. The positive number r is called the **radius** of the circle. Two circles with the same center are said to be **concentric.** Two circles with the same radius are said to be **congruent.** The set of all points of π whose distance from

P is less than r is called the **interior** of the circle. The set of all points of π whose distance from P is greater than r is called the **exterior** of the circle. A segment whose endpoints are points of the circle is said to be a **chord** of the circle. A chord which contains the center, P, is called a **diameter** of the circle. A line which contains a chord of a circle is called a **secant** of the circle. A line which lies in the plane of a circle and intersects the circle in a single point is called a **tangent** to the circle. The point of intersection of a circle and any line tangent to the circle is called the **point of tangency** or **point of contact** of the tangent. Two circles which are tangent to the same line at the same point are said to be **tangent** to each other. The segment determined by the center of a circle and any point of the circle is called a **radial segment** or, when there is no possibility of confusion with the earlier definition of radius, simply a **radius** of the circle.

From the point-plotting theorem, it is clear that any line which lies in the plane of a circle and passes through the center of the circle must intersect the circle in two points. Whether or not this is true of an arbitrary line which lies in the plane of the circle is determined by the following theorem:

Theorem 1. Let C be a circle with center P and radius r, let l be an arbitrary line in the plane of C, let F be the foot of the perpendicular to l which contains P, and let the perpendicular distance, PF, from P to l be p. Then

    (1) If $p < r$, the line l intersects the circle C in two points which are on opposite sides of, and equidistant from, F.

    (2) If $p = r$, the line l has only F in common with C, and l and C are tangent at F.

    (3) If $p > r$, every point of l lies in the exterior of C, and l and C have no points in common.

Proof. If there is a point, Q, on l which is also on the circle C, then necessarily $PQ = r$ (see Fig. 2.57), and by the Pythagorean theorem (or trivially if $Q = F$),

$$FQ = \sqrt{(PQ)^2 - (PF)^2} = \sqrt{r^2 - p^2}$$

If $p < r$, the quantity $r^2 - p^2$ is positive, and therefore $\sqrt{r^2 - p^2}$ is a positive number. But by the point-plotting theorem, there are on l exactly two points, Q_1 and Q_2, on opposite sides of F such that

$$FQ_1 = FQ_2 = \sqrt{r^2 - p^2}$$

Moreover, for these points it is clear that

$$PQ_1 = PQ_2 = \sqrt{p^2 + (\sqrt{r^2 - p^2})^2} = r$$

Hence Q_1 and Q_2 are, in fact, on the circle, as asserted. If $p = r$, then $r^2 - p^2 = 0$. Hence $FQ = \sqrt{r^2 - r^2} = 0$, and F itself is the

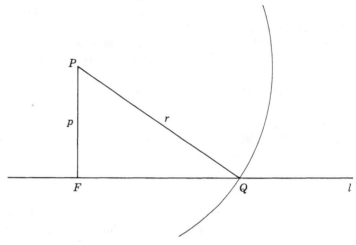

Fig. 2.57

only point on l whose distance from P is r. By definition, l is therefore tangent to C at F, as asserted. Finally, if $p > r$, the quantity $r^2 - p^2$ is negative, its square root does not exist (in the field of real numbers, with which we are exclusively concerned), and therefore there is no point on l whose distance from P is r; that is, l and C have no points in common. Moreover, by Corollary 1, Theorem 4, Sec. 2.11, the distance from P to any point of l is at least as great as the perpendicular distance, $PF = p > r$. Hence every point of l lies in the exterior of C, as asserted.

In Sec. 2.2, as one of our criticisms of the traditional presentation of euclidean geometry, we observed that Euclid assumed without proof that under certain conditions two circles have a point in common. It is now our purpose to investigate this question and establish conditions under which we can guarantee that two circles will or will not intersect.

Proceeding inductively, it appears from Fig. 2.58 that two circles, C_1 and C_2, with centers P_1 and P_2 and radii r_1 and r_2, will or will not intersect according as there does or does not exist a triangle with sides of length r_1, r_2, and $p = P_1P_2$. If we recall the triangle inequality

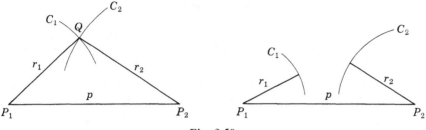

Fig. 2.58

(Theorem 6, Sec. 2.11), this suggests that two circles will intersect if simultaneously

$$r_1 + r_2 > p, \qquad r_1 + p > r_2, \qquad \text{and} \qquad r_2 + p > r_1\dagger$$

and will not intersect if

$$r_1 + r_2 < p, \qquad \text{or if} \quad r_1 + p < r_2, \qquad \text{or if} \quad r_2 + p < r_1$$

These observations we now make precise in the following theorem:

Theorem 2.   If C_1 and C_2 are two coplanar circles with centers P_1 and P_2 and radii r_1 and r_2, respectively, and if $P_1P_2 = p$, then

(1) If r_1, r_2, and p satisfy the inequalities

$$r_1 + r_2 > p, \qquad r_1 + p > r_2, \qquad r_2 + p > r_1$$

the circles have two and only two points in common.

(2) If r_1, r_2, and p satisfy one of the equalities

$$r_1 + r_2 = p, \qquad r_1 + p = r_2, \qquad r_2 + p = r_1$$

the circles have a single point in common and are tangent to each other at that point.

(3) If r_1, r_2, and p satisfy one of the inequalities

$$r_1 + r_2 < p, \qquad r_1 + p < r_2, \qquad r_2 + p < r_1$$

the circles do not intersect.

Proof.   Under the hypotheses of the theorem, we are given *two* circles, C_1 and C_2. Hence it is impossible that $P_1 = P_2$ and $r_1 = r_2$

---

† It is easy to show that for any positive numbers, r_1, r_2, and p, at least two of these inequalities must hold.

simultaneously.   Suppose first that $P_1 = P_2$ but $r_1 \neq r_2$.   Then $p = 0$ and the inequalities in (3) become simply

$$r_1 + r_2 < 0, \qquad r_1 < r_2, \qquad r_2 < r_1$$

The first of these cannot hold, since r_1 and r_2 are positive, but one and only one of the last two must hold.   Obviously this implies that one of the circles lies entirely in the interior of the other.   Hence they can have no point in common, and the assertion of the theorem is verified in this case.

Let us now suppose that $P_1 \neq P_2$, and let us consider first the conditions which must necessarily hold if the two circles are to intersect. Assuming then that Q is a point common to the two circles, let F be the foot of the perpendicular to $\overleftrightarrow{P_1P_2}$ which contains Q, let $QF = d$, let $P_1F = p_1$, and let $P_2F = p_2$ (see Fig. 2.59).   Then necessarily, by the Pythagorean theorem (or trivially if $Q = F$),

$$(P_1F)^2 + (QF)^2 = (QP_1)^2 \qquad \text{or} \qquad p_1^2 + d^2 = r_1^2$$

(1)   and

$$(P_2F)^2 + (QF)^2 = (QP_2)^2 \qquad \text{or} \qquad p_2^2 + d^2 = r_2^2$$

There are now five possibilities to consider:

(a) P_1 lies between F and P_2, and therefore $p_1 + p = p_2$.
(b) F lies between P_1 and P_2, and therefore $p_1 + p_2 = p$.
(c) P_2 lies between P_1 and F, and therefore $p_2 + p = p_1$.
(d) F coincides with P_1, and therefore $p_1 = 0$ and $p_2 = p$.
(e) F coincides with P_2, and therefore $p_1 = p$ and $p_2 = 0$.

Solving Eqs. (1) for p_1 and p_2 in terms of r_1, r_2, and p, we obtain in the respective cases

(1.1)

$$(a) \quad p_1 = \frac{r_2^2 - r_1^2 - p^2}{2p}, \qquad p_2 = \frac{r_2^2 - r_1^2 + p^2}{2p}$$

$$(b) \quad p_1 = \frac{r_2^2 - r_1^2 - p^2}{-2p}, \qquad p_2 = \frac{r_2^2 - r_1^2 + p^2}{2p}$$

$$(c) \quad p_1 = \frac{r_2^2 - r_1^2 - p^2}{-2p}, \qquad p_2 = \frac{r_2^2 - r_1^2 + p^2}{-2p}$$

$$(d) \quad p_1 = 0, \ p_2 = p, \text{ and, immediately, } p^2 + r_1^2 = r_2^2$$

$$(e) \quad p_1 = p, \ p_2 = 0, \text{ and, immediately, } p^2 + r_2^2 = r_1^2$$

Moreover, if we use Eqs. (1) and (1.1) to solve for d^2 in terms of

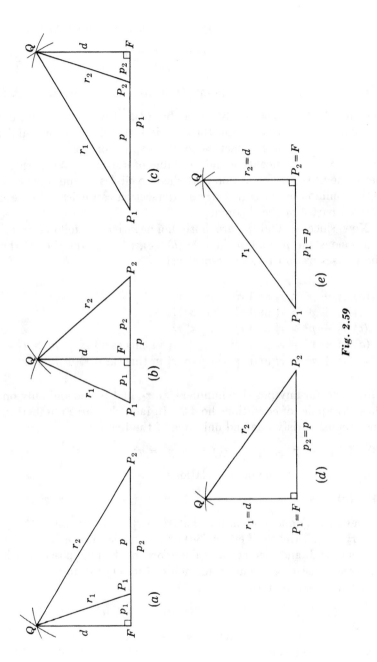

Fig. 2.59

r_1, r_2, and p, it is easy to verify that in each case we obtain

(2) $\quad d^2 = \dfrac{(r_1 + r_2 + p)(r_1 + r_2 - p)(r_1 + p - r_2)(r_2 + p - r_1)}{4p^2}$

Thus, *if C_1 and C_2 have a point, Q, in common, its distance, d, from the line $\overleftrightarrow{P_1 P_2}$ is necessarily given by* (2). However, if one of the inequalities in part 3 of the theorem is satisfied, then one (and perforce only one) of the factors in the expression for d^2 is negative. Hence d^2 itself is negative and no value of d exists. When this happens, the necessary condition that the two circles should have a point, Q, in common is not satisfied, no intersection can exist, and we have verified part 3 of the theorem.

Now since p_1 and p_2 are both nonnegative, it follows from the expressions for p_1 and p_2 in (a), (b), (c), (d), and (e) of (1.1) that in these cases we must have, respectively,

(*a*) $r_1^2 + p^2 < r_2^2$ and $r_2^2 + p^2 > r_1^2$
(*b*) $r_1^2 + p^2 > r_2^2$ and $r_2^2 + p^2 > r_1^2$
(*c*) $r_1^2 + p^2 > r_2^2$ and $r_2^2 + p^2 < r_1^2$
(*d*) $r_1^2 + p^2 = r_2^2$ and (since $r_2^2 > r_1^2$ in this case) $r_2^2 + p^2 > r_1^2$
(*e*) $r_2^2 + p^2 = r_1^2$ and (since $r_1^2 > r_2^2$ in this case) $r_1^2 + p^2 > r_2^2$

Moreover, for any positive numbers, r_1, r_2, and p, one and only one of these five pairs of conditions holds. In fact, the three numbers r_1, r_2, and p must satisfy one and only one of the relations

(3) $\quad$ (*a*) $r_1^2 + p^2 < r_2^2$, $\qquad$ (*b*) $r_1^2 + p^2 = r_2^2$, $\qquad$ (*c*) $r_1^2 + p^2 > r_2^2$

and one and only one of the relations

(4) $\quad$ (*a*) $r_2^2 + p^2 < r_1^2$, $\qquad$ (*b*) $r_2^2 + p^2 = r_1^2$, $\qquad$ (*c*) $r_2^2 + p^2 > r_1^2$

However, (3*a*) and (3*b*) imply that $r_2^2 > r_1^2$, and (4*a*) and (4*b*) imply that $r_2^2 < r_1^2$. Hence, if either (3*a*) or (3*b*) holds, then neither (4*a*) nor (4*b*) can hold, and vice versa. Therefore, of the nine possible pairs of relations consisting of one from each of the sets (3) and (4), the only ones which *can* hold are

$\qquad$ (3*a*,4*c*), $\qquad$ (3*c*,4*c*), $\qquad$ (3*c*,4*a*), $\qquad$ (3*b*,4*c*), $\qquad$ (3*c*,4*b*)

and one and only one of these *must* hold. Clearly, these are, respectively, the conditions characterizing the five cases (a), (b), (c), (d), and (e) above, and thus exactly one of these holds. Thus, since in each case, p_1 and p_2 and hence F are uniquely determined by Eqs. (1.1), it follows that *if the two circles intersect, their common points must all lie on the unique line in the plane of the circles which is perpendicu-*

lar to $\overleftrightarrow{P_1P_2}$ at the unique point F. Since, by Theorem 1, no line can have more than two points in common with a circle, it follows that two circles can intersect in at most two points.

Let us now investigate the sufficiency of the conditions in parts 1 and 2 of the theorem. In the first place, given any three positive numbers, r_1, r_2, and p, one and only one of the five possibilities (a), (b), (c), (d), and (e) must obtain. Furthermore, no matter which case we have, there is a unique pair of values, p_1 and p_2, satisfying Eqs. (1), and hence a unique point F is determined. Moreover, if the conditions of part 1 are satisfied, d^2 is positive and hence $d = QF$ exists. Therefore, on the line l which lies in the plane of the two circles and is perpendicular to $\overleftrightarrow{P_1P_2}$ at F, there are two and only two points, Q_1 and Q_2, such that $Q_1F = Q_2F = d$. Clearly, since Eqs. (1) are satisfied, each of these points is on C_1 and also on C_2. Hence they are points common to the two circles, and the first part of the theorem is verified. Finally, if the condition of part 2 is satisfied, $d^2 = 0$, and hence $d = QF = 0$. Thus on the line l, F itself, but no other point, lies at the distance $d = 0$ from F. Again, since Eqs. (1) are both satisfied, F is common to C_1 and C_2, and no other point has this property. Moreover, since $d = 0$ implies that $p_1 = r_1$ and $p_2 = r_2$, it follows from Theorem 1 that l is tangent to each circle at F, and therefore, by definition, the circles themselves are tangent at F. Thus the assertion of part 2 is verified, and the theorem is established.

Many of the important properties of circles involve the notion of **circular arcs** and angles related to these arcs:

Definition 2.　Any angle whose vertex is the center of a circle is called a central angle of the circle.

Definition 3.　If $\overline{AB}$ is a diameter of a circle and H is either of the halfplanes determined in the plane of the circle by the line $\overleftrightarrow{AB}$, the union of A, B, and all points of the circle which lie in H is called a semicircle.

Definition 4.　If A and B are two points which are not the ends of a diameter of a circle with center P, the union of A, B, and all points of the circle which lie in the interior of $\angle APB$ is called a minor arc of the circle. The union of A, B, and all points of the circle which lie in the exterior of $\angle APB$ is called a major arc of the circle.

Minor arcs, semicircles, and major arcs are referred to collectively simply as **arcs.** If X is any **interior point** of an arc, that is, any point of the arc except one of the endpoints, A and B, the arc is denoted by either of the symbols

$$\overset{\frown}{AXB} \quad \text{or} \quad \overset{\frown}{BXA}$$

When it is perfectly clear whether it is the minor or major arc with endpoints A and B which is intended, an arc is sometimes denoted by the simpler symbols

$$\overset{\frown}{AB} \quad \text{or} \quad \overset{\frown}{BA}$$

If $\overset{\frown}{AXB}$ is a minor arc of a circle and if Y is any point of the circle which is not a point of $\overset{\frown}{AXB}$, the major arc $\overset{\frown}{AYB}$ is said to be the major arc corresponding to $\overset{\frown}{AXB}$, and $\overset{\frown}{AXB}$ is said to be the minor arc corresponding to $\overset{\frown}{AYB}$.

By means of the following definition, it is possible to assign measures to arcs of all three kinds:

Definition 5.   The degree measure of any arc, $\overset{\frown}{AXB}$, is

   (1) The measure of the corresponding central angle if $\overset{\frown}{AXB}$ is a minor arc.

   (2) 180 if $\overset{\frown}{AXB}$ is a semicircle.

   (3) 360 minus the measure of the corresponding minor arc if $\overset{\frown}{AXB}$ is a major arc.

The degree measure of any arc, $\overset{\frown}{AXB}$,† we shall denote by the symbol

$$m\overset{\frown}{AXB}$$

Any arc, $\overset{\frown}{AXB}$, is clearly the union of two other arcs with endpoints (A,X) and (X,B), respectively. Hence it is natural to ask how the degree measure of $\overset{\frown}{AXB}$ is related to the degree measures of these two arcs. The following theorem justifies the obvious and expected answer to this question:

---

† For some purposes it is convenient to regard a circle as an arc, even though it has no endpoints; when this is done, the number 360 is assigned as its degree measure.

Theorem 3.* If $\overset{\frown}{AXB}$ and $\overset{\frown}{BYC}$ are two arcs of the same circle which have only the endpoint B in common, then

$$m\overset{\frown}{ABC} = m\overset{\frown}{AXB} + m\overset{\frown}{BYC}$$

Proof. Let us suppose first that $\overset{\frown}{AXB}$, $\overset{\frown}{BYC}$, and $\overset{\frown}{ABC}$ are all minor arcs on a circle with center P (see Fig. 2.60). Then B lies in

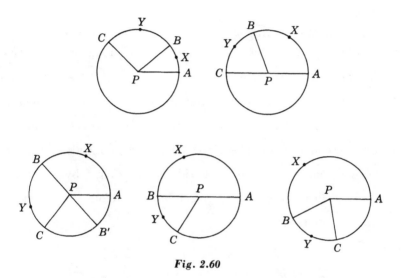

Fig. 2.60

the interior of $\angle APC$, and from the definition of the interior of an angle, it follows that $\overrightarrow{PB}$ is between $\overrightarrow{PA}$ and $\overrightarrow{PC}$. Furthermore, from the definition of the betweenness relation for rays, it follows from this that

$$m\angle APB + m\angle BPC = m\angle APC$$

But since $\overset{\frown}{AXB}$, $\overset{\frown}{BYC}$, and $\overset{\frown}{ABC}$ are all minor arcs, by hypothesis,

$$m\angle APB = m\overset{\frown}{AXB}, \quad m\angle BPC = m\overset{\frown}{BYC}, \quad \text{and } m\angle APC = m\overset{\frown}{ABC}$$

or, substituting,

$$m\overset{\frown}{AXB} + m\overset{\frown}{BYC} = m\overset{\frown}{ABC}$$

as asserted.

* To expedite the development, this theorem is taken as a postulate by the School Mathematics Study Group in its text, "Geometry with Coordinates."

Next, let us suppose that $\overset{\frown}{AXB}$ and $\overset{\frown}{BYC}$ are minor arcs but that $\overset{\frown}{ABC}$ is a semicircle.   Then obviously $\angle APB$ and $\angle BPC$ form a linear pair.   Hence

$$m\angle APB + m\angle BPC = 180$$

or $\qquad\qquad\quad m\overset{\frown}{AXB} + m\overset{\frown}{BYC} = m\overset{\frown}{ABC}$

as asserted.

Now suppose that $\overset{\frown}{AXB}$ and $\overset{\frown}{BYC}$ are minor arcs but that $\overset{\frown}{ABC}$ is a major arc.   Let B' be the other endpoint of the diameter of the given circle which has B as one endpoint.   Then since $\overset{\frown}{ABC}$ is a major arc, B and C are on opposite sides of $\overleftrightarrow{AP}$; hence B' and C are on the same side of $\overleftrightarrow{AP}$.   In other words, B' lies on the C side of $\overleftrightarrow{AP}$.   Similarly, B and A are on opposite sides of $\overleftrightarrow{CP}$.   Hence B' and A are on the same side of $\overleftrightarrow{CP}$; that is, B' is on the A side of $\overleftrightarrow{CP}$.   Thus, by Theorem 1, Sec. 2.7, B' lies in the interior of $\angle APC$.   Therefore $\overrightarrow{PB'}$ is between $\overrightarrow{PA}$ and $\overrightarrow{PC}$, and hence

$$m\angle APB' + m\angle B'PC = m\angle APC$$

But $\qquad\quad m\angle APB' = 180 - m\angle APB = 180 - m\overset{\frown}{AXB}$

$$m\angle B'PC = 180 - m\angle BPC = 180 - m\overset{\frown}{BYC}$$

$$m\angle APC = 360 - m\overset{\frown}{ABC}$$

Therefore, substituting,

$$(180 - m\overset{\frown}{AXB}) + (180 - m\overset{\frown}{BYC}) = 360 - m\overset{\frown}{ABC}$$

or $\qquad\qquad\quad m\overset{\frown}{ABC} = m\overset{\frown}{AXB} + m\overset{\frown}{BYC}$

as asserted.

If one of the component arcs, say $\overset{\frown}{AXB}$, is a semicircle and the other arc is a minor arc, then of course $\overset{\frown}{ABC}$ is a major arc.   In this case

$$m\overset{\frown}{AXB} = 180$$

$$m\overset{\frown}{BYC} = 180 - m\angle CPA$$

and hence, adding,

$$m\overarc{AXB} + m\overarc{BYC} = 180 + (180 - m\angle CPA)$$
$$= 360 - m\angle CPA$$
$$= m\overarc{ABC}$$

as asserted.

Finally, suppose that one of the component arcs, say $\overarc{AXB}$, is a major arc. Then, perforce, C lies in the interior of $\angle APB$ and therefore

$$m\angle APC + m\angle CPB = m\angle APB$$

But

$$m\angle APC = 360 - m\overarc{ABC}$$

$$m\angle CPB = m\overarc{BYC}$$

$$m\angle APB = 360 - m\overarc{AXB}$$

Hence, substituting,

$$(360 - m\overarc{ABC}) + m\overarc{BYC} = 360 - m\overarc{AXB}$$

or

$$m\overarc{AXB} + m\overarc{BYC} = m\overarc{ABC}$$

Thus in every case the assertion of the theorem is verified, and the theorem is established.

In addition to central angles, which provide the degree measures of arcs, other angles known as **inscribed angles** are also important in the study of circles:

Definition 6. An angle is said to be inscribed in an arc if and only if

(1) The vertex of the angle is an interior point of the arc.
(2) Each side of the angle contains an endpoint of the arc.

Definition 7. An angle is said to intercept an arc if and only if

(1) Each endpoint of the arc is a point of the angle.
(2) Each side of the angle contains at least one endpoint of the arc.
(3) Except for its endpoints, each point of the arc lies in the interior of the angle.

Clearly, on a given circle each inscribed angle intercepts a unique arc. Moreover, the measure of any inscribed angle is determined very simply by the degree measure of its intercepted arc:

Theorem 4.   The measure of an inscribed angle is equal to one half the angle measure of its intercepted arc.

Proof.   Let $\angle ABC$ be an inscribed angle in a circle with center P and let $\overset{\frown}{AXC}$ be the arc intercepted by $\angle ABC$. Let us suppose first that one side of the given angle, say $\overrightarrow{BC}$, contains a diameter of the circle (see Fig. 2.61).   Then, by the strong form of the exterior-angle

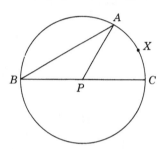

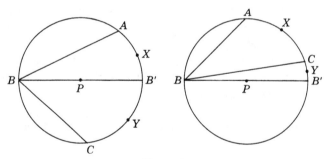

Fig. 2.61

theorem (Theorem 8, Sec. 2.11),

$$m\angle ABC + m\angle BAP = m\angle APC$$

Moreover $$m\angle ABC = m\angle BAP$$

since $\triangle ABP$ is isosceles, with congruent sides $\overline{AP}$ and $\overline{BP}$.   Therefore, substituting,

$$2m\angle ABC = m\angle APC = m\overset{\frown}{AXC}$$

or $$m\angle ABC = \tfrac{1}{2}m\overset{\frown}{AXC}$$

as asserted.

Next suppose that A and C are on opposite sides of the line which contains the diameter, $\overline{BB'}$, which passes through B. Let $A\stackrel{\frown}{X}B'$ be the arc intercepted by $\angle ABB'$ and let $\stackrel{\frown}{B'YC}$ be the arc intercepted by $\angle B'BC$. Then by the first part of our proof, we have

$$m\angle ABB' \doteq \tfrac{1}{2}m\stackrel{\frown}{AXB'} \qquad \text{and} \qquad m\angle B'BC = \tfrac{1}{2}m\stackrel{\frown}{B'YC}$$

Moreover, by hypothesis, $\overrightarrow{BB'}$ lies between $\overrightarrow{BA}$ and $\overrightarrow{BC}$, and $A\stackrel{\frown}{X}B'$ and $\stackrel{\frown}{B'YC}$ have only the point B' in common. Hence,

$$m\angle ABC = m\angle ABB' + m\angle B'BC = \tfrac{1}{2}m\stackrel{\frown}{AXB'} + \tfrac{1}{2}m\stackrel{\frown}{B'YC} = \tfrac{1}{2}m\stackrel{\frown}{AB'C}$$

Finally, suppose that A and C lie on the same side of the line containing the diameter, $\overline{BB'}$, which passes through B. Let $A\stackrel{\frown}{X}C$ be the arc intercepted by $\angle ABC$ and let $\stackrel{\frown}{CYB'}$ be the arc intercepted by $\angle CBB'$. Then if we suppose, for definiteness, that $\overrightarrow{BC}$ lies between $\overrightarrow{BA}$ and $\overrightarrow{BB'}$, $A\stackrel{\frown}{C}B'$ is the arc intercepted by $\angle ABB'$. Now by the first part of our proof, $m\angle ABB' = \tfrac{1}{2}m\stackrel{\frown}{ACB'}$ and $m\angle CBB' = \tfrac{1}{2}m\stackrel{\frown}{CYB'}$. Moreover, by hypothesis, $\overrightarrow{BC}$ lies between $\overrightarrow{BA}$ and $\overrightarrow{BB'}$, and $A\stackrel{\frown}{X}C$ and $\stackrel{\frown}{CYB'}$ have only the point C in common. Hence

$$
\begin{aligned}
m\angle ABC &= m\angle ABB' - m\angle CBB' \\
&= \tfrac{1}{2}m\stackrel{\frown}{ACB'} - \tfrac{1}{2}m\stackrel{\frown}{CYB'} \\
&= \tfrac{1}{2}(m\stackrel{\frown}{ACB'} - m\stackrel{\frown}{CYB'}) = \tfrac{1}{2}m\stackrel{\frown}{AXC}
\end{aligned}
$$

as asserted.

Corollary 1. An angle inscribed in a semicircle is a right angle.

Corollary 2. Angles inscribed in the same arc are congruent.

So far in this section, we have introduced and studied the angle measures, or more precisely the degree measures, of arcs, and it is now natural to ask if it is possible to develop a theory of length measures for arcs. The answer is "Yes," but to do so is much more difficult than we might at first expect, unless we are content to accept several key theorems as postulates. The reason for this is that our distance postulates provide us with no definition for the length of an arc of a circle until the sophisticated notion of a limit is introduced. We shall not attempt a complete discussion of the problem of defining and measuring arc length, but neither shall we take the easy way out and postulate the principal results.   Instead, we shall sketch the proofs of the fundamental theorems under conditions sufficiently general to illustrate the underlying ideas, yet sufficiently special to eliminate some of the major difficulties.

From an intuitive point of view, it is natural to approximate the length of an arc in the following way.   Let $\overset{\frown}{AXB}$ be a given arc and

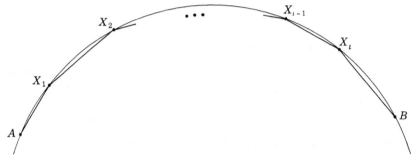

Fig. 2.62

let $X_1, X_2, \ldots, X_{i-1}, X_i$ be i interior points of the arc (see Fig. 2.62).   Then since the union of the $i + 1$ chords

$$\overline{AX_1}, \overline{X_1X_2}, \ldots, \overline{X_{i-1}X_i}, \overline{X_iB}$$

appears to be "close" to $\overset{\frown}{AXB}$, it seems that the sum of the lengths of these chords

$$AX_1 + X_1X_2 + \cdots + X_{i-1}X_i + X_iB$$

must be a good approximation to the length of the arc itself.

Though appealing, this procedure is illogical, since clearly we cannot speak of approximating something, such as the length of an arc, which has not yet been defined!   Once we realize this, however, our intuitive approach can be altered a little and used as the basis for *defining* what we shall mean by the length of an arc.   Specifically, let $X_1, X_2, \ldots,$ X_{i-1}, X_i be i interior points of an arc, $\overset{\frown}{AXB}$, and let

$$s_{i+1} = AX_1 + X_1X_2 + \cdots + X_{i-1}X_i + X_iB$$

be the sum of the lengths of the $i + 1$ chords $\overline{AX_1}, \overline{X_1X_2}, \ldots,$ $\overline{X_{i-1}X_i}, \overline{X_iB}$.   As i is made larger and larger in such a way that the length of each chord becomes smaller and smaller, it may be that the sum, s_{i+1}, will approach a limit, L.   If this happens, the number L is a natural definition of the length of $\overset{\frown}{AXB}$.

In the preceding discussion, no restriction was placed on the points X_1, X_2, . . . , X_{i-1}, X_i except, of course, that they be interior points of $\overset{\frown}{AXB}$ and that each of the lengths AX_1, X_1X_2, . . . , $X_{i-1}X_i$, X_iB should approach zero as i becomes infinite.   Unfortunately, this is too general a formulation for us to handle conveniently, and in our definition of the length of an arc we shall place certain restrictions on the choice of the points X_1, X_2, . . . , X_{i-1}, X_i.   In particular, we shall always take X_1, X_2, . . . , X_{i-1}, X_i to be consecutive vertices of a regular polygon; moreover, as a further simplification, we shall consider only regular polygons with

$$4, 8, 16, \ldots, 2^n \quad \text{sides}$$

With these restrictions, our definitions of the circumference of a circle and the length of a general arc become, respectively,

Definition 8.   The circumference of a circle is the limit as n becomes infinite of the perimeter of a regular polygon of 2^n sides inscribed in the circle.

Definition 9.   If $\overset{\frown}{AXB}$ is an arbitrary arc, if A is a vertex of a regular polygon of 2^n sides inscribed in the circle containing $\overset{\frown}{AXB}$, and if X_1, X_2, . . . , X_{k-1}, X_k are the other vertices of the inscribed polygon which are points of $\overset{\frown}{AXB}$, then the length of $\overset{\frown}{AXB}$ is the limit approached by the sum

$$AX_1 + X_1X_2 + \cdots + X_{k-1}X_k + X_kB$$

as n becomes infinite.

Since each of the central angles determined by two consecutive vertices of an inscribed regular polygon of 2^n sides is of measure $360/2^n$ and since $m\overset{\frown}{AXB}$ will in general not be evenly divisible by $360/2^n$, it is obvious that B will usually not be a vertex of the inscribed polygon which determines the points X_1, X_2, . . . , X_{k-1}, X_k (see Fig. 2.63). Clearly, however, if O is the center of the given circle, then

$$m\angle AOX_1 = m\angle X_1OX_2 = \cdots = m\angle X_{k-1}OX_k$$

$$= \frac{360}{2^n}, \quad m\angle X_kOB \leqq \frac{360}{2^n}$$

and $AX_1 = X_1X_2 = \cdots = X_{k-1}X_k, \quad X_kB \leqq AX_1$

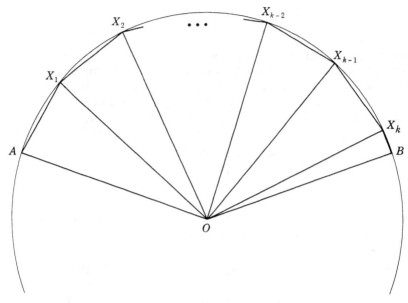

Fig. 2.63

Before these definitions can be of any use to us, we must, of course, establish the existence of the limits to which they refer:

Theorem 5.   The perimeter of a regular polygon of 2^n sides inscribed in a given circle approaches a limit as n becomes infinite.

Proof.   One of the most fundamental ways to establish the existence of the limit of a function, $f(n)$, as n becomes infinite is to show that

(1) $f(n)$ is monotonically increasing
(2) $f(n)$ is bounded from above

and this is the procedure we shall follow.   To do this, let C be a given circle with center O and let X_k and X_{k+1} be two consecutive vertices of a regular polygon, p_n, of 2^n sides inscribed in C.   The perimeter of the polygon p_n is then equal to

$$s_n = 2^n(X_kX_{k+1})$$

To obtain an inscribed regular polygon, p_{n+1}, of 2^{n+1} sides, it is clearly sufficient to locate on the circle between each pair of consecutive vertices of p_n a point, X_k', such that the ray $\overrightarrow{OX_k'}$ bisects $\angle X_kOX_{k+1}$, and by Corollary 1, Theorem 5, Sec. 2.6, this can always be done (see

Fig. 2.64). The perimeter of p_{n+1} is then

$$s_{n+1} = 2^n(X_k X'_k + X'_k X_{k+1})$$

However, by the triangle inequality,

$$X_k X'_k + X'_k X_{k+1} > X_k X_{k+1}$$

Hence, for any value of n,

$$s_{n+1} > s_n$$

which proves that the perimeter, s_n, of an inscribed regular polygon of 2^n sides increases monotonically with increasing n.

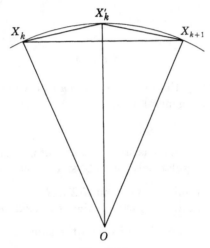

Fig. 2.64

To prove that the perimeter, s_n, of an inscribed regular polygon of 2^n sides is bounded from above for all values of n, it is convenient to consider the circumscribed regular polygon of 2^n sides, P_n, whose sides are tangent to the given circle at the vertices of the inscribed polygon, p_n (see Fig. 2.65). Using the triangle inequality again, it is easy to show by an argument just like the preceding one that for all values of n, the perimeter, S_n, of the circumscribed polygon, P_n, is greater than the perimeter, s_n, of the corresponding inscribed polygon, p_n, and moreover that the perimeter, S_n, of the circumscribed polygon, P_n, decreases monotonically as n increases. In other words, for all n,

$$S_n > S_{n+1} > s_{n+1} > s_n$$

Thus for all values of n, s_n is bounded by the value of S_n for any n. Hence we can conclude that s_n approaches a limit as n becomes infinite, since we have shown that it is a monotonically increasing function of n which is bounded from above.

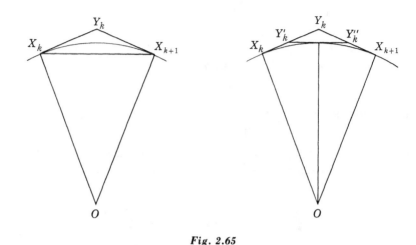

Fig. 2.65

In almost exactly the same way we can prove the important companion theorem for general arcs:

Theorem 6.   If $\overparen{AXB}$ is an arbitrary arc, if A is a vertex of a regular polygon of 2^n sides inscribed in the circle containing $\overparen{AXB}$, and if $X_1, X_2, \ldots, X_{k-1}, X_k$ are the other vertices of the inscribed polygon which are points of $\overparen{AXB}$, then the sum

$$AX_1 + X_1X_2 + \cdots + X_{k-1}X_k + X_kB$$

approaches a limit as n becomes infinite.

Using Definition 8 and certain of the simpler properties of limits, we can now prove the following fundamental theorem:

Theorem 7.   The ratio of the circumference to the length of any diameter is the same for all circles.

Proof.   Consider any two circles with centers O and O', radii r and r', and circumferences C and C', respectively, and in each let a regular polygon of 2^n sides be inscribed.   Let X_k and X_{k+1} be two consecutive vertices of the inscribed regular polygon in the first circle and let X'_k and X'_{k+1} be two consecutive vertices of the inscribed regular polygon in the second circle (see Fig. 2.66).   Now

$$m\angle X_kOX_{k+1} = m\angle X'_kO'X'_{k+1}$$

since each is equal to $360/2^n$. Hence, by Theorem 7, Sec. 2.12,

$$\triangle X_k O X_{k+1} \sim \triangle X_k' O' X_{k+1}'$$

Therefore $\dfrac{X_k X_{k+1}}{O X_k} = \dfrac{X_k' X_{k+1}'}{O' X_k'}$ or $\dfrac{X_k X_{k+1}}{r} = \dfrac{X_k' X_{k+1}'}{r'}$

or, multiplying each side of the last equality by $2^n/2$,

$$\frac{2^n (X_k X_{k+1})}{2r} = \frac{2^n (X_k' X_{k+1}')}{2r'}$$

Now $s_n = 2^n (X_k X_{k+1})$ and $s_n' = 2^n (X_k' X_{k+1}')$ are, respectively, the perimeters of the regular polygons inscribed in the two circles. Moreover, $2r$ and $2r'$ are, respectively, the lengths of the diameters, d and d',

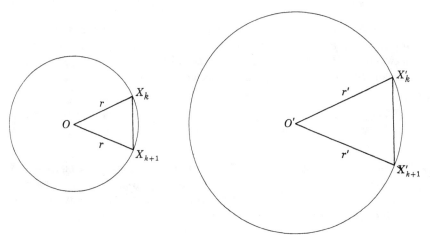

Fig. 2.66

of the two circles. Hence the last equation can be written

$$\frac{s_n}{d} = \frac{s_n'}{d'}$$

Now, taking the limit of each side as n becomes infinite, we have

$$\lim_{n \to \infty} \left(\frac{s_n}{d} \right) = \lim_{n \to \infty} \left(\frac{s_n'}{d'} \right)$$

or, since d and d' are constants, independent of n,

$$\frac{1}{d} \lim_{n \to \infty} s_n = \frac{1}{d'} \lim_{n \to \infty} s_n'$$

Finally, since $\lim\limits_{n \to \infty} s_n$ and $\lim\limits_{n \to \infty} s_n'$ are, by definition, the circumferences of the respective circles, we have

$$\frac{C}{d} = \frac{C'}{d'}$$

as asserted.

The name given to the number C/d, or $C/2r$, which we have just proved to be the same for all circles, is the familiar symbol π.   Hence we can restate Theorem 7 in the following form:

Theorem 8.   The circumference, C, of any circle of radius r is given by the formula $C = 2\pi r$.

Now that we have defined and established the existence of length measures of arcs, it is natural to ask what relation, if any, exists between the degree measure of an arc and the length of the arc.   The answer, though not easy to prove, is very simple:

Theorem 9.*   The lengths of arcs on congruent circles are proportional to the degree measures of the arcs.

Proof.   Let $\overparen{AXB}$ and $\overparen{A'X'B'}$ be two arcs on congruent circles, and in the circles let there be inscribed regular polygons of 2^n sides having A and A' as vertices.   From the properties of regular polygons and the fact that the two circles are congruent, it is clear that the sides of the two polygons all have the same length.   Similarly, it follows that the degree measure of the central angle determined by any side of either polygon is

$$\frac{360}{2^n}$$

Let q_n and r_n be, respectively, the quotient and remainder when $m\overparen{AXB}$ is divided by $360/2^n$, and let q'_n and r'_n be, similarly, the quotient and remainder when $m\overparen{A'X'B'}$ is divided by $360/2^n$.   Then

$$m\overparen{AXB} = \frac{360}{2^n}\, q_n + r_n \qquad \text{and} \qquad m\overparen{A'X'B'} = \frac{360}{2^n}\, q'_n + r'_n$$

---

* To simplify the development, this theorem is sometimes assumed without proof as an additional postulate.   In particular, this is done by the School Mathematics Study Group.

and, from these equations,

$$\frac{m\overset{\frown}{AXB}}{m\overset{\frown}{A'X'B'}} = \frac{(360/2^n)q_n + r_n}{(360/2^n)q_n' + r_n'} = \frac{(q_n/q_n') + (1/q_n')[r_n/(360/2^n)]}{1 + (1/q_n')[r_n'/(360/2^n)]}$$

Now from purely arithmetic considerations, it is clear that the remainders r_n and r_n' are both less than the divisor $360/2^n$ and that the quotients q_n and q_n' both become infinite as n becomes infinite. Hence, taking limits in the last expression, we have simply

$$\frac{m\overset{\frown}{AXB}}{m\overset{\frown}{A'X'B'}} = \lim_{n\to\infty} \left(\frac{q_n}{q_n'}\right)$$

Considering now the lengths of $\overset{\frown}{AXB}$ and $\overset{\frown}{A'X'B'}$, we observe first that the number of vertices of the inscribed regular polygons which are interior points of $\overset{\frown}{AXB}$ and $\overset{\frown}{A'X'B'}$ are, respectively, $k = q_n$ and $k' = q_n'$. Hence the sums whose limits define the lengths, L and L', of these arcs are

$$s_n = q_n(AX_1) + X_kB$$

and $$s_n' = q_n'(A'X_1') + X_{k'}'B' = q_n'(AX_1) + X_{k'}'B'$$

Therefore

$$\frac{s_n}{s_n'} = \frac{q_n(AX_1) + X_kB}{q_n'(AX_1) + X_{k'}'B'} = \frac{(q_n/q_n') + (1/q_n')(X_kB/AX_1)}{1 + (1/q_n')(X_{k'}'B'/AX_1)}$$

Finally, taking limits as n becomes infinite and observing that both X_kB and $X_{k'}'B'$ are less than or equal to AX_1, we have

$$\lim_{n\to\infty}\left(\frac{s_n}{s_n'}\right) = \frac{\lim\limits_{n\to\infty} s_n}{\lim\limits_{n\to\infty} s_n'} = \lim_{n\to\infty}\left(\frac{q_n}{q_n'}\right)$$

But $\lim\limits_{n\to\infty} s_n$ and $\lim\limits_{n\to\infty} s_n'$ are, respectively, the lengths, L and L', of $\overset{\frown}{AXB}$ and $\overset{\frown}{A'X'B'}$, and $\lim\limits_{n\to\infty}(q_n/q_n')$ we showed above to be equal to $m\overset{\frown}{AXB}/m\overset{\frown}{A'X'B'}$. Hence

$$\frac{L}{L'} = \frac{m\overset{\frown}{AXB}}{m\overset{\frown}{A'X'B'}}$$

as asserted.

Using Theorem 8, we can restate the last theorem in the following more explicit form:

Theorem 10.   The length of an arc of degree measure α on a circle of radius r is given by the formula $L = (\pi/180)\alpha r$.

Proof.   Let $\overset{\frown}{AXB}$ be an arc of degree measure α and length L on a circle of radius r.   Then, according to Theorem 9, we have

$$\frac{L}{\alpha} = \frac{L'}{m\overset{\frown}{A'X'B'}}$$

Hence the value of the ratio L/α can be found by evaluating the fraction on the right for any convenient arc on a circle of the same radius, r. Now the degree measure of an arc which is an entire circle is 360. Moreover, by Theorem 8, the circumference, that is, the length, of a circle of radius r is $2\pi r$.   Hence, substituting,

$$\frac{L}{\alpha} = \frac{2\pi r}{360} \qquad \text{or} \qquad L = \frac{\pi}{180}\alpha r$$

as asserted.

To conclude this section, we shall consider briefly the definition and measurement of the areas of certain regions associated with circles.

Definition 10.   The union of a circle and its interior is called a circular region.

Definition 11.   If $\overset{\frown}{AXB}$ is an arc of a circle, C, the union of all radial segments of C which have a point of $\overset{\frown}{AXB}$ as one endpoint is called a circular sector.

As in our definition of the length of circular arcs, so in our definition of the areas of circular regions and circular sectors the idea of a limit plays a fundamental role:

Definition 12.   The area of a circular region is the limit as n becomes infinite of the area of the polygonal region bounded by a regular polygon of 2^n sides inscribed in the circle.

Definition 13.   If $\overset{\frown}{AXB}$ is an arbitrary arc of a circle, C, with center O, if A is a vertex of a regular polygon of 2^n sides inscribed in C, and if $X_1, X_2, \ldots, X_{k-1}, X_k$

are the other vertices of the polygon which are points of $\overset{\frown}{AXB}$, then the area of the circular sector determined by $\overset{\frown}{AXB}$ is the limit approached by the sum of the areas of the triangular regions determined by the triangles

$$\triangle AOX_1, \; \triangle X_1OX_2, \; \ldots, \; \triangle X_{k-1}OX_k, \; \triangle X_kOB$$

as n becomes infinite.

Naturally, before these definitions can be used, we must demonstrate the existence of the limits to which they refer, and by arguments very much like the one we employed to prove Theorem 5, we can establish the following theorems:

Theorem 11. The area of the polygonal region bounded by a regular polygon of 2^n sides inscribed in a given circle approaches a limit as n becomes infinite.

Theorem 12. If $\overset{\frown}{AXB}$ is an arbitrary arc of a circle, C, with center O, if A is a vertex of a regular polygon of 2^n sides inscribed in C, and if $X_1, X_2, \ldots, X_{k-1}, X_k$ are the other vertices of the polygon which are points of $\overset{\frown}{AXB}$, then the sum of the areas of the triangular regions determined by the triangles

$$\triangle AOX_1, \; \triangle X_1OX_2, \; \ldots, \; \triangle X_{k-1}OX_k, \; \triangle X_kOB$$

approaches a limit as n becomes infinite.

Using Definitions 12 and 13 and Theorems 8 and 10, it is now a relatively simple matter to prove the following important results:

Theorem 13. The area, A, of a circular region of radius r is given by the formula

$$A = \pi r^2$$

Theorem 14. If $\overset{\frown}{AXB}$ is an arbitrary arc of degree measure α on a circle of radius r, then the area, A, of the circular sector determined by $\overset{\frown}{AXB}$ is given by the formula

$$A = \frac{\pi}{360} \, \alpha r^2$$

EXERCISES

1. Prove that no line is a circle.
2. Define the angle measure of an arc in terms of a general scale factor R.
3. If l is a line which does not intersect a given circle, C, prove that C lies entirely in one of the halfplanes determined by l.
4. If t is the tangent to a circle C at a point P, prove that with the exception of P, all points of C lie on the same side of t.
5. If s is a secant of a circle C, prove that there are points of C in each of the halfplanes determined by s.
6. The proof of Theorem 1 made use of the fact that if r_1, r_2, and p are any positive numbers, then at least two of the three inequalities

$$r_1 + r_2 > p \qquad r_1 + p > r_2 \qquad r_2 + p > r_1$$

   must hold.   Prove this statement.
7. Prove that if two circles have the property that the interior of each contains both interior and exterior points of the other, then the circles intersect in two distinct points.
8. Show that the circular arcs used in the usual construction of an equilateral triangle having a given segment as base actually intersect.
9. State and prove the converse of the triangle inequality, Theorem 6, Sec. 2.11.
10. If C is a circle with center O and radius r and if P and Q are two points on a ray having O as end point such that

$$(OP)(OQ) = r^2$$

   prove that the midpoint of the segment $\overline{PQ}$ is always a point in the exterior of C.   Using this fact, explain the fallacy in the "proof" presented in Exercise 4, Sec. 2.2.
11. If $\overset{\frown}{ABC}$ is a minor arc of a circle, prove that

$$AB < AC$$

12. Show that the perimeter of a regular polygon of n sides circumscribed about a circle is a monotonically decreasing function of n.

3

THE GEOMETRY
OF FOUR
DIMENSIONS

3.1 Introduction. When the theory of relativity burst upon the world in the early years of the twentieth century, it probably attracted as much attention outside the scientific community as it did within. To the physicist, it offered an explanation of paradoxical experiments that had perplexed him for more than a decade. To the mathematician, it offered exciting new opportunities for the application of differential geometry and the recently developed subject of tensor analysis. To the philosopher, it presented a direct challenge and a threat to an entire system of thought. And to the man in the street, it gave one of the most glittering and attractive intellectual toys he had ever known—the fourth dimension. In his epoch-making paper of 1905, Einstein made no explicit mention of a four-dimensional world in which the absolute character of space and time are lost in a new union of the two. But three years later, in 1908, the mathematician Minkowski made precisely this suggestion, to the considerable profit of theoretical physicists and the great excitement of nonscientists everywhere. Philosophers debated the existence of the fourth dimension. Theologians and mystics speculated about it as the abode of God, or at least the reservoir from whence come our dreams. Popularizers of all levels of competence rushed into print with explanations and interpretations, despite which the layman, finding himself unable to "visualize" anything more than the three dimensions of the familiar world through which he moved, quickly made of the fourth dimension a symbol of intellectual sophistication and abstruseness.

Looking back from the modest vantage point we achieved in Chap. 1, several things should be clear to us, even though we have never studied   **163** the geometry of more than three dimensions.   In the first place, the use of the definite article in the phrase *"the* fourth dimension" is probably unwarranted and misleading.   If a geometry is simply the totality of the logical consequences of a consistent set of axioms asserting properties of "geometric" objects, and if there are, in particular, different consistent plane geometries, surely it is to be expected that there are various four-dimensional geometries.   Second, such difficulties as the uninitiated encounter in their speculations about the possibility of a geometry of more than three dimensions are almost certainly perceptual rather than conceptual.   The layman, struggling to achieve an understanding of "the fourth dimension," is not thinking of geometry as we defined the term in Chap. 1 but, in fact, is struggling to visualize a vaguely defined model of a set of axioms he has never seen explicitly stated.   To use a figure of speech we employed earlier, he is trying to play a game with pieces he never saw before whose moves are governed by rules he does not know.

Conceptually, all geometries are alike as we first approach them. In each, as we now realize, we have a few undefined objects and relations, to which properties are ascribed by a set of axioms, and our task is simply to draw what conclusions we can from the axioms.   The motivation for the axioms of some particular geometry may be more obvious than the motivation for those of some other system.   The possible models of one geometry may be simpler or more familiar than those of another.   The technical difficulties in the study of one geometry may be less than those in the study of another.   But the logical organization of each is the same; each is as "real," and as abstract, as any other, and all exist in exactly the same sense as constructions of the inquiring human mind.

In this chapter we shall investigate certain of the simpler properties of four-dimensional euclidean space, or E_4, as it is sometimes called.[*] As we shall see in the next section, this will require a surprisingly modest change in the axiomatic system which defines the euclidean geometry of three dimensions—one new undefined term and three or four new postulates, not greatly different from those already in the system.   If we keep in mind that any perceptual difficulties we may seem to have are actually associated not with the axiomatic system itself but with its possible models and that, though real, they are nonetheless irrelevant, our exploration of the geometry of E_4 should be no harder, though more exciting, than the reappraisal of solid geometry which we undertook in the last chapter.   Intuition will be of little use to us at first, although in the recognition of analogies it should quickly

* More generally, a euclidean space of any number of dimensions, n, is often referred to simply as E_n.   In this notation, the spaces studied in plane geometry and in solid geometry are E_2 and E_3, respectively.

become helpful. As usual, we shall sometimes have to draw figures to remind us of the course of our proofs, but inevitably they will be so inadequate that the temptation to offer the evidence of a drawing in support of an argument will be much less than in the study of the geometry of two and three dimensions. Thus, to an unusual degree we shall be forced to work directly with ideas, one of the most important skills a person can learn.

EXERCISES

Discuss each of the following quotations:

1. "Among the splendid generalizations effected by modern mathematics, there is none more brilliant or more inspiring or more fruitful, and none more nearly commensurate with the limitless immensity of being itself, than that which has produced the great concept variously designated by such equivalent terms as hyperspace, multi-dimensional space, n-space, n-fold or n-dimensional space, and space of n-dimensions."

   C. J. Keyser, "The Human Worth of Rigorous Thinking," p. 101 [1].

2. "No concept that has come out of our heads or pens marked a greater forward step in our thinking, no idea of religion, philosophy, or science broke more sharply with tradition and commonly accepted knowledge than the idea of a fourth dimension."

   Kasner and Newman, "Mathematics and the Imagination," p. 131 [2].

3. "The problem of the fourth dimension is not merely a mathematical problem; it is a problem that affects our actual life, or at least the higher regions of our everyday life."

   Maurice Maeterlinck, "The Life of Space," p. 4 [3].

4. "The development of this idea [the notion of a fourth dimension] is as much due to our rather childish desire for consistency as to anything more profound. In this same striving after consistency and generality, mathematicians developed negative numbers, imaginaries, and transcendentals. It was not without a struggle that these now rather commonplace ideas were introduced into mathematics. The same struggle was repeated to introduce a fourth dimension, and there are still skeptics in the camp of the opposition."

   Kasner and Newman, "Mathematics and the Imagination," p. 117 [2]

5. "It is held by many, including perhaps the majority of mathematicians, that there are no hyperspaces of points and that n-dimensional geometries are, rightly speaking, not geometries at all, but that the facts dealt with in such so-called geometries are nothing but algebraic or analytic or numeric facts expressed in geometric language."

   C. J. Keyser, "The Human Worth of Rigorous Thinking," p. 235 [1].

6. "A great deal of nonsense has been written about what it would be like if we lived in a world of four spatial dimensions. Many writers have declared that in a world of four spatial dimensions people could eat an

egg without breaking the shell or leave a room without passing through the walls, floor, or ceiling. Now such speculations would be harmless if they did not give the impression that mathematicians actually believe in the real existence of a world of four spatial dimensions and hope some day to train our visual apparatus to perceive this world. No such belief is held nor is such a project contemplated."

Morris Kline, "Mathematics in Western Culture," p. 180 [4].

7. "Clearly to conceive a fourth dimension we should need other senses, a different brain, and a different body; in a word, we should need the power to emerge completely from our terrestrial envelop; and that means we should no longer be human."

Maurice Maeterlinck, "The Life of Space," p. 7 [3].

8. "When at last we have come to comprehend the fourth dimension, or are able to make use of it we shall be almost superhuman."

Maurice Maeterlinck, "The Life of Space," p. 115 [3].

9. "For its [the notion of hyperspace] fair understanding, for a lively sensibility to its manifold significance and quickening, a long and severe mathematical apprenticeship, however helpful it would be, is not demanded in preparation, but only the serious attention of a mature intelligence reasonably inured by discipline to the exaction of abstract thought and the austerities of the higher imagination."

C. J. Keyser, "The Human Worth of Rigorous Thinking," p. 101 [1].

10. "When we have said that we do not know precisely what the fourth dimension is we have said almost all that we really know of it. The rest is hypotheses, speculations, presentiments, and more or less reckless approximations. But these are useful as soundings which explore the incontestable unknown which one day we shall perhaps know."

Maurice Maeterlinck, "The Life of Space," p. 60 [3].

11. "A friend called at my study, and, finding me at work, asked 'What are you doing?' My reply was, 'I am trying to tell how a world which probably does not exist would look if it did.' I had been at work on a chapter of what is called four-dimensional geometry. The incident occurred ten years ago [1903]. The reply to my friend no longer represents my conviction. Subsequent reflection has convinced me that a space, S_n, of four or more dimensions has every kind of existence that may rightly be ascribed to the space, S_3, of ordinary geometry."

C. J. Keyser, "The Human Worth of Rigorous Thinking," p. 238 [1].

12. "The figures of four-dimensional geometry exist in the same sense as do figures in two and three dimensions."

Morris Kline, "Mathematics in Western Culture," p. 178 [4].

3.2 The Postulates of Four-dimensional Euclidean Geometry.

Instead of beginning with the postulates of three-dimensional euclidean geometry, as we did in Chap. 2, and developing plane geometry somewhat incidentally as a collection of results valid in each of the planes of space, we could, of course, have begun with a set of postulates for plane geometry and subsequently extended our study to space by modifying and adding postulates as necessary. Had we chosen to follow this procedure, the changes required to move from two to three dimensions would have been quite small. We would, of course, have had to introduce *plane* as a new undefined term; and to ensure that we were no longer restricted to a single plane, we would have had to postulate that not all points in our new system belonged to the same plane. Then, because of the existence of more than one plane, we would have needed several new postulates to tell us how planes are determined and to describe their intersection properties. But these, together with postulates for the measurement of volume,* are the only changes or additions we would have had to make in order to extend our investigations from two to three dimensions. The parallel postulate, the plane-separation postulate, the congruence postulate, and the postulates for the measurement of distance, angle, and area serve equally well for the study of plane or solid geometry. The essential distinction between the two geometries lies in the postulates of connection, and here the formal difference is small.

There is, of course, ample inductive evidence from the world of our experience to motivate the extension of geometric investigations from two to three dimensions but little or nothing to suggest the possibility of studying spaces of more than three dimensions. This, however, is completely irrelevant. We have acknowledged explicitly the debt geometry owes to the physical world, but we have also insisted that geometry is logically independent of the physical world. Whatever its original sources may have been, geometry, properly understood, is a creation of the human mind and only incidentally and fortuitously a reflection of the world of experience. If, motivated by the fact that we can pass from two to three dimensions by assuming little more than the existence of points not belonging to the plane of our original inquiry, we choose to explore the consequences of assuming the existence of points outside the space of our three-dimensional investigations, our intellectual curiosity is sufficient justification. Whether the results of such a study are in any sense "real" or "useful" is beside the point, although it is naturally a source of additional satisfaction if they are. The only limitation which really concerns us, however, is the logical requirement of consistency.

If we enlarge the space in which we propose to work by assuming that it includes points not belonging to the three-dimensional system

---

* In order to keep our development of three-dimensional euclidean geometry within reasonable bounds, we did not include the postulates for the measurement of volume in the last chapter.

we considered previously, we shall, presumably, create a space in which
there will exist many three-dimensional subspaces (analogous to the
planes in a three-dimensional space).   Hence it is necessary that our
new axiomatic system include, as an undefined term, some name for
these objects.   Various names are in use, including *solid, flat, three-
space*, and *prime*.   Of these, we shall use *prime* as our technical
undefined term, although for variety we shall occasionally use *three-
space* as an acceptable synonym.   In very much the same way that
we used parallelograms to suggest planes in our drawings in Chap. 2,
so in this chapter we shall represent primes by parallelepipeds, thus,

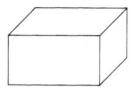

Of course, the use of this mode of representation does not imply that
a prime has "faces" or is in any way limited in extent, nor that some of
its points are "interior" points while others are "boundary" points.
   If, as we expect, our new space contains more than one prime, we
must, of course, introduce additional postulates telling us how primes
are determined and describing their intersection properties.   However, if, for brevity, we continue to exclude the measurement of volume
and "hypervolume" from our discussion, these postulates of connection
are the only new ones we shall need.   All the other postulates of our
three-dimensional system can be taken over without changes or additions.   For this reason, most of the definitions we introduced in the
last chapter are equally appropriate for the work we are about to
undertake and, except where explicitly noted, we shall use them without formal restatement.   There is, however, one new concept generalizing the ideas of collinear and coplanar points which must be defined
before our revised postulates of connection can be stated:

Definition 1.   The points of a set are said to be coprimal if and
            only if there is a prime which contains them all.

Although it would be entirely proper to present the postulates which
we shall use in our development of the euclidean geometry of four
dimensions without any formal reference to those we used in our study
of solid geometry, it is helpful to emphasize the similarity between the
two sets by listing the old and the new in parallel columns as shown on
pages 169 and 170.

Four-dimensional euclidean geometry	Three-dimensional euclidean geometry

Undefined terms

Point	•	Point	•
Line	————	Line	————
Plane		Plane	
Prime			

Postulates

1. Every line is a set of points and contains at least two points.

2. If P and Q are any two points, there is one and only one line which contains them.

3. Every plane is a set of points and contains at least three noncollinear points.

4. If P, Q, R are any three noncollinear points, there is one and only one plane which contains them.

5. Every prime is a set of points and contains at least four points which are neither collinear nor coplanar.

6. If P, Q, R, S are any four noncoplanar points, there is one and only one prime which contains them.

7. Space contains at least five points which are neither collinear, coplanar, nor coprimal.

8. If two points of a line lie in a plane, then every point of the line lies in the plane.

9. If three noncollinear points of a plane lie in a prime, then every point of the plane lies in the prime.

10. If a plane and a prime have a point in common, their intersection is a line.

1. Every line is a set of points and contains at least two points.

2. If P and Q are any two points, there is one and only one line which contains them.

3. Every plane is a set of points and contains at least three noncollinear points.

4. If P, Q, R are any three noncollinear points, there is one and only one plane which contains them.

7. Space (that is, the only prime in the system) contains at least four points which are neither collinear nor coplanar.

5. If two points of a line lie in a plane, then every point of the line lies in the plane.

6. If two planes (necessarily lying in the same prime) have a point in common, their intersection is a line.

The other postulates, namely,

The parallel postulate
The plane-separation postulate
The distance-measurement postulates
The angle-measurement postulates
The congruence postulate
The area-measurement postulates

are identical in the two systems and need not be formally restated.

We shall not consider in detail the question of the consistency of the postulates for four-dimensional euclidean geometry. However, to make plausible the fact that they are consistent and to illustrate some of the unexpected results we shall prove in later sections, it is interesting to consider briefly the simple finite system defined as follows:

Let there be given sixteen arbitrary objects, designated by the letters $A, B, C, \ldots, P$. These objects are the points of the system, and various sets of them, described explicitly in the tabulation on pages 171 and 172, are the lines, planes, and primes of the system.

It is an easy matter to verify that Postulates 1 to 10 and the parallel postulate are all satisfied in this system. The truth of Postulates 1, 3, and 5 can be checked immediately. Postulate 7 is clearly true since, for instance, there is no line, plane, or prime which contains the five points A, B, D, H, P. Postulate 2 is satisfied because, by definition, every possible pair of points is a line of the system. To check Postulate 4, we must consider every possible set of three noncollinear points and verify that there is one and only one plane which contains them. Though tedious, this can easily be done. Similarly, to verify Postulate 6, we must consider every possible set of four noncoplanar points and verify that there is one and only one prime which contains them. This, too, is a straightforward matter. The verification of Postulates 8, 9, and 10 likewise requires the checking of many particular cases, but again, though tedious, this is not difficult. Finally, that the parallel postulate is satisfied is almost obvious and easily checked by inspecting the table in which the lines of the system are displayed, since each line is parallel to every other line in the subtable in which it is listed and to no others. For instance, l_6 is the unique line which contains P_{11} and is parallel to l_1, since it is the only line which passes through P_{11}, lies in the plane determined by P_{11} and l_1, namely, π_{113}, and does not intersect

Points

$$P_1 = A \qquad P_5 = E \qquad P_9 = I \qquad P_{13} = M$$
$$P_2 = B \qquad P_6 = F \qquad P_{10} = J \qquad P_{14} = N$$
$$P_3 = C \qquad P_7 = G \qquad P_{11} = K \qquad P_{15} = O$$
$$P_4 = D \qquad P_8 = H \qquad P_{12} = L \qquad P_{16} = P$$

Lines

(arranged by eights in parallel families)

$l_1 = \{A,B\}$	$l_{25} = \{A,C\}$	$l_{49} = \{A,D\}$	$l_{73} = \{A,E\}$	$l_{97} = \{A,F\}$
$l_2 = \{C,H\}$	$l_{26} = \{B,H\}$	$l_{50} = \{B,G\}$	$l_{74} = \{B,F\}$	$l_{98} = \{B,E\}$
$l_3 = \{D,G\}$	$l_{27} = \{D,F\}$	$l_{51} = \{C,F\}$	$l_{75} = \{C,G\}$	$l_{99} = \{C,D\}$
$l_4 = \{E,F\}$	$l_{28} = \{E,G\}$	$l_{52} = \{E,H\}$	$l_{76} = \{D,H\}$	$l_{100} = \{G,H\}$
$l_5 = \{I,J\}$	$l_{29} = \{I,K\}$	$l_{53} = \{I,L\}$	$l_{77} = \{I,M\}$	$l_{101} = \{I,N\}$
$l_6 = \{K,P\}$	$l_{30} = \{J,P\}$	$l_{54} = \{J,O\}$	$l_{78} = \{J,N\}$	$l_{102} = \{J,M\}$
$l_7 = \{L,O\}$	$l_{31} = \{L,N\}$	$l_{55} = \{K,N\}$	$l_{79} = \{K,O\}$	$l_{103} = \{K,L\}$
$l_8 = \{M,N\}$	$l_{32} = \{M,O\}$	$l_{56} = \{M,P\}$	$l_{80} = \{L,P\}$	$l_{104} = \{O,P\}$
$l_9 = \{A,G\}$	$l_{33} = \{A,H\}$	$l_{57} = \{A,I\}$	$l_{81} = \{A,J\}$	$l_{105} = \{A,K\}$
$l_{10} = \{B,D\}$	$l_{34} = \{B,C\}$	$l_{58} = \{B,J\}$	$l_{82} = \{B,I\}$	$l_{106} = \{B,P\}$
$l_{11} = \{C,E\}$	$l_{35} = \{D,E\}$	$l_{59} = \{C,K\}$	$l_{83} = \{C,P\}$	$l_{107} = \{C,I\}$
$l_{12} = \{F,H\}$	$l_{36} = \{F,G\}$	$l_{60} = \{D,L\}$	$l_{84} = \{D,O\}$	$l_{108} = \{D,N\}$
$l_{13} = \{I,O\}$	$l_{37} = \{I,P\}$	$l_{61} = \{E,M\}$	$l_{85} = \{E,N\}$	$l_{109} = \{E,O\}$
$l_{14} = \{J,L\}$	$l_{38} = \{J,K\}$	$l_{62} = \{F,N\}$	$l_{86} = \{F,M\}$	$l_{110} = \{F,L\}$
$l_{15} = \{K,M\}$	$l_{39} = \{L,M\}$	$l_{63} = \{G,O\}$	$l_{87} = \{G,L\}$	$l_{111} = \{G,M\}$
$l_{16} = \{N,P\}$	$l_{40} = \{N,O\}$	$l_{64} = \{H,P\}$	$l_{88} = \{H,K\}$	$l_{112} = \{H,J\}$
$l_{17} = \{A,L\}$	$l_{41} = \{A,M\}$	$l_{65} = \{A,N\}$	$l_{89} = \{A,O\}$	$l_{113} = \{A,P\}$
$l_{18} = \{B,O\}$	$l_{42} = \{B,N\}$	$l_{66} = \{B,M\}$	$l_{90} = \{B,L\}$	$l_{114} = \{B,K\}$
$l_{19} = \{C,N\}$	$l_{43} = \{C,O\}$	$l_{67} = \{C,L\}$	$l_{91} = \{C,M\}$	$l_{115} = \{C,J\}$
$l_{20} = \{D,I\}$	$l_{44} = \{D,P\}$	$l_{68} = \{D,K\}$	$l_{92} = \{D,J\}$	$l_{116} = \{D,M\}$
$l_{21} = \{E,P\}$	$l_{45} = \{E,I\}$	$l_{69} = \{E,J\}$	$l_{93} = \{E,K\}$	$l_{117} = \{E,L\}$
$l_{22} = \{F,K\}$	$l_{46} = \{F,J\}$	$l_{70} = \{F,I\}$	$l_{94} = \{F,P\}$	$l_{118} = \{F,O\}$
$l_{23} = \{G,J\}$	$l_{47} = \{G,K\}$	$l_{71} = \{G,P\}$	$l_{95} = \{G,I\}$	$l_{119} = \{G,N\}$
$l_{24} = \{H,M\}$	$l_{48} = \{H,L\}$	$l_{72} = \{H,O\}$	$l_{96} = \{H,N\}$	$l_{120} = \{H,I\}$

Planes

(arranged by fours in parallel families)

$\pi_1 = \{A,B,C,H\}$	$\pi_{29} = \{A,B,D,G\}$	$\pi_{57} = \{A,B,E,F\}$	$\pi_{85} = \{A,B,I,J\}$	$\pi_{113} = \{A,B,K,P\}$
$\pi_2 = \{D,E,F,G\}$	$\pi_{30} = \{C,E,F,H\}$	$\pi_{58} = \{C,D,G,H\}$	$\pi_{86} = \{C,H,K,P\}$	$\pi_{114} = \{C,H,I,J\}$
$\pi_3 = \{I,J,K,P\}$	$\pi_{31} = \{I,J,L,O\}$	$\pi_{59} = \{I,J,M,N\}$	$\pi_{87} = \{D,G,L,O\}$	$\pi_{115} = \{D,G,M,N\}$
$\pi_4 = \{L,M,N,O\}$	$\pi_{32} = \{K,M,N,P\}$	$\pi_{60} = \{K,L,O,P\}$	$\pi_{88} = \{E,F,M,N\}$	$\pi_{116} = \{E,F,L,O\}$
$\pi_5 = \{A,B,L,O\}$	$\pi_{33} = \{A,B,M,N\}$	$\pi_{61} = \{A,C,D,F\}$	$\pi_{89} = \{A,C,E,G\}$	$\pi_{117} = \{A,C,I,K\}$
$\pi_6 = \{C,H,M,N\}$	$\pi_{34} = \{C,H,L,O\}$	$\pi_{62} = \{B,E,G,H\}$	$\pi_{90} = \{B,D,F,H\}$	$\pi_{118} = \{B,H,J,P\}$
$\pi_7 = \{D,G,I,J\}$	$\pi_{35} = \{D,G,K,P\}$	$\pi_{63} = \{IK,L,N\}$	$\pi_{91} = \{I,K,M,O\}$	$\pi_{119} = \{D,F,L,N\}$
$\pi_8 = \{E,F,K,P\}$	$\pi_{36} = \{E,F,I,J\}$	$\pi_{64} = \{J,M,O,P\}$	$\pi_{92} = \{J,L,N,P\}$	$\pi_{120} = \{E,G,M,O\}$
$\pi_9 = \{A,C,J,P\}$	$\pi_{37} = \{A,C,L,N\}$	$\pi_{65} = \{A,C,M,O\}$	$\pi_{93} = \{A,D,E,H\}$	$\pi_{121} = \{A,D,I,L\}$
$\pi_{10} = \{B,H,I,K\}$	$\pi_{38} = \{B,H,M,O\}$	$\pi_{66} = \{B,H,L,N\}$	$\pi_{94} = \{B,C,F,G\}$	$\pi_{122} = \{B,G,J,O\}$
$\pi_{11} = \{D,F,M,O\}$	$\pi_{39} = \{D,F,I,K\}$	$\pi_{67} = \{D,F,J,P\}$	$\pi_{95} = \{I,L,M,P\}$	$\pi_{123} = \{C,F,K,N\}$
$\pi_{12} = \{E,G,L,N\}$	$\pi_{40} = \{E,G,J,P\}$	$\pi_{68} = \{E,G,I,K\}$	$\pi_{96} = \{J,K,N,O\}$	$\pi_{124} = \{E,H,M,P\}$
$\pi_{13} = \{A,D,J,O\}$	$\pi_{41} = \{A,D,K,N\}$	$\pi_{69} = \{A,D,M,P\}$	$\pi_{97} = \{A,F,G,H\}$	$\pi_{125} = \{A,E,I,M\}$
$\pi_{14} = \{B,G,I,L\}$	$\pi_{42} = \{B,G,M,P\}$	$\pi_{70} = \{B,G,K,N\}$	$\pi_{98} = \{B,C,D,E\}$	$\pi_{126} = \{B,F,J,N\}$
$\pi_{15} = \{C,F,M,P\}$	$\pi_{43} = \{C,F,I,L\}$	$\pi_{71} = \{C,F,J,O\}$	$\pi_{99} = \{I,N,O,P\}$	$\pi_{127} = \{C,G,K,O\}$
$\pi_{16} = \{E,H,K,N\}$	$\pi_{44} = \{E,H,J,O\}$	$\pi_{72} = \{E,H,I,L\}$	$\pi_{100} = \{J,K,L,M\}$	$\pi_{128} = \{D,H,L,P\}$
$\pi_{17} = \{A,E,J,N\}$	$\pi_{45} = \{A,E,K,O\}$	$\pi_{73} = \{A,E,L,P\}$	$\pi_{101} = \{A,F,I,N\}$	$\pi_{129} = \{A,F,J,M\}$
$\pi_{18} = \{B,F,I,M\}$	$\pi_{46} = \{B,F,L,P\}$	$\pi_{74} = \{B,F,K,O\}$	$\pi_{102} = \{B,E,J,M\}$	$\pi_{130} = \{B,E,I,N\}$
$\pi_{19} = \{C,G,L,P\}$	$\pi_{47} = \{C,G,I,M\}$	$\pi_{75} = \{C,G,J,N\}$	$\pi_{103} = \{C,D,K,L\}$	$\pi_{131} = \{C,D,O,P\}$
$\pi_{20} = \{D,H,K,O\}$	$\pi_{48} = \{D,H,J,N\}$	$\pi_{76} = \{D,H,I,M\}$	$\pi_{104} = \{G,H,O,P\}$	$\pi_{132} = \{G,H,K,L\}$
$\pi_{21} = \{A,F,K,L\}$	$\pi_{49} = \{A,F,O,P\}$	$\pi_{77} = \{A,G,I,O\}$	$\pi_{105} = \{A,G,J,L\}$	$\pi_{133} = \{A,G,K,M\}$
$\pi_{22} = \{B,E,O,P\}$	$\pi_{50} = \{B,E,K,L\}$	$\pi_{78} = \{B,D,J,L\}$	$\pi_{106} = \{B,D,I,O\}$	$\pi_{134} = \{B,D,N,P\}$
$\pi_{23} = \{C,D,I,N\}$	$\pi_{51} = \{C,D,J,M\}$	$\pi_{79} = \{C,E,K,M\}$	$\pi_{107} = \{C,E,N,P\}$	$\pi_{135} = \{C,E,I,O\}$
$\pi_{24} = \{G,H,J,M\}$	$\pi_{52} = \{G,H,I,N\}$	$\pi_{80} = \{F,H,N,P\}$	$\pi_{108} = \{F,H,K,M\}$	$\pi_{136} = \{F,H,J,L\}$
$\pi_{25} = \{A,G,N,P\}$	$\pi_{53} = \{A,H,I,P\}$	$\pi_{81} = \{A,H,J,K\}$	$\pi_{109} = \{A,H,L,M\}$	$\pi_{137} = \{A,H,N,O\}$
$\pi_{26} = \{B,D,K,M\}$	$\pi_{54} = \{B,C,J,K\}$	$\pi_{82} = \{B,C,I,P\}$	$\pi_{110} = \{B,C,N,O\}$	$\pi_{138} = \{B,C,L,M\}$
$\pi_{27} = \{C,E,J,L\}$	$\pi_{55} = \{D,E,L,M\}$	$\pi_{83} = \{D,E,N,O\}$	$\pi_{111} = \{D,E,I,P\}$	$\pi_{139} = \{D,E,J,K\}$
$\pi_{28} = \{F,H,I,O\}$	$\pi_{56} = \{F,G,N,O\}$	$\pi_{84} = \{F,G,L,M\}$	$\pi_{112} = \{F,G,J,K\}$	$\pi_{140} = \{F,G,I,P\}$

Primes

(arranged by pairs in parallel families)

$\Sigma_1 = \{A,B,C,H,I,J,K,P\}$	$\Sigma_{11} = \{A,B,D,G,I,J,L,O\}$	$\Sigma_{21} = \{A,B,E,F,I,J,M,N\}$
$\Sigma_2 = \{D,E,F,G,L,M,N,O\}$	$\Sigma_{12} = \{C,E,F,H,K,M,N,P\}$	$\Sigma_{22} = \{C,D,G,H,K,L,O,P\}$
$\Sigma_3 = \{A,B,C,H,L,M,N,O\}$	$\Sigma_{13} = \{A,B,D,G,K,M,N,P\}$	$\Sigma_{23} = \{A,B,E,F,K,L,O,P\}$
$\Sigma_4 = \{D,E,F,G,I,J,K,P\}$	$\Sigma_{14} = \{C,E,F,H,I,J,L,O\}$	$\Sigma_{24} = \{C,D,G,H,I,J,M,N\}$
$\Sigma_5 = \{A,C,D,F,I,K,L,N\}$	$\Sigma_{15} = \{A,C,E,G,I,K,M,O\}$	$\Sigma_{25} = \{A,D,E,H,I,L,M,P\}$
$\Sigma_6 = \{B,E,G,H,J,M,O,P\}$	$\Sigma_{16} = \{B,D,F,H,J,L,N,P\}$	$\Sigma_{26} = \{B,C,F,G,J,K,N,O\}$
$\Sigma_7 = \{A,C,D,F,J,M,O,P\}$	$\Sigma_{17} = \{A,C,E,G,J,L,N,P\}$	$\Sigma_{27} = \{A,D,E,H,J,K,N,O\}$
$\Sigma_8 = \{B,E,G,H,I,K,L,N\}$	$\Sigma_{18} = \{B,D,F,H,I,K,M,O\}$	$\Sigma_{28} = \{B,C,F,G,I,L,M,P\}$
$\Sigma_9 = \{A,F,G,H,I,N,O,P\}$	$\Sigma_{19} = \{A,F,G,H,J,K,L,M\}$	$\Sigma_{29} = \{A,B,C,D,E,F,G,H\}$
$\Sigma_{10} = \{B,C,D,E,J,K,L,M\}$	$\Sigma_{20} = \{B,C,D,E,I,N,O,P\}$	$\Sigma_{30} = \{I,J,K,L,M,N,O,P\}$

l_1. On the other hand, although l_6 and l_9 have no point in common, they are skew and not parallel, since there is no plane which contains them both.

Thus, assuming that the necessary details have been checked, we have a specific model in which Postulates 1 to 10 and the parallel postulate are satisfied. Hence these eleven postulates, which represent the essential difference between three-dimensional and four-dimensional euclidean geometry, form a subsystem which is absolutely consistent.

Various interesting results in the geometry of four dimensions are illustrated in the preceding finite example. For instance, in four dimensions it is possible for two planes to have a single point in common, as is the case with π_1 and π_{13}, which have only the point P_1 in common. Furthermore, in four dimensions nonintersection is not a sufficient condition for two planes to be parallel, since there is now the additional requirement that the planes must lie in the same three-space.* For example, even though they do not intersect, π_1 and π_{36} are not parallel since there is no prime which contains them both. Also, in four dimensions there is a new relation which can exist between a line and a plane. In solid geometry a given line either lies in a given plane, intersects the plane in a single point, or is parallel to the plane. In four dimensions, however, in addition to these possibilities, the line and the plane may be skew, that is, they may have no point in common yet still not be parallel because they do not lie in the same three-space. For instance, although l_4 and π_{77} have no point in common, they are not parallel since there is no prime which contains them both.

At this point it is of course not clear whether the unusual features we have just observed in our finite model are peculiar to the model or are consequences of the postulates and hence true in every model. Actually, they follow from the postulates and are not special properties of this particular model, and in the next section we shall prove them as theorems.

EXERCISES

1. In the finite four-dimensional geometry discussed in the text, determine
    (a) The lines which contain the point P_1 and are parallel to the plane π_2
    (b) The lines which contain the point P_1 and are parallel to, that is, do not intersect, the prime Σ_2

---

* This, of course, is analogous to the requirement, which appears when we pass from plane to solid geometry, that parallel lines, in addition to having no point in common, must lie in the same plane.

  (*c*) The lines which contain the point P_1 and are skew to the plane π_2

  (*d*) The planes which contain the point P_1 and are parallel to, that is, do not intersect, the prime Σ_2

  (*e*) The planes which contain the point P_1 and are skew to the plane π_2

  (*f*) The planes which contain the point P_1 and meet π_2 in a single point

 2. Prove that in any finite four-dimensional geometry satisfying the postulates introduced in this section and the euclidean parallel postulate, if there exists one line which contains exactly n points, then

  (*a*) Every line contains exactly n points.

  (*b*) Every plane contains exactly n^2 points.

  (*c*) Every prime contains exactly n^3 points.

  (*d*) Space contains exactly n^4 points.

 3. In any finite four-dimensional geometry satisfying the postulates introduced in this section and the euclidean parallel postulate,

  (*a*) How many lines are there?

  (*b*) How many planes are there?

  (*c*) How many primes are there?

 4. What new undefined terms, if any, would be required in an axiomatic development of five-dimensional euclidean geometry? six-dimensional euclidean geometry? n-dimensional euclidean geometry?

 5. What changes in or additions to the postulates of extension and connection do you think would be required in order to pass from four-dimensional to five-dimensional euclidean geometry?

3.3   Connection Theorems.   Among the consequences of the connection postulates are a number of results whose proofs are immediate but which nonetheless are of sufficient interest to warrant explicit mention.   The first two are familiar properties of solid geometry which hold equally well in four dimensions.   The others all involve primes and hence are meaningful only in E_4.

 Theorem 1. There is a unique plane which contains a given line and a point not on the line.

 Theorem 2. There is a unique plane which contains each of two intersecting lines.

 Theorem 3. If two points of a line lie in a prime, then every point of the line lies in the prime.

 Proof. Let l be a line and let P and Q be two points of l which lie in a prime, Σ.   By Postulate 5, there is at least one point, R, in Σ which is not collinear with P and Q.   By Postulate 4, there is a unique plane, π, which contains P, Q, and R; and, by Postulate 8, this plane contains every point of l. Finally, by Postulate 9, π lies entirely in Σ. Hence l, which lies in π, must lie in Σ, as asserted.

Theorem 4.   There is a unique prime which contains a given plane and a point not in the plane.

Theorem 5.   There is a unique prime which contains a given plane and a line intersecting the plane in a single point.

Proof.   Let π be a plane and let l be a line which has a single point, P, in common with π. By Postulate 1, l contains at least one point, Q, distinct from P, and, by hypothesis, Q does not lie in π. Hence, by Theorem 4, there is a unique prime, Σ, which contains π and Q (see Fig. 3.1). Moreover, by Theorem 3, l lies entirely in Σ since two of its

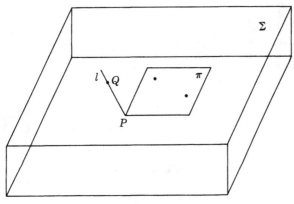

Fig. 3.1

points, P and Q, lie in Σ. Finally, Σ is the only prime with this property, for any prime, Σ', which contains π and l necessarily has four noncoplanar points in common with Σ and hence, by Postulate 6, must coincide with Σ. Thus there is one and only one prime containing π and l, as asserted.

Theorem 6.   There is a unique prime which contains each of two planes which have a line in common.

Theorem 7.   There is a unique prime which contains each of two skew lines.

Theorem 8.   There is a unique prime which contains each of three concurrent noncoplanar lines.

In the last section we saw that in one particular model in which the connection postulates of four-dimensional euclidean geometry are satisfied it is possible for two planes to have a single point in common.   At

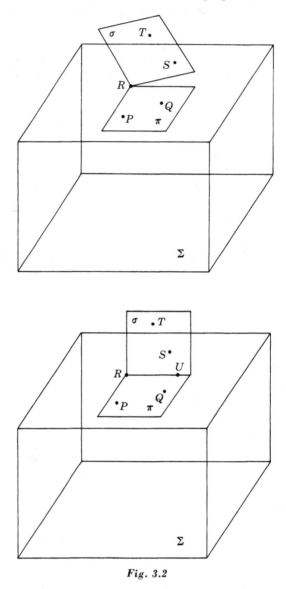

Fig. 3.2

that time we were unable to say whether this novel property was peculiar to the particular model or was a logical consequence of the postulates and hence true in every model.   That the latter is in fact the case is guaranteed by the following theorem:

Theorem 9.   There exist pairs of planes which have a single point in common.

Proof.   Let Σ be an arbitrary prime and let P, Q, R, S be four non-collinear noncoplanar points of Σ, the existence of which is guaranteed by Postulate 5. By Postulate 7, there exists a point, T, which is not a point of Σ, and, by Theorem 3, T is not collinear with any two of the points P, Q, R, S. Hence, by Postulate 4, the points P, Q, R determine a plane, π, and the points R, S, T determine a plane, σ. Clearly, these two planes have the point R in common; moreover, they can have no other common point (see Fig. 3.2). To prove this, let us assume the contrary and suppose that π and σ have a second point, U, in common. The point U cannot lie on the line $\overleftrightarrow{RS}$ because if this were the case, then by Postulate 8, $\overleftrightarrow{RS}$ would lie entirely in π, contrary to the hypothesis that S is not coplanar with P, Q, and R. Hence σ has three noncollinear points, R, S, U, in common with Σ, and therefore by Postulate 9, every point of σ must lie in Σ. But this is impossible, since by hypothesis, T, which is a point of σ, does not lie in Σ. Thus π and σ have only the one point P in common, as asserted.

The postulates of E_4 give us no explicit information about the intersection of two primes, but from them it is possible to prove the following important result:

Theorem 10.   If two primes have a point in common, their intersection is a plane.

Proof.   Let Σ_1 and Σ_2 be two primes having a point, P, in common (see Fig. 3.3).

1. In Σ_1 there are at least two points, A and A', which are not collinear with P.	1. Postulate 5.
2. A, A', and P determine a plane, α.	2. Postulate 4.
3. α lies in Σ_1.	3. Postulate 9.
4. The intersection of α and Σ_2 is a line, a, which contains P.	4. Postulate 10.
5. In Σ_2 there are at least two points, B and B', which are not coplanar with a.	5. Postulate 5.

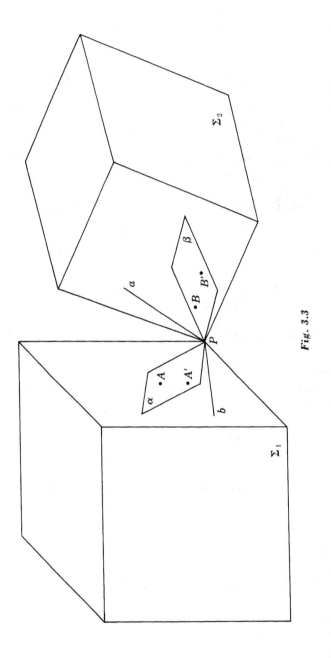

Fig. 3.3

6. B and B' are not collinear with P.	**6.** If they were, $\overleftrightarrow{BB'}$ would be coplanar with a, contrary to step 5.
7. B, B', P determine a plane, β, lying in Σ_2.	**7.** Postulate 4, Postulate 9.
8. β does not contain a.	**8.** Step 5.
9. The intersection of β and Σ_1 is a line, b, which contains P.	**9.** Postulate 10.
10. a lies in both Σ_1 and Σ_2.	**10.** Steps 3 and 4.
11. b lies in both Σ_1 and Σ_2.	**11.** Steps 7 and 9.
12. a and b are distinct lines.	**12.** b lies in β (step 9) but a does not (step 8).
13. a and b determine a plane, π.	**13.** Theorem 2.
14. π lies in both Σ_1 and Σ_2.	**14.** Postulate 9.
15. Σ_1 and Σ_2 can have nothing more than π in common.	**15.** Postulate 6.

Q.E.D.

We are now in a position to answer the following fundamental question: In four-dimensional euclidean geometry, is the geometry in each prime identical with euclidean solid geometry? Reasoning by analogy, we would expect this to be the case, because we know that the geometry in each plane in three-dimensional euclidean geometry is identical with euclidean plane geometry. However, although plausible, this result must be proved, because the postulates of E_4 do not include all the postulates of solid geometry. Specifically, Postulate 6 of three-dimensional euclidean geometry is not included among the postulates for E_4, and conceivably, its omission could mean that certain results, provable by its use in solid geometry, could not be proved in four dimensions. Fortunately, this is not the case, because in E_4, Postulate 6 is provable as a theorem:

Theorem 11. If two planes which lie in the same prime have a point in common, their intersection is a line.

Proof. Let π and σ be two planes which lie in a prime, Σ, and have a point, P, in common (see Fig. 3.4).

1. There is a point, Q, which is not in Σ and hence not in σ.	**1.** Postulate 7.
2. Q and σ determine a prime, Σ_1.	**2.** Theorem 4.

3. The intersection of Σ and Σ_1 is σ.

   3. Theorem 10.

4. π and Σ_1 have a line, l, in common.

   4. Postulate 10.

5. l lies in Σ.

   5. l lies in π (step 4) and π lies in Σ (hypothesis).

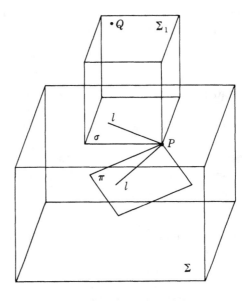

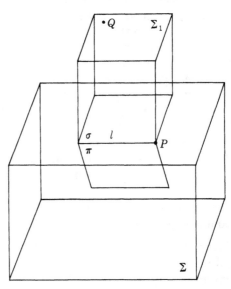

Fig. 3.4

6. l lies in the intersection of Σ
   and Σ_1.
7. l lies in σ.
8. The intersection of π and σ
   is l.

6. l lies in Σ (step 5) and also in
   Σ_1 (step 4).
7. Step 6 and step 3.
8. l lies in π (step 4) and also
   in σ (step 7).

Q.E.D.

With Theorem 11 established, every postulate of solid geometry is true in four dimensions, either as a postulate of E_4 or as a theorem in E_4. Hence every theorem of solid geometry holds true in any three-space in E_4, just as every theorem of plane geometry holds true in any plane in E_3. For instance, although we shall see in Sec. 3.5 that the familiar assertion

> There is a unique line which passes through a given point in a given plane and is perpendicular to the given plane.

is false in E_4, the following statement is true:

> *In any prime containing a given plane,* there is a unique line which passes through a given point in the given plane and is perpendicular to the plane.

Moreover, theorems such as those dealing with the congruence and similarity of triangles, which depend upon postulates that are identical in three- and four-dimensional geometry, are true without restriction in E_4.

With identical information available in each case, the theorems of solid geometry can be proved in any prime in E_4 in exactly the same way as they were established in E_3. Hence there is no reason to state and prove these results again in this chapter. Accordingly, we shall say little or nothing about the measurement of distance, angle, and area, the congruence and similarity of triangles, or circles and their properties. The essentially new problems we have to investigate are those associated with configurations which cannot be contained in a single prime. These involve primarily the notions of perpendicularity and parallelism, and to these we shall devote our attention in the rest of this chapter, using without further proof any results from solid geometry we find convenient.

EXERCISES

1. Prove Theorem 1.
3. Prove Theorem 4.
5. Prove Theorem 7.

2. Prove Theorem 2.
4. Prove Theorem 6.
6. Prove Theorem 8.

7. Prove that there exist nonintersecting planes which do not lie in the same prime.

8. Do there exist nonintersecting, nonparallel lines which do not lie in the same prime?

9. Prove that the lines which pass through an arbitrary point in E_4 do not all lie in the same prime.

10. Prove that for every plane there exists at least one line which does not intersect the plane and does not lie in the same prime with the plane.

11. Prove that it is unnecessary to assume in Postulate 5 that the points of an arbitrary prime are noncollinear.

12. State and prove the four-dimensional generalization of the space-separation theorem (Theorem 9, Sec. 2.5).

13. If l_1 and l_2 are two skew lines, does there exist a line which passes through an arbitrary point and intersects both l_1 and l_2?

14. If l_1, l_2, and l_3 are three lines which are mutually skew and do not all lie in the same prime, how many lines, if any, can intersect l_1, l_2, and l_3?

15. If three primes have a point in common, do the three planes in which they intersect by twos lie in the same prime?

3.4 Parallel Relations in E_4. In plane geometry, parallel lines are defined by the single condition that they be nonintersecting. In solid geometry, however, this is insufficient, and the additional requirement that they be coplanar must be added. By analogy, although in solid geometry a line and a plane or two planes are parallel if they are nonintersecting, it is to be expected that in four-dimensional geometry the additional requirement that they be coprimal is necessary, and this is indeed the case:

Definition 1.   A line and a plane which are coprimal and do not intersect are said to be parallel.

Definition 2.   A line and a plane which are not coprimal and do not intersect are said to be skew.

Definition 3.   Two planes which are coprimal and do not intersect are said to be parallel.

Definition 4.   Two planes which are not coprimal and do not intersect are said to be skew.

Definition 5.   Two planes which intersect in a single point are said to be semiskew.

The existence of semiskew planes is guaranteed by Theorem 9, Sec. 3.3. The existence of lines and planes which are skew to each other and the existence of skew planes follows from the results of Exercises 10 and 7, Sec. 3.3, respectively.

Definition 6.   A line and a prime which do not intersect are said to be parallel.

Definition 7.   A plane and a prime which do not intersect are said to be parallel.

Definition 8.   Two primes which do not intersect are said to be parallel.

The existence of lines, planes, and primes parallel to a given prime is guaranteed by theorems to which we shall shortly turn our attention.

Among the simpler theorems involving parallel relations in E_4, we have the following, which are little more than restatements of familiar results from solid geometry:

Theorem 1.   Two lines which are parallel to the same line are parallel to each other.

Proof.   Let l_1, l_2, and l_3 be three lines such that $l_1 \| l_2$ and $l_2 \| l_3$. From the definition of parallel lines, l_1 and l_2 lie in a plane, π, and l_2 and l_3 lie in a plane, σ. Moreover, since π and σ have l_2 in common, it follows from Theorem 6, Sec. 3.3, that there is a unique prime which contains these planes and hence contains l_1, l_2, and l_3. Thus we are actually concerned with a configuration which lies in a single prime and to which, therefore, the familiar results of solid geometry are applicable. Since we have already proved that in three dimensions two lines which are parallel to the same line are parallel to each other, the present theorem is also established.

The proof of Theorem 1 is also essentially the proof of the following useful result:

Theorem 2.   There is a unique prime which contains each of three noncoplanar parallel lines.

Theorem 3.   If l_1 and l_1' are two lines which intersect in a point P_1, and if l_2 and l_2' are two lines which intersect in a point P_2 and are respectively parallel to l_1 and l_1', then the plane determined by l_2 and l_2' is parallel to the plane determined by l_1 and l_1'.

Proof.   Since $l_1 \| l_2$, l_1 and l_2 determine a plane π; and since $l_1' \| l_2'$, l_1' and l_2' determine a plane π' (see Fig. 3.5).   Moreover, by Theorem 6, Sec. 3.3, since π and π' have the line $\overleftrightarrow{P_1 P_2}$ in common, there is a unique

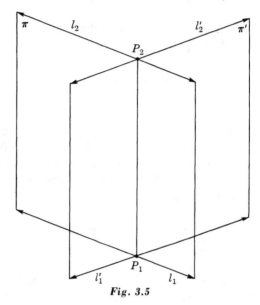

Fig. 3.5

prime which contains them both and, perforce, contains l_1, l_1', l_2, and l_2'. The assertion of the theorem is thus an assertion about a configuration existing in a single prime and as such is decided by the facts of solid geometry.   In particular, in Exercise 11, Sec. 2.15, the result in question was shown to be true in three dimensions; hence the present theorem is established.

Using Theorems 1 and 3, it is now easy to prove the transitive property for the parallel relation for planes:

Theorem 4.   Two planes which are parallel to the same plane are parallel to each other.

Proof.   Let π_1, π_2, and π_3 be three planes such that $\pi_1 \| \pi_2$ and $\pi_2 \| \pi_3$ (see Fig. 3.6).

| 1. | π_1 and π_2 are coprimal, and π_2 and π_3 are coprimal. | 1. | Definition of parallel planes. |

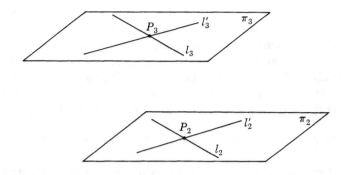

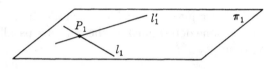

Fig. 3.6

2. In π_2 there are two intersecting lines, l_2 and l_2'.	**2.** Postulate 3.
3. Through an arbitrary point, P_1, in π_1 there is a line, l_1, lying in π_1 and parallel to l_2; through an arbitrary point, P_3, in π_3 there is a line, l_3, lying in π_3 and parallel to l_2.	**3.** Step 1 and the facts of solid geometry.
4. $l_1 \| l_3$.	**4.** Theorem 1.
5. Through P_1 there is a line, l_1', in π_1 and parallel to l_2'; through P_3 there is a line, l_3', lying in π_3 and parallel to l_2'.	**5.** Step 1 and the facts of solid geometry.
6. $l_1' \| l_3'$.	**6.** Theorem 1.
7. $\pi_1 \| \pi_3$.	**7.** Steps 4 and 6 and Theorem 3.
	Q.E.D.

The next four theorems establish the existence of lines, planes, and primes parallel to a given prime.

> **Theorem 5.** If l is a line in a prime, Σ, then any line which is parallel to l and contains a point which is not in Σ is parallel to Σ.

Proof. Let l be a line in a prime, Σ, and let l' be a line which is parallel to l and contains a point, P', which is not in Σ. To prove the theorem, let us assume the contrary and suppose that l' intersects Σ in a point, P. Obviously P is not a point of l, since l and l' are parallel. Therefore P and l determine a plane, π, which by Postulate 9 lies entirely in Σ. Moreover, since parallel lines are necessarily coplanar, l' must lie in π and hence in Σ. But this is impossible, since l' contains P', which is not a point of Σ. Thus l' is parallel to Σ, as asserted.

By an almost identical argument, we can establish the following:

Theorem 6. If π is a plane in a prime, Σ, then any plane which is parallel to π and contains a point which is not in Σ is parallel to Σ.

Theorem 7. If each of two intersecting lines is parallel to a prime, then the plane determined by these lines is parallel to the prime.

Proof. Let l_2 and l_2' be two intersecting lines, each parallel to a prime, Σ; let P' be their point of intersection; and let π_2 be the plane they determine. Then through an arbitrary point, P, of Σ there is a unique line, l_1, parallel to l_2 and a unique line, l_1', parallel to l_2'. Moreover, by Theorem 3, the plane, π_2, containing l_2 and l_2' is parallel to the plane, π_1, containing l_1 and l_1'. But if this is the case, then by Theorem 6, π_2 is parallel to Σ, as asserted.

From the last three theorems it should be clear that through a point not in a given prime there passes more than one line which is parallel to the prime and also more than one plane which is parallel to the prime. What the situation is with respect to primes which are parallel to primes we must now determine. The parallel postulate, of course, guarantees that through a given point not on a given line there passes one and only one line which is parallel to the given line. Similarly, the theorems of solid geometry guarantee that through a given point not in a given plane there passes one and only one plane which is parallel to a given plane. In fact, if there is such a parallel plane, it must, by definition, lie in the prime determined by the given point and the given plane; and in this prime, the existence and uniqueness of the parallel plane is one of the familiar properties of the geometry of three dimensions. By analogy with the foregoing, it is natural to expect that there is a unique prime which is parallel to a given prime and contains a given point which is not in the given prime. The following theorem assures us that this is the case:

Theorem 8.   There is one and only one prime which is parallel to a given prime and contains a given point not in the given prime.

Proof.   Let Σ be a given prime and let P' be a given point which is not in Σ (see Fig. 3.7).   Let P be an arbitrary point in Σ and let r, s,

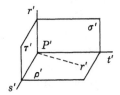

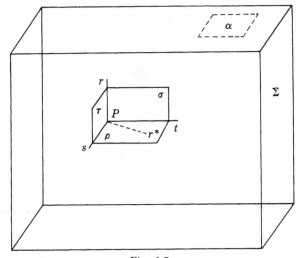

Fig. 3.7

and t be three noncoplanar lines through P in Σ.   Then through P' there are, by the parallel postulate, unique lines, r', s', and t', respectively, parallel to r, s, and t.   We hope to show that the lines r', s', and t' determine a prime which is the unique prime parallel to Σ through P'.   Hence we must first prove that r', s', and t' are noncoplanar, so that they do, indeed, determine a prime.   Now s' and t' determine a plane, ρ', which by Theorem 3 is parallel to the plane, ρ, determined by s and t.   If r' were contained in ρ', then in the prime containing the parallel planes, ρ and ρ', the intersection of ρ and the plane determined by r' and P would be a line, r^*, passing through P and parallel to r'.   Moreover, r and r^* are different lines, since, by hypothesis, r does not lie in the plane, ρ, determined by s and t.

Hence, through P there are two lines, r and r^*, each parallel to r'. Since this contradicts the parallel postulate, the assumption that r' lies in the plane, ρ', containing s' and t' must be abandoned. Therefore r', s', and t' are noncoplanar and hence, by Theorem 8, Sec. 3.3, there is a unique prime, Σ', which contains them, and this prime is parallel to Σ. To prove this, we observe first that by Theorem 7, the planes ρ', σ', and τ' determined by the respective pairs of lines (s',t'), (r',t'), (r',s') are all parallel to Σ. Next we note that if Σ and Σ' are not parallel, then by Theorem 10, Sec. 3.3, their intersection is a plane, α. Finally, since ρ', σ', and τ' are parallel to Σ, each is parallel to α. But this is impossible because, from solid geometry, we know that in a prime such as Σ', there is a unique plane which is parallel to a given plane, that is, α, and contains a given point, that is, P', which is not in the given plane. Thus the assumption that Σ and Σ' intersect must be abandoned, and the two primes are parallel. Finally, let us suppose that besides Σ' there is a prime Σ'' which also contains P' and is parallel to Σ. Then the plane determined by r and P', which of course contains r', intersects Σ'' in a line which passes through P' and is parallel to r. But since there is a unique parallel to r through P', this line and r' must be the same; that is, Σ' and Σ'' have r' in common. Similarly, Σ' and Σ'' must have s' and t' in common, and therefore, by Theorem 8, Sec. 3.3, they must coincide. Thus the prime which is parallel to Σ through P' is unique, as asserted.

Using Theorem 8, it is now an easy matter to prove the following theorem:

Theorem 9.   Two primes which are parallel to the same prime are parallel to each other.

We now turn our attention to a number of intersection theorems which involve parallel lines, planes, and primes. The first is very much like Theorem 9, Sec. 2.15, planes in E_3 playing in many respects the role of primes in E_4.

Theorem 10.   If a line intersects one of two parallel primes in a single point, it intersects the other in a single point also.

Proof. Let Σ_1 and Σ_2 be parallel primes and let l be a line which has a single point, P_1, in common with Σ_1 (see Fig. 3.8). To prove the theorem, let us assume the contrary and suppose that l does not intersect Σ_2. Now l and an arbitrary point, P_2, in Σ_2 determine a plane, π. Moreover, since π has P_1 in common with Σ_1 and P_2 in common with

Σ_2, it intersects Σ_1 in a line, l_1, and Σ_2 in a line, l_2. Clearly, since Σ_1 and Σ_2 are parallel, l_1 and l_2 can have no point in common. Furthermore, l can have no point in common with l_2, since l_2 lies in Σ_2 and our

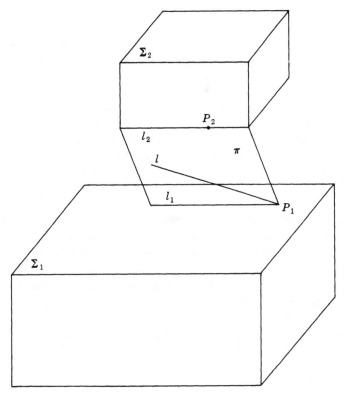

Fig. 3.8

working hypothesis is that l does not intersect Σ_2. Moreover, l and l_1 are distinct lines, since l_1 lies in Σ_1 and l does not. Hence in the plane π there are two lines, l and l_1, each passing through P_1 and each having no point in common with l_2. Since this violates the parallel postulate, the assumption that l does not intersect Σ_2 must be abandoned, and the theorem is established.

The following related result is proved in very much the same way:

Theorem 11. If a prime intersects one of two parallel lines in a single point, it intersects the other in a single point also.

For planes instead of lines we have the following result, which is analogous to Theorem 15, Sec. 2.15:

Theorem 12.   If a plane intersects one of two parallel primes in a
line, it intersects the other in a line also, and the
two lines of intersection are parallel.

Proof.   Let Σ_1 and Σ_2 be two parallel primes and let π be a plane
which intersects Σ_1 in a line, l_1.   To prove, first, that π intersects Σ_2,
let us assume the contrary and suppose that π has no point in common
with Σ_2.   Then π and an arbitrary point, P, of Σ_2 determine a prime,
Σ_3, which, by Theorem 10, Sec. 3.3, intersects Σ_1 in a plane, π_1, and
intersects Σ_2 in a plane, π_2.   Moreover, π_1 cannot intersect π_2, since
Σ_1 and Σ_2 are parallel.   Furthermore, under our working hypothesis,
π does not intersect Σ_2; hence π cannot intersect π_2, since π_2 lies in Σ_2.
Finally, π and π_1 are distinct planes since the latter lies in Σ_1 and the
former does not.   Therefore, in the prime Σ_3 there are two planes,
π and π_1, each passing through P and each having no point in common
with π_2.   But this contradicts Theorem 13, Sec. 2.15.   Hence the
assumption that π does not intersect Σ_2 must be abandoned, and
therefore, by Postulate 10, π and Σ_2 have a line, l_2, in common.   Moreover, if l_1 and l_2, both of which lie in π, had a point in common, this
point would be common to Σ_1 and Σ_2, contrary to the hypothesis that
these primes are parallel.   Hence the lines of intersection, l_1 and l_2,
are parallel, as asserted.

In very much the same way, the following theorem can be established:

Theorem 13.   If a prime intersects one of two parallel planes in a
line, it intersects the other in a line also, and the
two lines of intersection are parallel.

The next intersection theorem is an immediate consequence of
Theorem 8.

Theorem 14.   If a prime intersects one of two parallel primes in a
plane, it intersects the other in a plane also, and the
two planes of intersection are parallel.

The last of this group of intersection theorems follows at once from
Postulate 10 and Theorem 9, Sec. 2.15:

Theorem 15.   If a plane intersects one of two parallel planes in a
single point, it intersects the other in a single point
also.

EXERCISES

1. Prove Theorem 6.
2. Prove Theorem 9.
3. Prove Theorem 11.
4. Prove Theorem 13.
5. Prove Theorem 14.
6. Prove Theorem 15.
7. Show that the lines which contain a given point and are parallel to a given plane, π, all lie in a plane which is parallel to π.
8. If a line, l, and a plane, π, are skew, does there exist a plane which contains l and is parallel to π?
9. Show that the lines which pass through a given point and are parallel to a given prime all lie in the prime which contains the given point and is parallel to the given prime.
10. Show that the planes which pass through a given point and are parallel to a given prime all lie in the prime which contains the given point and is parallel to the given prime.
11. If l_1 and l_2 are two skew lines and if P is a point which does not lie in the prime determined by l_1 and l_2, show that there is a unique plane which contains P and is parallel to both l_1 and l_2.
12. If π_1 and π_2 are two skew planes and if P is a point which does not lie in either π_1 or π_2, show that there is a unique line which contains P and is parallel to both π_1 and π_2.
13. If π_1 and π_2 are two skew planes and if P is a point which does not lie in either π_1 or π_2, show that there is a unique prime which contains P and is parallel to both π_1 and π_2.
14. If a plane intersects one of two parallel planes in a line, does it necessarily intersect the other plane?
15. Can a line be parallel to each of two skew lines?
16. If a plane is skew to one of two parallel planes, is it necessarily skew to the other?
17. Can a plane be skew to each of (a) Two parallel planes? (b) Two planes which have a line in common? (c) Two planes which have a point in common? (d) Two skew planes?
18. Can a prime be parallel to each of two skew lines?
19. Prove that two planes which contain a point, P, and are each parallel to a line, l, intersect in a line which is parallel to l.
20. Prove that two primes which contain a point, P, and are each parallel to a plane, π, intersect in a plane which is parallel to π.
21. Prove that two primes which contain a point, P, and are each parallel to a line, l, intersect in a plane which is parallel to l.
22. If a plane and a prime which does not contain the plane are each parallel to a line, l, and have a point in common, prove that they intersect in a line which is parallel to l.
23. If π_1 and π_2 are two semiskew planes and if l is a line in π_1, can π_2 contain a line which is parallel to l?
24. If π_1 and π_2 are two skew planes and if l is a line in π_1, can π_2 contain a line which is parallel to l? Does it necessarily contain a line which is parallel to l?

25.   Prove that all lines which are parallel to two skew planes are parallel to each other.

3.5   Perpendicularity Relations in E_4.   It is a fundamental fact of plane geometry that there is a unique perpendicular to a given line at a given point of that line.   In other words, to put it somewhat artificially,

> The perpendiculars to a given line at a given point of that line all lie in a line.

This statement is, of course, false in solid geometry, since there the perpendicular to a given line at a given point is not unique, and the correct assertion is that

> The perpendiculars to a given line at a given point of that line all lie in a plane.

Our knowledge of plane and solid geometry thus suggests that the last statement probably becomes false when we move from three to four dimensions, just as the first statement became false when we passed from plane to solid geometry, and furthermore that in E_4 the correct statement is probably

> The perpendiculars to a given line at a given point of that line all lie in a prime.

Our first four theorems deal with this conjecture and together establish its truth.

> **Theorem 1.**   The lines which are perpendicular to a given line at a given point of that line do not all lie in the same plane.

Proof.   Let p be an arbitrary line, let P be any point on p, and let Σ be any prime containing p (see Fig. 3.9).   Then, from the familiar properties of solid geometry, we know that in Σ there is a unique plane, π, which is perpendicular to p at P and which, perforce, contains all the lines of Σ which are perpendicular to p at P.   Now let Q be any point which is not in Σ.   Clearly, Q cannot be a point of p; hence Q and p determine a plane, σ.   Now we know from the properties of plane geometry that in σ there is a unique line, q, which is perpendicular to p at P.   Moreover, since σ and Σ intersect only in p, it is evident that q cannot lie in π.   Thus we have shown that the perpendiculars to a given line at a given point do not all lie in the same plane, as asserted.

Theorem 1 assures us that there are at least three noncoplanar perpendiculars to a given line, p, at a given one of its points, P, but we do not yet know that the prime determined by three such perpendiculars contains every perpendicular to p at P nor, conversely, that every

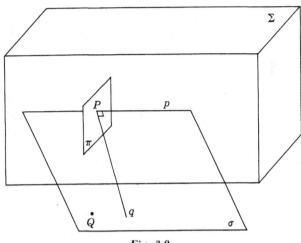

Fig. 3.9

line through P in such a prime is perpendicular to p.   The next two theorems settle these important questions.

Theorem 2.   A line which is perpendicular to each of three concurrent noncoplanar lines at their common point is perpendicular to every line which passes through their intersection and lies in the prime they determine.

Proof.   Let a, b, and c be three noncoplanar lines each of which is perpendicular to a line, p, at the point P, and let Σ be the prime determined by a, b, and c (see Fig. 3.10).   Clearly, p cannot lie in Σ.

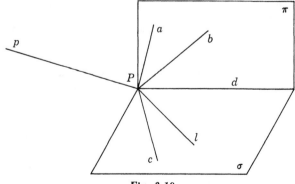

Fig. 3.10

Let l be any line in Σ which passes through P and is distinct from a, b, and c. Now a and b intersect in P and hence determine a plane, π. Likewise c and l intersect in P and hence determine a plane, σ. Now π and σ both lie in the prime Σ and have P in common. Hence, by Theorem 11, Sec. 3.3, they must intersect in a line, d, which passes through P and is necessarily distinct from c. Furthermore, since π contains two lines, a and b, which are perpendicular to p, every line in π which passes through P is perpendicular to p. Therefore d, which of course lies in π, is perpendicular to p at P and hence σ contains two lines, c and d, each perpendicular to p at P. This, in turn, means that every line in σ which contains P is perpendicular to p. Finally, since l lies in σ and passes through P, l is perpendicular to p at P, as asserted.

Theorem 3.   The lines which are perpendicular to a given line at a given point of the line all lie in the same prime.

Proof.   Let p be an arbitrary line and let a, b, and c be three non-coplanar lines each perpendicular to p at a point P (see Fig. 3.11).

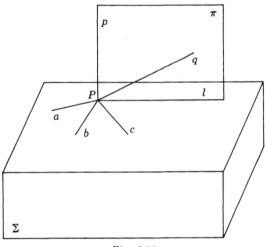

Fig. 3.11

By the last theorem, a, b, and c determine a prime, Σ, with the property that every line in Σ which passes through P is perpendicular to p at P. Now suppose that there is a line, q, which is perpendicular to p at P but does not lie in Σ. The lines p and q determine a plane, π, and this plane, having P in common with Σ, intersects Σ in a line, l, obviously distinct from q, which does not lie in Σ. By the last theorem, l is perpendicular to p at P, and by hypothesis, q is perpendicular to p at P. Hence in the plane π we have two lines, q and l, each perpendicular to

p at P. Since this is impossible, we must abandon the assumption that there is a line, q, which is perpendicular to p at P and does not lie in Σ. Hence the theorem is established.

Introducing the obvious definition, we may restate the preceding results more concisely as a single theorem:

Definition 1. A prime and a line are said to be perpendicular to each other if and only if their intersection is a point and every line in the prime which passes through this point is perpendicular to the line.

Theorem 4. At any point of a given line there is a unique prime which is perpendicular to the given line.

Using Theorem 4 it is now an easy matter to establish its expected counterpart:

Theorem 5. At every point of a given prime there is a unique line which is perpendicular to the given prime.

Proof. Let P be an arbitrary point of an arbitrary prime, Σ, and let a, b, and c be three noncoplanar lines of Σ each of which contains P. At P there is a unique prime, Σ_a, perpendicular to a, a unique prime, Σ_b, perpendicular to b, and a unique prime, Σ_c, perpendicular to c. Moreover, Σ_a and Σ_b, having P in common, have at least a plane, π, in common. Furthermore π and Σ_c, having P in common, have at least a line, p, in common. Since p passes through P and lies in Σ_a, Σ_b, and Σ_c, it is simultaneously perpendicular to a, b, and c. Hence, by Theorem 2, it is perpendicular to Σ at P. Now suppose that there is a second line, p', perpendicular to Σ at P. The plane of p and p', having P in common with Σ, must intersect Σ in a line, s, to which both p and p' must be perpendicular. But this is impossible, since in a given plane there is only one line perpendicular to a given line at a given point. Hence the assumption of a second line perpendicular to Σ at P must be abandoned, and the theorem is established.

By analogy with the theorem of solid geometry which asserts that there is a unique plane which passes through a given point not on a given line and is perpendicular to the given line, there is reason to believe that in E_4 there is a unique prime which passes through a given point not on a given line and is perpendicular to the given line, and we do, in fact, have the following extension of Theorem 5:

Theorem 6.   There is a unique prime which passes through a given point and is perpendicular to a given line.

Proof.   Let p be a given line and let P be a given point. If P is a point of p, the existence of a unique prime which contains P and is perpendicular to p is simply a restatement of Theorem 4. Suppose, therefore, that P is not a point of p. From P there is then a unique perpendicular to p, say l. Let F be the foot of this perpendicular. Now at F there is, by Theorem 4, a unique prime perpendicular to p, and since this prime contains every line which is perpendicular to p at P, it contains l and hence passes through P. Thus there is at least one prime which contains P and is perpendicular to p. Suppose now that there were a second prime perpendicular to p at a point, F', necessarily distinct from F since there is only one prime perpendicular to a given line at a given point. In the plane determined by p and P we now have two lines, $\overleftrightarrow{PF}$ and $\overleftrightarrow{PF'}$, each perpendicular to p and passing through P. This is impossible, however. Hence the assumption of a second prime containing P and perpendicular to p must be rejected, and the theorem is established.

We now turn our attention to certain perpendicularity relations involving the planes of E_4:

Theorem 7.   There is more than one line which is perpendicular to a given plane at a given point of the plane.

Proof.   Let P be an arbitrary point of an arbitrary plane, π, and let Σ_1 be a prime containing π. Then, from the results of solid geometry, there is in Σ_1 a unique line, p_1, which passes through P and is perpendicular to π. Let Q be an arbitrary point which is not in Σ_1. The point Q and the plane π determine a second prime, Σ_2, whose intersection with Σ_1 is of course π. In Σ_2 there is also a unique line, p_2, perpendicular to π at P. Clearly, p_1 and p_2 are distinct, since p_1 lies in Σ_1 and p_2 does not. Hence the theorem is established.

Theorem 8.   The lines which are perpendicular to a given plane at a given point all lie in a plane.

Proof.   Let P be an arbitrary point of an arbitrary plane, π, and let a and b be two lines in π which intersect at P. Now any line which is perpendicular to π at P must be perpendicular to a and hence must lie in the prime, Σ_a, which is perpendicular to a at P. Similarly, any line perpendicular to π at P must also be perpendicular to b and hence must lie in the prime, Σ_b, which is perpendicular to b at P. Moreover, the only lines which are perpendicular to a at P are the lines which pass through P and lie in Σ_a, and the only lines which are perpendicular to b at P are the lines which pass through P and lie in Σ_b. Therefore, the only lines which are simultaneously perpendicular to a and to b and hence to π are the lines which pass through P and lie in both Σ_a and Σ_b. Now Σ_a and Σ_b are distinct, for otherwise there would be two lines, a and b, each perpendicular to the same prime, $\Sigma_a = \Sigma_b$, at the same point, which is impossible by Theorem 5. Hence Σ_a and Σ_b intersect in a plane, σ, and a line is perpendicular to π at P if and only if it passes through P and lies in σ, as asserted. Moreover, not only is every line through P in σ perpendicular to π but, conversely, every line through P in π is perpendicular to σ. Hence π is also the union of all the lines which are perpendicular to σ at P.

Definition 2.   Two planes which intersect in a single point and have the property that every line in either plane which passes through their common point is perpendicular to the other plane are said to be absolutely perpendicular.

As a matter of notation, we shall indicate that the planes π and σ are absolutely perpendicular by writing

$$\pi \underset{\text{a}}{\perp} \sigma$$

Theorem 8 assures us that there is a unique plane which is absolutely perpendicular to a given plane at a given point of that plane. It is also easy to show, by an argument very much like the one we used to prove Theorem 6, that there is a unique plane which passes through a given point not in a given plane and is absolutely perpendicular to that plane. Hence, combining these results, we have the following theorem:

Theorem 9.   There is a unique plane which contains a given point and is absolutely perpendicular to a given plane.

From the proof of Theorem 8, it is clear that if a plane, σ, contains two lines which are perpendicular to a given plane, π, at a given point, P, then σ is absolutely perpendicular to π at P, and every line through P in σ is perpendicular to π. On the other hand, there are planes which are semiskew to a given plane, π, at a point, P, and contain either just one or none of the lines which are perpendicular to π at P. The latter is the usual relation which holds between two semiskew

planes.   The former, like absolute perpendicularity, is a special property which does not hold in general and for which we need a special
name:

Definition 3.   A plane which is semiskew to a given plane at a
given point and contains one and only one perpendicular to the given plane at the given point is said
to be semiperpendicular to the given plane.

We shall indicate that a plane, σ, is semiperpendicular to a plane, π, by
writing

$$\sigma \underset{s}{\perp} \pi$$

Theorem 10.   If a plane, σ, is semiperpendicular to a plane, π,
then π is semiperpendicular to σ.

Proof.   Let σ be a plane which is semiperpendicular to a plane π at a
point P; let l be the perpendicular to π which lies in σ; and let l' be any
line other than l which lies in σ and passes through P (see Fig. 3.12).

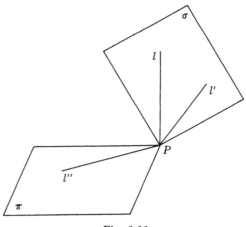

Fig. 3.12

At P there is a unique prime which is perpendicular to l'; and this prime
intersects π in a line, l'', which is perpendicular to both l and l'.   Thus
l'' is perpendicular to σ.   Moreover, no other line in π can be perpendicular to σ because, if such were the case, π would be absolutely
perpendicular to σ, and therefore σ would be absolutely perpendicular
to π, contrary to the hypothesis that σ is semiperpendicular to π.
Thus π, which is semiskew to σ, by hypothesis, contains one and only

one line which is perpendicular to σ at P and is therefore semiperpendicular to σ, as asserted.

In general, of course, two semiskew planes are neither absolutely perpendicular nor semiperpendicular; that is, neither contains even one line which is perpendicular to the other. We do, however, have the following interesting result for general semiskew planes:

Theorem 11. If two semiskew planes are neither absolutely perpendicular nor semiperpendicular, then any line in either plane which passes through the point common to the two planes is perpendicular to one and only one line in the other plane.

Proof. Let π and σ be two semiskew planes which are neither absolutely perpendicular nor semiperpendicular, let P be their common point, and let l be an arbitrary line through P in one or the other of the planes, say π. Then at P there is a unique prime which is perpendicular to l, and this prime intersects σ in a line l' which is perpendicular to l. Moreover, there can be no other line in σ which is perpendicular to l, for if there were, π would be at least semiperpendicular to σ, contrary to hypothesis. Hence there is one and only one line in either plane which is perpendicular to an arbitrary line which lies in the other plane and passes through the intersection of the two planes.

EXERCISES

1. If π_1 and π_2 are two planes which intersect in a line, l, and if σ_1 and σ_2 are, respectively, the planes which are absolutely perpendicular to π_1 and π_2 at a point of l, show that σ_1 and σ_2 have a line in common.
2. If π_1 and π_2 are two planes which intersect in a line, l, show that the lines which are perpendicular to both π_1 and π_2 at the points of l all lie in a plane, σ. Can π_1, π_2, and σ be coprimal?
3. Prove that if three planes are absolutely perpendicular to a fourth plane at three collinear points, the three planes are coprimal.
4. If l is a line in a plane, π, prove that the lines which are perpendicular to π at the points of l are all coprimal.
5. If l is a line in a prime, Σ, prove that the lines which are perpendicular to Σ at the points of l are all coplanar.
6. A plane and a prime are said to be perpendicular if their intersection is a line and if every line in the plane which is perpendicular to the line of intersection is perpendicular to the prime and conversely. Prove that if l is a line which is perpendicular to a prime, Σ, then any plane which contains l is perpendicular to Σ.

7. Can two semiskew planes be perpendicular to the same prime?
8. If a plane, σ, is absolutely perpendicular to a plane, π, prove that any prime which contains σ is perpendicular to π.
9. Prove that if two intersecting primes are perpendicular to the same plane, π, their intersection is a plane which is absolutely perpendicular to π.
10. If O is any point and r is any positive number, the set of all points, P, such that $OP = r$ is called a **hypersphere**. Any line which is perpendicular to a radius $\overline{OP}$ at P is said to be **tangent** to the hypersphere at the point P. Can a tangent to a hypersphere have two points in common with the hypersphere? What is the union of all the tangents to a hypersphere at a given point of the hypersphere?

3.6   Relations Involving Both Perpendicularity and Parallelism in E_4.   In the last two sections we have investigated configurations involving, respectively, only parallel relations and only perpendicularity relations in E_4. In this section we shall conclude our study of four-dimensional euclidean geometry by examining certain configurations in which both parallelism and perpendicularity relations are involved. Our first theorems, dealing with lines and primes in E_4, bear a striking resemblance to corresponding results in solid geometry dealing with lines and planes.

> **Theorem 1.**   Two primes which are perpendicular to the same line are parallel.

Proof.   Let Σ_1 and Σ_2 be two primes each perpendicular to a line, p, at the points P_1 and P_2, respectively. By Theorem 4, Sec. 3.5, P_1 and P_2 are distinct points. Now suppose that Σ_1 and Σ_2 have a point, Q, in common. Then the distinct lines $\overleftrightarrow{QP_1}$ and $\overleftrightarrow{QP_2}$ each pass through Q and are perpendicular to p. Since this is impossible, the assumption that Σ_1 and Σ_2 intersect must be rejected, and the two primes are parallel, as asserted.

> **Theorem 2.**   Two lines which are perpendicular to the same prime are parallel.

Proof.   Let l_1 and l_2 be two lines each perpendicular to a prime, Σ, at the points P_1 and P_2, respectively (see Fig. 3.13). If l_1 and l_2 had a point, Q, in common, $\overleftrightarrow{QP_1}$ and $\overleftrightarrow{QP_2}$ would be two distinct lines each passing through Q and each perpendicular to $\overleftrightarrow{P_1P_2}$, which is impossible. Hence l_1 and l_2 cannot intersect, and to prove the theorem we need only show that they are coplanar. To do this, let us assume the contrary, that is, let us assume that l_1 and l_2 are skew. Then, by Theorem 7, Sec. 3.3, l_1 and l_2 determine a prime, Σ', which, having $\overleftrightarrow{P_1P_2}$ in common with Σ, must intersect Σ in a plane, π. Now by

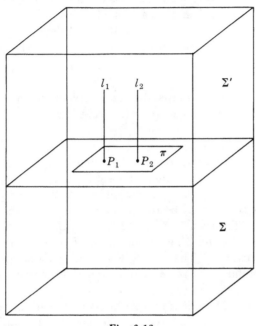

Fig. 3.13

Theorem 2, Sec. 3.5, l_1 and l_2 are both perpendicular to π. Hence since l_1, l_2, and π all lie in the same prime, Σ', it follows from Theorem 7, Sec. 2.15, that they are parallel, contrary to the assumption that they are skew. This contradiction thus establishes the theorem.

It is interesting to compare the simple proof of the last theorem with the relatively complicated proof required for its counterpart in solid geometry, namely, Theorem 7, Sec. 2.15. Of course we had to use Theorem 7, Sec. 2.15, to prove the last theorem, but the elaborate use of congruence relations was required explicitly only in the transition from two to three dimensions. In some respects, then, the "going" is easier once one gets beyond three dimensions!

> **Theorem 3.**   A prime which is perpendicular to one of two parallel lines is perpendicular to the other also.

Proof. Let l_1 and l_2 be two parallel lines, and let Σ be a prime which is perpendicular to one of these lines, say l_1, at a point, P_1. By

Theorem 11, Sec. 3.4, Σ must intersect l_2 in a point, P_2. If l_2 is perpendicular to Σ at P_2, there is nothing more to prove. Therefore, let us suppose that l_2 is not perpendicular to Σ. Then, by Theorem 5, Sec. 3.5, there is a line, p, distinct from l_2, which is perpendicular to Σ at P_2. Moreover, by Theorem 2, p is parallel to l_1. Therefore, there are two distinct lines, l_2 and p, each of which contains P_2 and is parallel to l_1. Since this contradicts the parallel postulate, we must reject the assumption that l_2 is not perpendicular to Σ, and the theorem is established.

Theorem 4.   A line perpendicular to one of two parallel primes is perpendicular to the other also.

Proof.   Let Σ_1 and Σ_2 be two parallel primes, and let l be a line which is perpendicular to one of the primes, say Σ_1, at a point, P_1. By Theorem 10, Sec. 3.4, l must intersect Σ_2 in a point, P_2. Now if l is not perpendicular to Σ_2 at P_2, then of course Σ_2 is not perpendicular to l. In this case, let Σ_2' be the prime which is perpendicular to l at P_2. Then, by Theorem 1, Σ_1 and Σ_2' are parallel, and therefore there are two primes, Σ_2 and Σ_2', each containing P_2 and each parallel to Σ_1. Since this contradicts Theorem 8, Sec. 3.4, the assumption that l is not perpendicular to Σ_2 is untenable, and the theorem is established.

Theorem 5.   There is a unique line which contains a given point not in a given prime and is perpendicular to the given prime.

Proof.   Let P be a given point which does not lie in a given prime, Σ, and let Q be an arbitrary point in Σ. Then, by Theorem 5, Sec. 3.5, there is a unique line, q, which is perpendicular to Σ at Q. If q passes through P, it is a line which contains P and is perpendicular to Σ. If q does not pass through P, let p be the line which contains P and is parallel to q. Then, by Theorem 3, p is perpendicular to Σ. That p is unique follows at once from the observation that if there were two lines which passed through P and were perpendicular to Σ, the plane they determined would intersect Σ in a line, l, to which there would be two perpendiculars from a point not on l. Since this is impossible, the uniqueness of the perpendicular from P to Σ is established.

The last theorem and Theorem 5, Sec. 3.5, are often combined into a single inclusive theorem:

Theorem 6.   There is a unique line which passes through a given point and is perpendicular to a given prime.

The next two theorems deal with absolutely perpendicular planes and have no counterparts in solid geometry.

Theorem 7.   Two planes which are absolutely perpendicular to the same plane are parallel.

Proof.   Let π_1 and π_2 be two planes absolutely perpendicular to a third plane, σ, at points P_1 and P_2, respectively (see Fig. 3.14). By

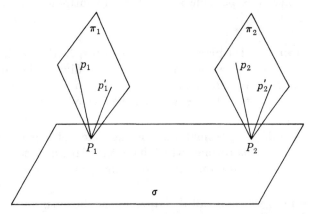

Fig. 3.14

Theorem 9, Sec. 3.5, π_1 and π_2 can have no point in common. Let Σ be any prime containing σ. Then Σ intersects π_1 and π_2 in lines p_1 and p_2, respectively. Moreover, in Σ, p_1 and p_2 are both perpendicular to σ; hence, from the well-known results of solid geometry, they are parallel. Similarly, a second prime, Σ', containing σ will intersect π_1 and π_2 in lines p_1' and p_2', which are necessarily distinct from p_1 and p_2 and parallel to each other. Hence π_1 contains two intersecting lines, p_1 and p_1', which are, respectively, parallel to two intersecting lines, p_2 and p_2', in π_2. Therefore, by Theorem 3, Sec. 3.4, π_1 and π_2 are parallel, as asserted.

Theorem 8.   If a plane is absolutely perpendicular to one of two parallel planes, it is absolutely perpendicular to the other plane also.

Proof.   Let π_1 and π_2 be two parallel planes, and let σ be a plane which is absolutely perpendicular to one of these planes, say π_1, at the point P_1. Then, by Theorem 15, Sec. 3.4, σ must intersect π_2 in a single point, P_2. Now if σ is not absolutely perpendicular to π_2, then of course π_2 is not absolutely perpendicular to σ. In this case, let π_2' be the plane which is absolutely perpendicular to σ at P_2. Then by the last theorem, π_2' is parallel to π_1. Thus there are two planes, π_2 and π_2', each containing P_2 and each parallel to π_1. Since this is impossible, it follows that σ must be absolutely perpendicular to π_2, as asserted.

As a final topic we shall indicate the four-dimensional generalization of the notion of dihedral angles. In the first place, we observe that the space-separation theorem of solid geometry is valid in any prime in E_4. Hence it is possible and natural to introduce the following definitions:

Definition 1.   If a plane, π, lies in a prime, Σ, then either of the two convex sets into which π divides the points of Σ which are not points of π is called a halfprime, and the plane π is called the face of either halfprime.

Definition 2.   A diprimal angle is the union of a plane and two halfprimes which have the plane as a common face and do not lie in the same prime.

The halfprimes we shall call the *sides* of the diprimal angle, and the plane which is their common face we shall call the *face* of the angle. We shall denote a diprimal angle by a symbol of the form

$$\angle A\text{-}BCD\text{-}E$$

where A and E are points in the respective halfprimes and B, C, and D are three noncollinear points in the common face of the two halfprimes.

Definition 3.   The angle formed by two concurrent rays which are perpendicular to the face of a diprimal angle and extend into the respective sides of the diprimal angle is called a plane angle of the diprimal angle.

It is important to know whether or not the plane angles of a diprimal angle are all congruent, because if they are, their common measure is a natural measure of the diprimal angle itself. This question is answered in the affirmative by the following theorem:

Theorem 9.   All plane angles of a diprimal angle have the same measure.

Proof. Let $\angle A\text{-}BCD\text{-}E$ be an arbitrary diprimal angle, and let $\angle P_1QP_2$ and $\angle P_1'Q'P_2'$ be any two of its plane angles, P_1 and P_1', lying in one side of the diprimal angle and P_2 and P'_2 lying in the other (see Fig. 3.15).   Then from the definition of a plane angle of a diprimal angle, $\overrightarrow{QP_1}$ and $\overrightarrow{Q'P'_1}$ are perpendicular to the same plane in the same

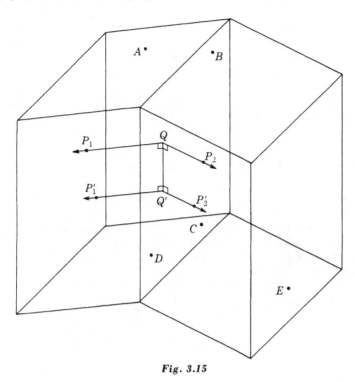

Fig. 3.15

halfprime and hence are parallel.   Similarly, $\overrightarrow{QP_2}$ and $\overrightarrow{Q'P'_2}$ are perpendicular to the same plane in the same halfprime and so are also parallel.   Moreover, each of these four rays is perpendicular to $\overleftrightarrow{QQ'}$. Therefore, $\angle P_1QP_2$ and $\angle P'_1Q'P'_2$ are plane angles of the dihedral angle $\angle P_1\text{-}QQ'\text{-}P'_2$ and hence, by Theorem 17, Sec. 2.15, are congruent, as asserted.

EXERCISES

1. If two planes are perpendicular to the same line, are they necessarily parallel?
2. Is there always at least one plane absolutely perpendicular to each of two parallel planes?   How many are there?
3. If two absolutely perpendicular planes are intersected by a prime, can the lines of intersection be parallel?

4. If two planes which are perpendicular to the same prime have no point in common, are they necessarily parallel?
5. If two planes are each perpendicular to the same prime, are their lines of intersection with the prime necessarily parallel?
6. Two primes are said to be **perpendicular** if their union contains a right diprimal angle. Prove that if two primes are perpendicular, then any line in either of the primes which is perpendicular to their plane of intersection is perpendicular to the other.
7. If two primes are perpendicular to the same prime, are they necessarily parallel?
8. If each of two parallel primes, Σ_1 and Σ_2, is perpendicular to a prime, Σ_3, prove that the planes in which Σ_1 and Σ_2 intersect Σ_3 are parallel.

CREDITS FOR QUOTATIONS

1. Cassius J. Keyser, "The Human Worth of Rigorous Thinking," Columbia University Press, New York, 1916.
2. Edward Kasner and James Newman, "Mathematics and the Imagination," Simon and Schuster, Inc., New York, 1941.
3. Maurice Maeterlinck, "The Life of Space," translated by Bernard Miall, Dodd, Mead & Company, Inc. 1928.
4. Morris Kline, "Mathematics in Western Culture." Copyright 1953 by Oxford University Press, Fair Lawn, N.J.

4

PLANE
HYPERBOLIC
GEOMETRY

4.1 Introduction. In our general discussion of the axiomatic method in Chap. 1, we observed that, subject to the one requirement of consistency, postulates are essentially arbitrary and certainly need not be self-evident. Euclid appears to have understood this, but his contemporaries as well as many of his successors did not. As a consequence, Euclid's parallel postulate,* which is much less obvious than his other assumptions, was viewed with suspicion from the very beginning. The ancient Greeks were familiar with the hyperbola and well aware that its branches approach the corresponding asymptotes yet never meet them. With this evidence before them, they saw no reason to believe that two straight lines might not exhibit the same behavior. Hence, far from being self-evident, Euclid's parallel postulate seemed to them as likely to be false as true, and therefore they tried to settle the matter by finding a proof for it.

For over two thousand years the search continued, and many "proofs" were forthcoming. But in each case, the argument rested upon some alternative hypothesis, usually more obvious than the parallel postulate, but no easier to establish. Among the propositions logically equivalent to Euclid's parallel postulate are the following:

1. There exists a triangle for which the sum of the measures of the angles is 180.

---

* "If a straight line falling on two straight lines make the interior angles on the same side less than two right angles, the two straight lines, if produced indefinitely, will meet on that side on which are the angles less than two right angles."

2. The sum of the measures of the angles of a triangle is the same for all triangles.
3. There exists a triangle similar, but not congruent, to any given triangle.
4. There exist lines which are everywhere equidistant.
5. There exists a rectangle.
6. There exists a circle passing through any three noncollinear points.
7. Through any point in the interior of an angle there is at least one line which intersects both sides of the angle.
8. If a line lying in the plane of two parallel lines intersects one of the lines, it intersects the other also.
9. There is a unique line which passes through a given point not on a given line and is parallel to the given line.

From each of these, Euclid's parallel postulate can be deduced, and many of the purported proofs of the postulate involved relatively short derivations from one or another of these assumptions.

Especially noteworthy among the attempts to prove the parallel postulate is one of a somewhat different sort, undertaken by the Jesuit, Gerolamo Saccheri (1667–1733), a professor at the University of Pavia. In his book, "Euclides ab omni naevo vindicatus," he adopted a contrary hypothesis* and explored its consequences, hoping that he would thereby be led to a *reductio ad absurdum* proof of the parallel postulate. However, the necessary contradiction did not appear, and he evolved a long chain of results, consistent among themselves but so different from those of euclidean geometry that he finally concluded his hypothesis had to be abandoned and that, indeed, "Euclid was vindicated of all flaws." Without realizing it, Saccheri was on the threshold of the discovery of noneuclidean geometry, and the theorems which he finally rejected are among the central results of what is now called hyperbolic geometry. Had he not held, with the rest of the scientific world, the certainty that euclidean geometry is the only possible geometry, he, rather than Bolyai, and Lobachevsky, and Riemann, a century later, would have been the founder of noneuclidean geometry.

It was not until 1763, thirty years after the publication of Saccheri's work, that the possibility that the parallel postulate might, indeed, be unprovable was seriously suggested. And it was not until Gauss did so in 1816 that anyone expressed the conviction that, in fact, it was undemonstrable. And, of course, until at least a few mathematicians shared this belief, noneuclidean geometry could not be said to exist as an object of serious investigation.

---

* Actually, Saccheri investigated each of two alternative hypotheses, one of which he rejected very quickly.

Gauss himself worked extensively in noneuclidean geometry, but chose to leave his results unpublished, perhaps from a desire to avoid controversy with those who did not share his views. The first published contributions to the new field came independently from the Russian mathematician Nicholaus Ivanovich Lobachevsky (1793–1856), a professor and later rector of the University of Kazan, and from the Hungarian mathematician Johann Bolyai (1802–1860). In 1829, in the Kazan *Messenger*, Lobachevsky published an account of his researches in noneuclidean geometry; and in 1831, in a brief appendix to a two-volume work on geometry by his father, Wolfgang, Johann Bolyai published the results of his investigations. Both Lobachevsky and Bolyai replaced the parallel postulate of euclidean geometry with the contradictory assertion that *two* parallels can be drawn to a given line through a given point and proceeded to develop what the great German mathematician Felix Klein (1849–1925) later called **hyperbolic geometry.** Not until 1854, in the inaugural lecture* of G. B. F. Riemann (1826–1866) at the University of Gottingen, was the alternative hypothesis that there are *no* parallels to a given line through a given point systematically explored. The geometry stemming from this assumption, together with the rejection of the euclidean supposition that a line is of infinite length, is known in Klein's terminology as **elliptic geometry,** the second of the so-called classical noneuclidean geometries. Despite the work of Lobachevsky, Bolyai, Riemann, and others, the impossibility of proving the parallel postulate was not shown until 1868, when the Italian mathematician Eugenio Beltrami (1835–1900) exhibited a particular model within euclidean geometry in which the postulates of hyperbolic geometry were satisfied, thereby proving that these were at least relatively consistent.

In the foregoing summary of the origins of noneuclidean geometry we would be grossly unfair if we left the impression that it was developed exclusively by the four or five men we have chosen to mention. Like all branches of mathematics, it has been enriched by the labors of many men; and like all branches of mathematics, its history is fascinating to retrace. Our concern, however, is not with history but with geometry, and so we now turn our attention to an axiomatic development of the more elementary portions of hyperbolic geometry.

4.2 The Parallel Postulate for Hyperbolic Geometry. The chief difference between the axiomatic structure of plane hyperbolic geometry and plane euclidean geometry is, of course, in the respective parallel postulates. However, there are several other postulates which

---

* "On the Hypotheses Which Underlie the Foundations of Geometry."

must be deleted from the list we introduced in Chap. 2 before we have a set appropriate to our present purposes.   In the first place, since the role of Postulates 4, 5, 6, and 7 is to ensure that we are *not* restricted to a single plane, they certainly are unnecessary in the study of any kind of *plane* geometry.   Furthermore, since rectangles do not exist in hyperbolic geometry, Postulates 19, 20, 21, and 22 have no place in the work of this chapter, and a different basis for the measurement of area will have to be established.

With the exceptions we have just noted, the postulates of euclidean geometry and hyperbolic geometry are identical.   As a consequence, there are many theorems which hold equally well in both geometries. In fact, all theorems except those depending directly or indirectly on a particular parallel postulate are common to the two geometries, and without repeating their proofs we shall employ them, as necessary, in the work of this chapter.   Specifically, we shall make frequent use of the point-plotting theorem, the angle-construction theorem, the vertical-angle theorem, the weak form of the exterior-angle theorem, the plane-separation theorem, Pasch's theorem and its corollary, and the various congruence theorems.   On the other hand, we cannot use any of the theorems on the angles formed by two parallel lines and a transversal, the strong form of the exterior-angle theorem, the angle-sum theorem, the various similarity theorems, or the euclidean area formulas, since these depend on postulates which are not available in hyperbolic geometry.

Our work in hyperbolic geometry begins, then, with the substitution of the following assertion for the euclidean parallel postulate, Postulate 17, Sec. 2.10:

Postulate 17$_H$.   A given point not on a given line is the endpoint of exactly two noncollinear rays which do not intersect the line but are such that every ray between them does intersect the line.

Definition 1.   Each of the two rays whose existence is guaranteed by Postulate 17$_H$ is said to be parallel to the given line at the given point.

Definition 2.   A line is said to be parallel to a given line at a given point if it contains a ray which is parallel to the given line at the given point.

Definition 3.   One ray is said to be parallel to another ray at a given point if the first ray is parallel at the given

point to the line determined by the second ray and
if the two rays lie on the same side of the line determined by their endpoints.

If P is a point not on a line, l, we indicate that $\overrightarrow{PA}$ and $\overrightarrow{PB}$ are the
rays parallel to l at P by writing

$$\overrightarrow{PA} \,\|l \quad \text{and} \quad \overrightarrow{PB} \,\|l$$

The line determined by P and A is parallel to l at the point P because
it contains the ray $\overrightarrow{PA}$ which is parallel to l at P. We indicate this
fact by writing $\overleftrightarrow{PA} \,\|l$ but *not* $\overleftrightarrow{AP} \,\|l$. Similarly, we indicate that the
line determined by P and B is parallel to l at the point P by writing
$\overleftrightarrow{PB} \,\|l$ but *not* $\overleftrightarrow{BP} \,\|l$. To indicate that $\overrightarrow{PA}$ is parallel to a ray, r, at the
point P, we write $\overrightarrow{PA} \,\|r$.

If P is a point which is not on a line, l, Postulate 17_H identifies a
certain set of rays which have P as endpoint and intersect l, namely,
those between the rays, $\overrightarrow{PA}$ and $\overrightarrow{PB}$, which are parallel to l at P. On
the other hand, it does not say that these are the only rays originating
at P which intersect l. That this is the case, however, is guaranteed
by the following theorem:

> ***Theorem 1.*** If P is a point which is not on a line, l, the only rays
> originating at P which intersect l are the rays between
> those which are parallel to l at P.

Proof. Let $\overrightarrow{PA}$ and $\overrightarrow{PB}$ be the rays which are parallel to a line, l,
at a point, P, which is not on l. Let $\overrightarrow{PA'}$ and $\overrightarrow{PB'}$ be, respectively,
the rays opposite to $\overrightarrow{PA}$ and $\overrightarrow{PB}$. To prove the theorem, we must
show that none of the following rays can intersect l:

1. The rays between $\overrightarrow{PA'}$ and $\overrightarrow{PB'}$
2. The rays between $\overrightarrow{PA}$ and $\overrightarrow{PB'}$
3. The rays between $\overrightarrow{PA'}$ and $\overrightarrow{PB}$
4. The rays $\overrightarrow{PA'}$ and $\overrightarrow{PB'}$

Suppose, first, that a ray, $\overrightarrow{PC'}$, between $\overrightarrow{PA'}$ and $\overrightarrow{PB'}$ intersected l, say in the point Q' (Fig. 4.1a). Now by Exercise 1, Sec. 2.7, the ray, $\overrightarrow{PC}$, which is opposite to $\overrightarrow{PC'}$ lies between $\overrightarrow{PA}$ and $\overrightarrow{PB}$ and hence, by our new parallel postulate, Postulate 17$_H$, must intersect l in a point Q,

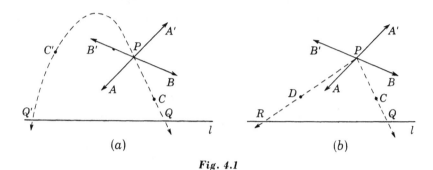

(a)   (b)

Fig. 4.1

necessarily distinct from Q', since it lies on the opposite side of $\overleftrightarrow{AA'}$ from Q'. Thus the line $\overleftrightarrow{CC'}$ has two distinct intersections with l. Since this is impossible, we conclude that it is impossible for a ray between $\overrightarrow{PA'}$ and $\overrightarrow{PB'}$ to intersect l. Now consider the possibility that a ray, $\overrightarrow{PD}$, between $\overrightarrow{PA}$ and $\overrightarrow{PB'}$ intersects l, say in the point R (Fig. 4.1b). Let $\overrightarrow{PC}$ be any ray between $\overrightarrow{PA}$ and $\overrightarrow{PB}$, necessarily intersecting l in a point, Q. Then by Exercise 2, Sec. 2.7, $\overrightarrow{PA}$ is between $\overrightarrow{PD}$ and $\overrightarrow{PC}$. Hence, by Corollary 1, Theorem 1, Sec. 2.11, $\overrightarrow{PA}$ must intersect the segment $\overline{QR}$, that is, $\overrightarrow{PA}$ must intersect l. But this is impossible, since $\overrightarrow{PA}\|l$. Hence no ray between $\overrightarrow{PA}$ and $\overrightarrow{PB'}$ can intersect l. In exactly the same way we can prove that no ray between $\overrightarrow{PA'}$ and $\overrightarrow{PB}$ can intersect l and finally that neither $\overrightarrow{PA'}$ nor $\overrightarrow{PB'}$ can intersect l. Hence the theorem is established.

> **Corollary 1.** If $\overrightarrow{PA}$ is a ray which is parallel to a line l at the point P and if $\overrightarrow{PC}$ is a ray which intersects l, then any ray between $\overrightarrow{PA}$ and $\overrightarrow{PC}$ intersects l.

From Postulate 17$_H$ and from Theorem 1, we know that a ray originating at P intersects l if and only if it lies between the two rays which are parallel to l at P. By Theorem 5, Sec. 2.9, which depends only on postulates common to euclidean and hyperbolic geometry and hence is true in each geometry, there is a unique ray originating at P and perpendicular to l. Since this ray obviously intersects l, it must lie between the rays which are parallel to l at P. It is natural now to conjecture from considerations of symmetry that the angles formed by the perpendicular ray and the respective parallel rays are congruent, and this is in fact the case, as the following theorem shows:

Theorem 2. If P is a point which is not on a line, l, if F is the foot of the perpendicular from P to l, and if $\overrightarrow{PA}$ and $\overrightarrow{PB}$ are the two rays which are parallel to l from P, then $\angle APF \cong \angle BPF$.

Proof. If $\angle APF$ is not congruent to $\angle BPF$, then the measure of one of the angles is necessarily greater than the measure of the other. Suppose for definiteness that $m\angle APF > m\angle BPF$ (see Fig. 4.2). Then

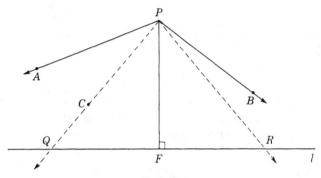

Fig. 4.2

in the halfplane determined by $\overleftrightarrow{PF}$ which does not contain B there exists a ray, $\overrightarrow{PC}$, such that $\angle CPF \cong \angle BPF$. Moreover, since

$$m\angle CPF = m\angle BPF < m\angle APF$$

it follows that $\overrightarrow{PC}$ is between $\overrightarrow{PF}$ and $\overrightarrow{PA}$. Hence $\overrightarrow{PC}$ must intersect l, say at Q. Let R be the point on the ray opposite to $\overrightarrow{FQ}$ such that $FR = FQ$. Then $\triangle PFQ \cong \triangle PFR$, since they are right triangles with congruent legs. Hence $\angle FPR \cong \angle FPQ$. But $\angle FPQ \cong \angle FPB$; hence $\angle FPR \cong \angle FPB$, and therefore, by the angle-construction theorem, $\overrightarrow{PR} = \overrightarrow{PB}$. But this is impossible, since $\overrightarrow{PR}$ intersects l, while $\overrightarrow{PB}$, being parallel to l, does not. This contradiction overthrows the possibility that $\angle APF$ and $\angle BPF$ are not congruent, and the theorem is established.

As immediate consequences of Theorem 2, we have the following results:

Corollary 1. If $\overrightarrow{PA}$ and $\overrightarrow{PB}$ are the rays which are parallel to a line, l, from the point P and if F is the foot of the perpendicular to l from P, then $\angle APF$ and $\angle BPF$ are both acute.

Corollary 2. The ray $\overrightarrow{PA}$ is parallel to a line, l, from P if and only if $\overrightarrow{PA}$ does not intersect l but every ray between $\overrightarrow{PA}$ and $\overrightarrow{PF}$, where F is any point on l, does intersect l.

Corollary 3. If $l = \overleftrightarrow{LM}$ is an arbitrary line and P is any point which is not on l, there is a unique ray, $\overrightarrow{PA}$, such that $\overrightarrow{PA}$ is parallel to l from P and A lies on the same side of $\overleftrightarrow{PL}$ as M.

Corollary 4. If each of two noncollinear rays, $\overrightarrow{PA}$ and $\overrightarrow{PB}$, is parallel to a line, l, from the point P, then the bisector of $\angle APB$ is perpendicular to l.

Corollary 5. If $\overrightarrow{PA} \,\|l$ and $\overrightarrow{P'A'}\|l'$ and if the perpendicular distances, PF and $P'F'$, from P and P' to l and l', respectively, are equal, then $m\angle APF = m\angle A'P'F'$.

In view of Theorem 2 and its corollaries, it is now appropriate to speak, at least informally, of the right parallel ray, or line, and the left parallel ray, or line, to a given line from a given point and to interpret Postulate 17_H to mean that there is a unique right parallel ray and a unique left parallel ray to a given line from a given point. Moreover, since Corollary 5 assures us that the measure of the acute angle determined by the perpendicular ray and either parallel ray to a given line from a given point is uniquely determined by the perpendicular distance from the point to the line, it is natural to introduce the following definitions:

Definition 4. If $\overrightarrow{PA}\,\|l$ and if F is the foot of the perpendicular from P to l, the measure of the acute angle, $\angle APF$,

is called the amplitude of parallelism, $\pi(d)$, for the
distance $PF = d$.

Definition 5.   The distances d_1 and d_2 are said to be complementary if the sum of their amplitudes of parallelism
is 90, that is, if $\pi(d_1) + \pi(d_2) = 90$.

To indicate that the distances d_1 and d_2 are complementary, we write
either $d_2 = \bar{d}_1$ or $d_1 = \bar{d}_2$.

The fact that we have used the word "parallel" to describe the rays
referred to in Postulate 17_H does not justify, without proof, the
assumption that the relation of parallelism in hyperbolic geometry
has all or even any of the properties of parallelism in euclidean geom-
etry.   For instance, if $\overrightarrow{PA} \| l$ and Q is a point between P and A, is it
true that $\overrightarrow{QA} \| l$?   Or to put it more generally, if a line is the right
parallel to a given line from one of its points, is it necessarily the right
parallel to the given line from each of its points?   With the adjective
"right" dropped, this result is, of course, true in euclidean geometry,
and for both right and left parallels it turns out to be true in hyperbolic
geometry, as the following theorem assures us:

Theorem 3.   A line which is parallel to a given line from a given
point in a given direction is at each of its points the
parallel to the given line in the given direction.

Proof.   Let l be an arbitrary line, let P be an arbitrary point which
is not on l, and let $p = \overleftrightarrow{PQ}$ be parallel to l through P.   To prove the
theorem we must show that if P' is any other point on p, then p is also
the parallel to l in the same direction from P'.   To do this, suppose
first that P' is any point other than P on the ray $\overrightarrow{PQ}$ and, for convenience of notation, let us assume that Q is chosen such that P' is
between P and Q (Fig. 4.3a).   Let F be an arbitrary point on l and let
$\overrightarrow{P'R}$ be any ray between $\overrightarrow{P'F}$ and $\overrightarrow{P'Q}$, R being on the same side of l as
P'.   It is easy to show that R lies in the interior of $\angle FPQ$; hence $\overrightarrow{PR}$
lies between $\overrightarrow{PF}$ and $\overrightarrow{PQ}$ and therefore must intersect l, say at S.   Now
by Pasch's theorem, since $\overleftrightarrow{P'R}$ intersects one side, $\overline{PS}$, of $\triangle PSF$, it
must intersect a second side.   Clearly, the ray opposite to $\overrightarrow{P'R}$ cannot
intersect either $\overline{PF}$ or $\overline{FS}$ since, except for the vertex P and the end-

point P', $\triangle PSF$ and the ray opposite to $\overrightarrow{P'R}$ lie on opposite sides of p. Moreover, $\overrightarrow{P'R}$ cannot intersect $\overline{PF}$ since, except for the endpoints P' and F, the ray $\overrightarrow{P'R}$ and the segment $\overline{PF}$ lie on opposite sides of $\overleftrightarrow{P'F}$.

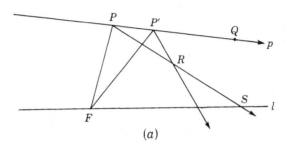

(a)

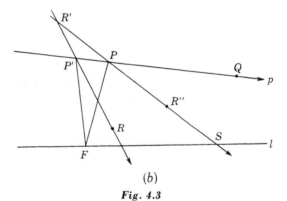

(b)

Fig. 4.3

Therefore $\overrightarrow{P'R}$ must intersect $\overline{FS}$, that is, it must intersect l. Now $\overrightarrow{P'Q}$ cannot intersect l since it is a subset of $\overrightarrow{PQ}$ which, by hypothesis, does not intersect l. On the other hand, we have just shown that any ray, such as $\overrightarrow{P'R}$, between $\overrightarrow{P'Q}$ and an arbitrary intersecting ray, $\overrightarrow{P'F}$, does intersect l. Hence, by Corollary 2, Theorem 2, $\overrightarrow{P'Q}$ is parallel to l; that is, $\overleftrightarrow{PQ}$ is parallel to l from every point, P', on $\overrightarrow{PQ}$.

Now suppose that P' is an arbitrary point, distinct from P, on the ray opposite to $\overrightarrow{PQ}$ (Fig. 4.3b). As before, let F be an arbitrary point on l, let $\overrightarrow{P'R}$ be any ray between $\overrightarrow{P'F}$ and $\overrightarrow{P'Q}$, R being on the same side of l as P', let R' be any point on the ray opposite to $\overrightarrow{P'R}$, and let R'' be any point on the ray opposite to $\overrightarrow{PR'}$. It is easy to show that R''

lies in the interior of $\angle FPQ$. Hence $\overrightarrow{PR''}$ lies between $\overrightarrow{PF}$ and $\overrightarrow{PQ}$ and therefore must intersect l, say at S. Now by the corollary to Pasch's theorem, $\overleftrightarrow{P'R}$ must intersect $\overline{PF}$. Hence, by Pasch's theorem, $\overleftrightarrow{P'R}$ must intersect one other side of $\triangle PFS$. Clearly, $\overleftrightarrow{P'R}$ cannot intersect either $\overline{FS}$ or $\overline{PS}$, and $\overrightarrow{P'R}$ cannot intersect $\overline{PS}$. Therefore, $\overrightarrow{P'R}$ must intersect $\overline{FS}$, that is, $\overrightarrow{P'R}$ must intersect l. As before, $\overrightarrow{P'Q}$ cannot intersect l, but we have now shown that every ray between $\overrightarrow{P'Q}$ and the arbitrary intersecting ray, $\overrightarrow{P'F}$, does intersect l. Hence $\overleftrightarrow{PQ}$ is also parallel to l from any point on the ray opposite to $\overrightarrow{PQ}$. Thus, combining the two parts of our proof, $\overleftrightarrow{PQ}$ is the right parallel to l from any point on $\overleftrightarrow{PQ}$, as asserted. In exactly the same way the theorem can be established for left parallels.

In view of Theorem 3, it is now correct to speak of a line, $\overleftrightarrow{PQ}$, being right-parallel or left-parallel to a line, l, without singling out for special attention any particular point on $\overleftrightarrow{PQ}$.

Another result whose truth we would certainly anticipate from our knowledge of euclidean geometry is the assertion of the following theorem that the relation of parallelism is a symmetric one:

Theorem 4. If $\overleftrightarrow{AB}$ is parallel to $\overleftrightarrow{CD}$ and if B and D are on the same side of $\overleftrightarrow{AC}$, then $\overleftrightarrow{CD}$ is parallel to $\overleftrightarrow{AB}$.

Proof. Let $\overleftrightarrow{AB}$ be parallel to the line $c = \overleftrightarrow{CD}$ where, for convenience, C is assumed to be the foot of the perpendicular from A to c (see Fig. 4.4). Now let $\overrightarrow{CF}$ be any ray between $\overrightarrow{CD}$ and $\overrightarrow{CA}$. If $\overrightarrow{CF}$ intersects $\overleftrightarrow{AB}$, our theorem is immediately established by Corollary 2, Theorem 2. Hence we shall suppose that $\overrightarrow{CF}$ does not intersect $\overleftrightarrow{AB}$, which means, in other words, that $\overrightarrow{CF}$, except for its endpoint, C, lies in the interior of $\angle CAB$. Let G be the foot of the perpendicular from A to $\overleftrightarrow{CF}$. By the exterior-angle theorem, G must be a point of the ray $\overrightarrow{CF}$, distinct from C. Hence G must lie in the interior of $\angle CAB$, and

$\overrightarrow{AG}$ must lie between $\overrightarrow{AC}$ and $\overrightarrow{AB}$. Moreover, since $m\angle ACG < 90$, while $m\angle AGC = 90$, it follows, from Theorem 4, Sec. 2.11, that $AG < AC$. Thus there is a point, H, between A and C such that

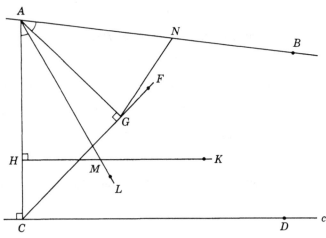

Fig. 4.4

$AH = AG$. Now let $\overrightarrow{HK}$ be perpendicular to $\overleftrightarrow{AC}$ on the same side of $\overleftrightarrow{AC}$ as B and D. Since $m\angle GAB < m\angle HAB$, we can determine a ray, $\overrightarrow{AL}$, between $\overrightarrow{AH}$ and $\overrightarrow{AB}$ such that $m\angle HAL = m\angle GAB$. Moreover, $\overrightarrow{AL}$ must intersect $\overrightarrow{HK}$, say in M, since $\overrightarrow{AL}$ must in fact intersect $\overleftrightarrow{CD}$, which is on the opposite side of $\overleftrightarrow{HK}$ from A. Now locate N on $\overrightarrow{AB}$ such that $AN = AM$. Then $\triangle HAM \cong \triangle GAN$, since they have two pairs of corresponding sides, $\overline{AH}$, $\overline{AG}$ and $\overline{AM}$, $\overline{AN}$, and the included angles, $\angle HAM$ and $\angle GAN$, congruent. Hence

$$\angle AGN \cong \angle AHM$$

that is, $m\angle AGN = 90$. Thus $\overrightarrow{GN}$ and $\overrightarrow{GF}$ are both perpendicular to $\overleftrightarrow{AG}$ at G. Since they lie on the same side of $\overleftrightarrow{AG}$, they must therefore coincide, which implies that $\overrightarrow{CF}$ intersects $\overrightarrow{AB}$, contrary to our initial supposition. This contradiction establishes the theorem.

Under our original definition, the unsymmetrical parallel relation

$$\overleftrightarrow{PQ} \parallel l$$

though it involves a direction on the line $\overleftrightarrow{PQ}$, does not involve a direction on l, and if $l = \overleftrightarrow{LM}$, either $\overleftrightarrow{PQ} \parallel \overleftrightarrow{LM}$ or $\overleftrightarrow{PQ} \parallel \overleftrightarrow{ML}$ is equivalent to $\overleftrightarrow{PQ} \parallel l$. However, with the symmetrical character of the parallel relation now established, it is desirable that our notation be modified or restricted so that it will indicate the appropriate direction on each of the lines in question. If $p = \overleftrightarrow{PQ}$ and $l = \overleftrightarrow{LM}$, we therefore agree that henceforth when we write

$$\overleftrightarrow{PQ} \parallel \overleftrightarrow{LM}$$

we assert both $\overleftrightarrow{PQ} \parallel l$ and $\overleftrightarrow{LM} \parallel p$. Hence, if $\overleftrightarrow{PQ} \parallel \overleftrightarrow{LM}$ is a true statement, then $\overleftrightarrow{PQ} \parallel \overleftrightarrow{ML}$, $\overleftrightarrow{QP} \parallel \overleftrightarrow{LM}$, and $\overleftrightarrow{QP} \parallel \overleftrightarrow{ML}$ are all false.

Another familiar property of parallel lines which also holds in hyperbolic geometry is the property of transitivity. Specifically, we have the following theorem:

Theorem 5. If $\overleftrightarrow{AB}$, $\overleftrightarrow{CD}$, and $\overleftrightarrow{EF}$ are three lines such that $\overleftrightarrow{AB}$ and $\overleftrightarrow{EF}$ are both parallel to $\overleftrightarrow{CD}$, then $\overleftrightarrow{AB}$ and $\overleftrightarrow{EF}$ are parallel.

Proof. Suppose first that $\overleftrightarrow{AB}$ and $\overleftrightarrow{EF}$ lie on opposite sides of $\overleftrightarrow{CD}$, so that $\overline{AE}$ intersects $\overleftrightarrow{CD}$, say in G (Fig. 4.5a). Let $\overrightarrow{AH}$ be any ray between $\overrightarrow{AE}$ and $\overrightarrow{AB}$. Then since $\overrightarrow{AB}\|\overrightarrow{CD}$, it follows that $\overrightarrow{AH}$ must intersect $\overrightarrow{GD}$, say in K. Clearly, the ray opposite to $\overrightarrow{KA}$, say $\overrightarrow{KA'}$,

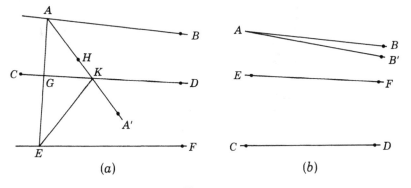

(a)   (b)

Fig. 4.5

lies between $\overrightarrow{KE}$ and $\overrightarrow{KD}$.   Hence, since $\overleftrightarrow{CD} \parallel \overleftrightarrow{EF}$, it follows that $\overrightarrow{KA'}$ intersects $\overrightarrow{EF}$.   In other words, any ray, such as $\overrightarrow{AK}$, between $\overrightarrow{AE}$ and $\overrightarrow{AB}$ intersects $\overleftrightarrow{EF}$.   Since $\overleftrightarrow{AB}$ and $\overleftrightarrow{EF}$ do not intersect, because they lie on opposite sides of $\overleftrightarrow{CD}$, it follows, from Corollary 2, Theorem 2, that $\overleftrightarrow{AB} \parallel \overleftrightarrow{EF}$.

Now suppose that $\overleftrightarrow{AB}$ and $\overleftrightarrow{EF}$ are on the same side of $\overleftrightarrow{CD}$ (Fig. 4.5b). Clearly $\overleftrightarrow{AB}$ and $\overleftrightarrow{EF}$ cannot intersect, for if they had a point, P, in common, then through P there would be two lines parallel to $\overleftrightarrow{CD}$ in the same direction, which is impossible.   Hence of the three lines, either $\overleftrightarrow{AB}$ or $\overleftrightarrow{EF}$, say $\overleftrightarrow{EF}$, is between the other two.   Now if $\overleftrightarrow{AB}$ and $\overleftrightarrow{EF}$ are not parallel, there is a line, $\overleftrightarrow{AB'}$, which is parallel to $\overleftrightarrow{EF}$ through A. Then $\overleftrightarrow{EF}$ lies between $\overleftrightarrow{AB'}$ and $\overleftrightarrow{CD}$ and is parallel to both.   Hence, by the first part of our proof, $\overleftrightarrow{AB'}$ is parallel to $\overleftrightarrow{CD}$.   Thus through A there are two lines, $\overleftrightarrow{AB}$ and $\overleftrightarrow{AB'}$, each parallel to $\overleftrightarrow{CD}$, which is impossible, since the parallel to a given line from a given point in a given direction is unique.   Hence in every case, $\overleftrightarrow{AB}$ is parallel to $\overleftrightarrow{EF}$, as asserted.

Definition 6.   Three lines, $l = \overleftrightarrow{L_1L_2}$, $m = \overleftrightarrow{M_1M_2}$, $n = \overleftrightarrow{N_1N_2}$, are said to be parallel in the same sense if

$$\overleftrightarrow{L_1L_2} \parallel \overleftrightarrow{M_1M_2} \parallel \overleftrightarrow{N_1N_2}$$

The last several theorems dealt with properties of parallel lines which hold in both hyperbolic and euclidean geometry.   As we should certainly expect, however, there are numerous properties of parallel lines which hold in euclidean geometry but do not hold in hyperbolic geometry.   The following theorem asserts an important result of this contradictory character:

Theorem 6.   Corresponding angles formed by two parallel lines and a transversal cannot be congruent.

Proof.   Let $\overleftrightarrow{AB}$ and $\overleftrightarrow{CD}$ be two parallel lines cut by the transversal $\overleftrightarrow{AC}$.   Let C' be a point on the ray opposite to $\overrightarrow{AC}$, and let us assume that $\angle C'AB \cong \angle ACD$.   Let M be the midpoint of $\overline{AC}$ and let F and

G be, respectively, the feet of the perpendiculars from M to $\overleftrightarrow{AB}$ and $\overleftrightarrow{CD}$. By Corollary 1, Theorem 2, the congruent angles $\angle C'AB$ and $\angle ACD$ cannot be right angles. If $\angle C'AB$ and $\angle ACD$ are acute (Fig. 4.6a), we have a contradiction of the exterior-angle theorem unless G is a point of the ray $\overrightarrow{CD}$ and F is a point of the ray opposite to $\overrightarrow{AB}$.

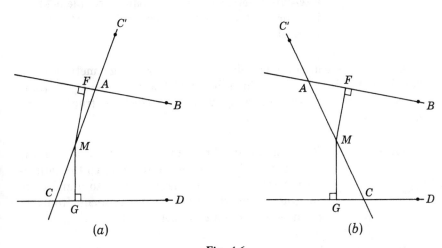

(a)   (b)

Fig. 4.6

Similarly, if $\angle C'AB$ and $\angle ACD$ are obtuse (Fig. 4.6b), we have a contradiction of the exterior-angle theorem unless F is a point of the ray $\overrightarrow{AB}$ and G is a point of the ray opposite to $\overrightarrow{CD}$. Thus in every case, F and G lie on opposite sides of the line $\overleftrightarrow{AC}$. Now by the vertical-angle theorem, if $\angle C'AB$ and $\angle ACD$ are acute, or by Theorem 7, Sec. 2.7, if $\angle C'AB$ and $\angle ACD$ are obtuse, it follows that

$$\angle FAM \cong \angle GCM$$

Hence, by Exercise 3, Sec. 2.9, $\triangle AMF \cong \triangle CMG$. Therefore

$$\angle AMF \cong \angle CMG$$

and hence, by Exercise 12, Sec. 2.7, $\overrightarrow{MF}$ and $\overrightarrow{MG}$ are collinear. But this contradicts Corollary 1, Theorem 2. Hence the assumption that $\angle C'AB$ and $\angle ACD$ are congruent must be abandoned, and the theorem is established.

    Corollary 1.   Alternate interior angles formed by two parallel lines and a transversal cannot be congruent.

Corollary 2.   The sum of the measures of the interior angles on the same side of a transversal of two parallel lines as the direction of parallelism is always less than 180.

Corollary 3.   If a transversal forms with two lines a pair of congruent corresponding angles, or a pair of congruent alternate interior angles, or a pair of supplementary interior angles, the lines neither intersect nor are parallel.

As we shall see later on, parallel lines in hyperbolic geometry are not equidistant.   However, for three lines which are parallel in the same sense we do have the following property, which is true in both geometries:

Theorem 7.   If a, b, and c are three lines which are parallel in the same sense, with b between a and c, and if the perpendicular segments from one point of b to a and c are congruent, then the perpendicular segments from any point of b to a and c are congruent.

Proof.   Let B be a point of b, let A and C be, respectively, the feet of the perpendiculars from B to a and c, and let $AB = CB$ (see Fig. 4.7).   Then if B' is any other point of b, it follows that

$$m\angle ABB' = m\angle CBB'$$

since each is either the amplitude of parallelism for the distance $AB = CB$, if $\overleftrightarrow{BB'}$ is parallel to a and c, or 180 minus the amplitude of

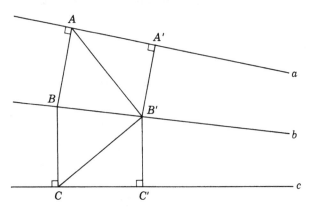

Fig. 4.7

parallelism for this distance, if $\overleftrightarrow{B'B}$ is parallel to a and c.   Therefore, $\triangle ABB' \cong \triangle CBB'$ and $AB' = CB'$.   Now let A' and C' be the feet of the perpendiculars from B' to a and c, respectively.   Then

$$\angle B'AA' \cong \angle B'CC'$$

since they are complements of the congruent angles $\angle BAB'$ and $\angle BCB'$.   Hence $\triangle AA'B' \cong \triangle CC'B'$, and therefore $A'B' = C'B'$, as asserted.

EXERCISES

1. Complete the proof of Theorem 1 by showing in detail that no ray between $\overrightarrow{PA'}$ and $\overrightarrow{PB}$ can intersect l and that neither $\overrightarrow{PA'}$ nor $\overrightarrow{PB'}$ can intersect l.
2. Prove Corollary 1, Theorem 1.
3. Prove (a) Corollary 1, Theorem 2; (b) Corollary 2, Theorem 2; (c) Corollary 3, Theorem 2; (d) Corollary 4, Theorem 2; (e) Corollary 5, Theorem 2.
4. Show that the point R in Fig. 4.3a lies in the interior of $\angle FPP'$, as asserted in the first part of the proof of Theorem 3.
5. Show that the point R'' in Fig. 4.3b lies in the interior of $\angle FPQ$, as asserted in the second part of the proof of Theorem 3.
6. In what respect, if any, is the proof of Theorem 4 changed if the ray $\overrightarrow{AL}$ in Fig. 4.4 is between $\overrightarrow{AG}$ and $\overrightarrow{AN}$ instead of between $\overrightarrow{AH}$ and $\overrightarrow{AG}$?
7. Prove (a) Corollary 1, Theorem 6; (b) Corollary 2, Theorem 6; (c) Corollary 3, Theorem 6.
8. If $p\|r$ and $q\|r$, is it necessarily true that $p\|q$?   If $p\|q$, $q\|r$, and $r\|p$, are p, q, and r necessarily parallel in the same sense?   If not, what other possibilities are there?
9. If $\overrightarrow{PP'} \parallel \overrightarrow{QQ'}$, prove that the bisectors of $\angle QPP'$ and $\angle PQQ'$ intersect. If P' and Q' are on the same side of $\overleftrightarrow{PQ}$ but if $\overrightarrow{PP'}$ and $\overrightarrow{QQ'}$ are not parallel, is this necessarily the case?
10. In Exercise 9, if A is the intersection of the two angle bisectors, prove that the perpendicular distances from A to $\overleftrightarrow{PP'}$, from A to $\overleftrightarrow{QQ'}$, and from A to $\overleftrightarrow{PQ}$ are all equal.

4.3 Asymptotic Triangles.   We have already seen that certain fundamental properties of triangles hold equally in hyperbolic and euclidean geometry.   In particular, the weak form of the exterior-angle theorem and Pasch's theorem and its corollary hold in both geometries.   Moreover, in hyperbolic geometry there are certain

other configurations, called **asymptotic triangles,** which also possess
these properties, and to these we now turn our attention.

> **Definition 1.**   If $\overleftrightarrow{AB} \parallel \overleftrightarrow{CD}$, the union of $\overrightarrow{AB}, \overrightarrow{CD}$, and $\overline{AC}$ is called an
> asymptotic triangle.

The asymptotic triangle determined by the parallel rays $\overrightarrow{AB}$ and $\overrightarrow{CD}$
(see Fig. 4.8) is denoted by the symbol

$$\triangle(AB,CD)$$

The points A and C are called the **vertices,** and $\overrightarrow{AB}, \overrightarrow{CD}$, and $\overline{AC}$ are

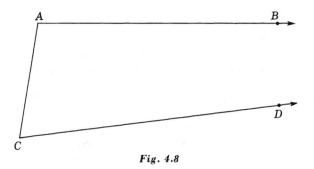

Fig. 4.8

called the **sides** of the asymptotic triangle.   The angles $\angle ACD$ and
$\angle CAB$ are called the **angles** of the asymptotic triangle.   The angle
formed by $\overrightarrow{AB}$ and the ray opposite to $\overrightarrow{AC}$ and the angle formed by
$\overrightarrow{CD}$ and the ray opposite to $\overrightarrow{CA}$ are called the **exterior angles** of the
asymptotic triangle.   The **interior** of an asymptotic triangle is the
intersection of the interiors of its two angles.

In the asymptotic triangle $\triangle(AB,CD)$ the sides $\overrightarrow{AB}$ and $\overrightarrow{CD}$ are
clearly of infinite length, and therefore it is meaningless to speak of
them as being either equal or unequal.   However, it is occasionally
useful to have the following definition, provided we are careful to read
into it no implied comparison of infinite lengths:

> **Definition 2.**   If the angles of an asymptotic triangle are con-
> gruent, the asymptotic triangle is said to be isosceles.

The simplest theorem involving asymptotic triangles is the following,
which in essence is an extension of the corollary of Pasch's theorem:

Theorem 1.   If P is a point in the interior of $\triangle(AB,CD)$, then

   (1) $\overrightarrow{AP}$ intersects $\overrightarrow{CD}$.

   (2) $\overrightarrow{CP}$ intersects $\overrightarrow{AB}$.

   (3) The line through P which is parallel to $\overleftrightarrow{AB}$ and $\overleftrightarrow{CD}$ intersects $\overline{AC}$.

Proof.   Since P lies in the interior of $\angle CAB$ and $\overleftrightarrow{AB} \parallel \overleftrightarrow{CD}$, it follows from Corollary 1, Theorem 1, Sec. 4.2, that $\overrightarrow{AP}$ must intersect $\overrightarrow{CD}$, say at Q (see Fig. 4.9).   Similarly, $\overrightarrow{CP}$ must intersect $\overrightarrow{AB}$.   Finally, the

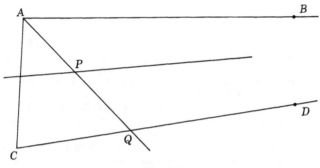

Fig. 4.9

line through P which is parallel to $\overleftrightarrow{AB}$ and $\overleftrightarrow{CD}$ intersects $\overline{AQ}$ in P. Therefore, since it cannot intersect $\overline{CQ}$, it must, by Pasch's theorem, intersect the side $\overline{AC}$ of $\triangle ACQ$, as asserted.

Pasch's theorem itself is valid for asymptotic triangles.   Specifically, we have the following theorem:

Theorem 2.   If a line which does not pass through either vertex of an asymptotic triangle, $\triangle(AB,CD)$, and is not parallel to $\overleftrightarrow{AB}$ and $\overleftrightarrow{CD}$ intersects one side of the asymptotic triangle, then it intersects one and only one of the other two sides.

Proof.   Let $\triangle(AB,CD)$ be an arbitrary asymptotic triangle and let l be a line which does not pass through A or C and is not parallel to $\overleftrightarrow{AB}$ and $\overleftrightarrow{CD}$.   Suppose, first, that l intersects $\overline{AC}$, say in E (Fig. 4.10a).

Then through E there is a unique line, $\overleftrightarrow{EF}$, such that $\overleftrightarrow{AB}$, $\overleftrightarrow{CD}$, and $\overleftrightarrow{EF}$ are parallel in the same sense.   Since l must extend into the interior of

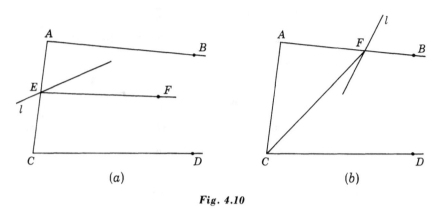

(a)   (b)

Fig. 4.10

either $\angle FEA$ or $\angle FEC$, but not both, it must, by the preceding theorem, intersect either $\overrightarrow{AB}$ or $\overrightarrow{CD}$, but not both.   On the other hand, if l intersects either $\overrightarrow{AB}$ or $\overrightarrow{CD}$, say $\overrightarrow{AB}$, in a point, F (Fig. 4.10b), then l must either extend into the interior of $\angle CFB$ and hence, by the preceding theorem, must intersect $\overrightarrow{CD}$, or else it must extend into the interior of $\triangle FAC$ and hence, by the corollary of Pasch's theorem, must intersect $\overline{AC}$.   Thus the assertion of the theorem is verified in all cases.

As the counterpart of the exterior-angle theorem for ordinary triangles, we have the following result:

Theorem 3.   In any asymptotic triangle, $\triangle(AB,CD)$, the measure of the exterior angle at A (or at C) is greater than the measure of the interior angle at C (or A).

Proof.   Let C' be a point on the ray opposite to $\overrightarrow{AC}$ (see Fig. 4.11). Then by the angle-construction theorem, there is a ray, $\overrightarrow{AF}$, extending into the halfplane determined by $\overleftrightarrow{AC}$ and B such that

$$m\angle C'AF = m\angle ACD$$

Since $\overleftrightarrow{AB} \parallel \overleftrightarrow{CD}$, it follows from Theorem 6, Sec. 4.2, that $\overrightarrow{AF}$ and $\overrightarrow{AB}$ are not the same ray.   Furthermore, by Postulate 17$_H$, if $\overrightarrow{AF}$ is between

$\overrightarrow{AB}$ and $\overrightarrow{AC}$, it must intersect $\overrightarrow{CD}$, thereby forming a triangle in which the measure of the exterior angle at A is equal to the measure of the

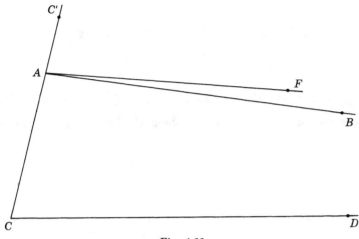

Fig. 4.11

interior angle at C.   Since this contradicts the exterior-angle theorem, it follows that $\overrightarrow{AF}$ cannot lie between $\overrightarrow{AB}$ and $\overrightarrow{AC}$.   Hence

$$m\angle C'AB > m\angle C'AF$$

and therefore $\qquad m\angle C'AB > m\angle ACD$

as asserted.

Interestingly enough, asymptotic triangles also have certain congruence properties suggestive of, though stronger than, the congruence properties of ordinary triangles.   These are described in the next three theorems.

Theorem 4.   If two asymptotic triangles have their finite sides congruent and an angle of one congruent to an angle of the other, then their second angles are congruent.

Proof.   Let $\triangle(AB,CD)$ and $\triangle(A'B',C'D')$ be two asymptotic triangles with $\overline{AC} \cong \overline{A'C'}$ and $\angle A \cong \angle A'$ (see Fig. 4.12).   If $\angle C$ is not congruent to $\angle C'$, then one of these angles, say $\angle C$, has the larger measure.   By the protractor postulate, there is a ray, $\overrightarrow{CE}$, between $\overrightarrow{CA}$ and $\overrightarrow{CD}$ such that

$$\angle ACE \cong \angle A'C'D'$$

Since $\overleftrightarrow{CD} \parallel \overleftrightarrow{AB}$, it follows that $\overrightarrow{CE}$ must intersect $\overrightarrow{AB}$, say in F. Now on $\overrightarrow{A'B'}$ there is a unique point, F', such that

$$\overline{A'F'} \cong \overline{AF}$$

Then, since $\angle A' \cong \angle A$ and $\overline{A'C'} \cong \overline{AC}$, it follows that

$$\triangle A'C'F' \cong \triangle ACF$$

Therefore $$m\angle A'C'F' = m\angle ACF$$

But $m\angle ACF = m\angle A'C'D'$. Hence $m\angle A'C'F' = m\angle A'C'D'$, which is impossible since $\overrightarrow{C'F'}$ and $\overrightarrow{C'D'}$ are obviously distinct rays. Therefore,

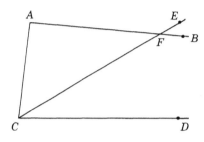 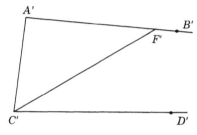

Fig. 4.12

we must reject the supposition that $\angle C$ and $\angle C'$ are not congruent, and thus the theorem is established.

Theorem 5. If two isosceles asymptotic triangles have their finite sides congruent, then the angles of one asymptotic triangle are congruent to the angles of the other.

Proof. Let $\triangle(AB,CD)$ and $\triangle(A'B',C'D')$ be two asymptotic triangles such that $\angle A \cong \angle C$, $\angle A' \cong \angle C'$, and $\overline{AC} \cong \overline{A'C'}$. If

$$m\angle A \neq m\angle A'$$

(and hence $m\angle C \neq m\angle C'$), suppose, for definiteness, that

$$m\angle A > m\angle A'$$

(and hence $m\angle C > m\angle C'$). Then there is a ray, $\overrightarrow{AE}$, between $\overrightarrow{AB}$ and $\overrightarrow{AC}$ such that $\angle CAE \cong \angle C'A'B'$ and a ray, $\overrightarrow{CF}$, between $\overrightarrow{CA}$ and $\overrightarrow{CD}$ such that $\angle ACF \cong \angle A'C'D'$ (see Fig. 4.13). Since $\overleftrightarrow{AB} \parallel \overleftrightarrow{CD}$, it follows that $\overrightarrow{AE}$ must intersect $\overrightarrow{CD}$, say in G. Hence $\overrightarrow{CF}$ must intersect $\overline{AG}$, say in H. Now on $\overrightarrow{A'B'}$ there is a point, H', such that $\overline{A'H'} \cong \overline{AH}$. Hence, since $\overline{A'C'} \cong \overline{AC}$ and $\angle C'A'B' \cong \angle CAE$, it follows that

$$\triangle C'A'H' \cong \triangle CAH$$

Therefore $\angle A'C'H' \cong \angle ACH$

But $\angle ACH \cong \angle A'C'D'$

Hence $\angle A'C'H' \cong \angle A'C'D'$

which is impossible since, clearly, $\overrightarrow{C'H'}$ and $\overrightarrow{C'D'}$ are distinct rays.

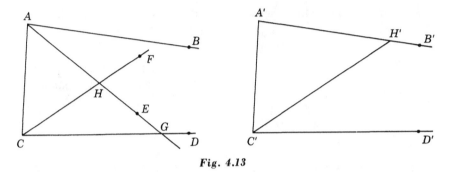

Fig. 4.13

Thus we must abandon the assumption that $m\angle A \neq m\angle A'$ (and $m\angle C \neq m\angle C'$), and the theorem is established.

Theorem 6. If corresponding angles of two asymptotic triangles are congruent, then the finite sides of the two asymptotic triangles are congruent.

Proof. Let $\triangle(AB,CD)$ and $\triangle(A'B',C'D')$ be two asymptotic triangles in which $\angle A \cong \angle A'$ and $\angle C \cong \angle C'$ (see Fig. 4.14). If $AC \neq A'C'$, suppose, for definiteness, that $AC > A'C'$. Then on $\overline{AC}$

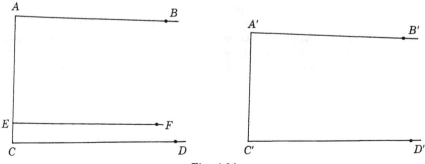

Fig. 4.14

there is a point, E, between A and C such that $AE = A'C'$.   Let $\overrightarrow{EF}$
be the ray parallel to $\overrightarrow{AB}$ through E.   Then, by Theorem 4,

$$\angle AEF \cong \angle A'C'D'$$
But, by hypothesis, $\quad \angle A'C'D' \cong \angle ACD$
Hence $\qquad\qquad\qquad \angle AEF \cong \angle ACD$

However, this contradicts Theorem 3 applied to the asymptotic triangle $\triangle(EF,CD)$.   Therefore the lengths of $\overline{AC}$ and $\overline{A'C'}$ cannot be different, as we tentatively supposed, and the theorem is established.

Using the preceding theorems it is now easy to prove the following result:

Theorem 7.   If $\overrightarrow{AB}$ and $\overrightarrow{CD}$ are two rays such that $\angle BAC$, $\angle DCA$, and $\overline{AC}$ are congruent to the corresponding parts of an asymptotic triangle, $\triangle(A'B',C'D')$, then $\overleftrightarrow{AB}$ and $\overleftrightarrow{CD}$ are parallel.

EXERCISES

1.   Show that the sum of the measures of the angles of an asymptotic triangle is less than 180.
2.   If $\overrightarrow{AB}$ and $\overrightarrow{CD}$ are nonintersecting rays such that B and D are on the same side of $\overleftrightarrow{AC}$, does the exterior-angle theorem hold for the union of $\overrightarrow{AB}$, $\overrightarrow{CD}$, and $\overline{AC}$?   Does Pasch's theorem hold?
3.   Prove Theorem 7.
4.   If A' and C' are two interior points of the segment $\overline{AC}$ and if B, D, B', and D' are points on the same side of $\overleftrightarrow{AC}$ such that

$$\overrightarrow{AB} \parallel \overrightarrow{CD} \parallel \overrightarrow{A'B'} \parallel \overrightarrow{C'D'}$$

how does the sum of the measures of the angles of $\triangle(AB,CD)$ compare with the sum of the measures of the angles of $\triangle(A'B',C'D')$?
5.   Prove that the sum of the measures of the angles of $\triangle(AB,CD)$ is less than the sum of the measures of the angles of $\triangle(ABC)$.
6.   If E is any point in the interior of $\triangle(AB,CD)$, prove that the sum of the measures of the angles of $\triangle(AB,CD)$ is less than the sum of the measures of the angles of $\triangle ACE$.
7.   If $P, Q, R,$ and S are four points such that $\overrightarrow{PS} \parallel \overrightarrow{QR}$ and $\overrightarrow{QP} \parallel \overrightarrow{RS}$, prove that

$$m\angle PSR > m\angle PQR$$

8.   In Exercise 7, is it possible that simultaneously $PS = QR$ and $QP = RS$?

4.4 Saccheri Quadrilaterals. One of the most important configurations in hyperbolic geometry is the **Saccheri quadrilateral**, so called because Saccheri made extensive use of it in his unsuccessful attempt to prove the euclidean parallel postulate:

> **Definition 1.** A quadrilateral $ABCD$ in which A and B are right angles and $AD = BC$ is called a Saccheri quadrilateral.

The fundamental properties of Saccheri quadrilaterals are set forth in the following theorem:

> **Theorem 1.** If $ABCD$ is a Saccheri quadrilateral with right angles at A and at B, then
>
> (1) The perpendicular bisector of $\overline{AB}$ is also the perpendicular bisector of $\overline{CD}$.
> (2) $\angle C$ and $\angle D$ are congruent acute angles.

Proof. Let $ABCD$ be a Saccheri quadrilateral with right angles at A and at B, and $AD = BC$ (see Fig. 4.15). Let E and F be, respec-

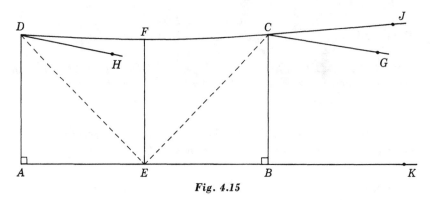

Fig. 4.15

tively, the midpoints of $\overline{AB}$ and $\overline{CD}$. Then since

$$AD = BC, \qquad AE = EB, \qquad \text{and } m\angle DAE = m\angle CBE \ (= 90)$$

it follows that

$$\triangle DAE \cong \triangle CBE$$

Hence $DE = CE$. Therefore, since $DF = CF$, it follows that

$$\triangle DFE \cong \triangle CFE$$

Therefore $\angle DFE \cong \angle CFE$

Since these congruent angles form a linear pair, each must be a right angle, which proves that $\overline{EF}$ is perpendicular to $\overline{DC}$. Also, since $\angle AED \cong \angle BEC$, from the first pair of congruent triangles, and $\angle DEF \cong \angle CEF$, from the second pair of congruent triangles, it follows that $\angle AEF \cong \angle BEF$. Hence each of these angles is a right angle, and we also have $\overline{FE} \perp \overline{AB}$. Furthermore, since $\angle ADE \cong \angle BCE$ and $\angle EDF \cong \angle ECF$, it follows that $\angle ADF \cong \angle BCF$. To complete the proof of the theorem, it thus remains to prove that $\angle ADF$ and $\angle BCF$ are acute. To do this, let J and K be, respectively, points on the rays opposite to $\overrightarrow{CF}$ and $\overrightarrow{BE}$, and consider the rays $\overrightarrow{CG}$ and $\overrightarrow{DH}$ which are parallel to $\overleftrightarrow{AB}$ through C and D, respectively. By the first part of the theorem, $\overleftrightarrow{AB}$ and $\overleftrightarrow{DC}$ have a common perpendicular, namely $\overleftrightarrow{EF}$; hence they must be nonparallel nonintersecting lines. Therefore, $\overrightarrow{DH}$ lies between $\overrightarrow{DA}$ and $\overrightarrow{DF}$, and $\overrightarrow{CG}$ lies between $\overrightarrow{CB}$ and $\overrightarrow{CJ}$. Now, by Theorem 3, Sec. 4.3, applied to the asymptotic triangle $\triangle(DH,CG)$, we have

$$m\angle JCG > m\angle CDH$$

Also, from Theorem 4, Sec. 4.3, applied to $\triangle(AE,DH)$ and $\triangle(BK,CG)$, we have

$$m\angle GCB = m\angle HDA$$

Hence, adding, $m\angle JCB > m\angle CDA$

But $\angle CDA \cong \angle DCB$; hence $m\angle JCB > m\angle DCB$. In other words, of the supplementary angles, $\angle JCB$ and $\angle DCB$, the latter has the smaller measure and hence is acute, as asserted.

For birectangular quadrilaterals which are not Saccheri quadrilaterals, we have the following theorem and its converse:

Theorem 2. If in a quadrilateral $ABCD$, $\angle A$ and $\angle B$ are right angles and $AD > BC$, then $m\angle D < m\angle C$.

Proof. Since $BC < AD$, there is a unique point, E, between A and D such that $AE = BC$ (see Fig. 4.16). Then $ABCE$ is a Saccheri quadrilateral and

$$m\angle AEC = m\angle BCE$$

Furthermore, from the exterior-angle theorem,

$$m\angle ADC < m\angle AEC$$

Moreover, since $\overrightarrow{CE}$ is between $\overrightarrow{CB}$ and $\overrightarrow{CD}$, we have

$$m\angle BCE < m\angle BCD$$

Hence, combining the last three relations,

$$m\angle ADC < m\angle BCD$$

as asserted.

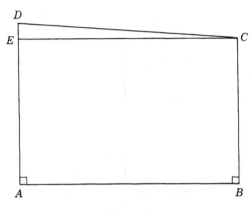

Fig. 4.16

Corollary 1. If in a quadrilateral $ABCD$, $\angle A$ and $\angle B$ are right angles and $\angle C \cong \angle D$, then $AD = BC$.

Theorem 3. If in a quadrilateral $ABCD$, $\angle A$ and $\angle B$ are right angles and $m\angle D < m\angle C$, then $AD > BC$.

Proof. Clearly we must have either

$$AD = BC \quad\text{or}\quad AD < BC \quad\text{or}\quad AD > BC$$

The first of these possibilities must be rejected because if $AD = BC$, then $ABCD$ is a Saccheri quadrilateral and, perforce, $m\angle D = m\angle C$, contrary to hypothesis. Similarly, if $AD < BC$, then by Theorem 2, $m\angle D > m\angle C$, which again contradicts our hypothesis. Hence the remaining possibility, $AD > BC$, must be the correct one, as asserted.

Related to the Saccheri quadrilateral is the so-called **Lambert quadrilateral*** which is a quadrilateral with three and only three right angles. Clearly, such a configuration cannot exist in euclidean

---

* Named for the German mathematician Johann Heinrich Lambert (1728–1777), who discovered many of the theorems of hyperbolic geometry in the course of his investigations of the euclidean parallel postulate.

geometry, but its existence in hyperbolic geometry is guaranteed by the following theorem:

Theorem 4.   If three angles of a quadrilateral are right angles, the fourth angle is acute.

Proof.   Let $ABCD$ be a quadrilateral with right angles at A, B, and C (see Fig. 4.17). Then on the ray opposite to $\overrightarrow{CD}$ there is a

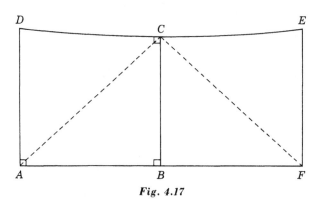

Fig. 4.17

point, E, such that $CD = CE$, and on the ray opposite to $\overrightarrow{BA}$ there is a point, F, such that $BA = BF$. Since $\angle ABC$ and $\angle FBC$ are both right angles, it follows that

$$\triangle ABC \cong \triangle FBC$$

and hence that

$$m\angle ACB = m\angle FCB \qquad \text{and} \qquad m\angle BAC = m\angle BFC$$

Moreover, $\overrightarrow{CA}$ lies between $\overrightarrow{CB}$ and $\overrightarrow{CD}$, and $\overrightarrow{CF}$ lies between $\overrightarrow{CB}$ and $\overrightarrow{CE}$. Therefore, since $\angle BCD$ and $\angle BCE$ are both right angles, it follows that

$$\angle ACD \cong \angle FCE$$

Furthermore, from the congruent triangles $\triangle ABC$ and $\triangle FBC$, we also have

$$AC = FC$$

Hence $$\triangle ADC \cong \triangle FEC$$
and therefore $$AD = FE$$
Also, $$m\angle CAD = m\angle CFE$$
and hence

$$m\angle BFC + m\angle CFE = m\angle BAC + m\angle CAD = m\angle BAD = 90$$

by hypothesis.   Thus $AFED$ is a Saccheri quadrilateral, and therefore $\angle ADC$ is acute, as asserted.

It is now easy to establish the following useful results:

Theorem 5.   If the acute angles of two Lambert quadrilaterals are congruent and if a side of one quadrilateral is congruent to the corresponding side of the other, then all pairs of corresponding sides of the two quadrilaterals are congruent.

Theorem 6.   If two sides of a Lambert quadrilateral are congruent to the corresponding sides of a second Lambert quadrilateral, then all pairs of corresponding parts of the two quadrilaterals are congruent.

Definition 2.   Two quadrilaterals are said to be congruent if there exists a correspondence between their vertices such that corresponding sides and corresponding angles of the two quadrilaterals are congruent.

EXERCISES

1. If $ABCD$ is a quadrilateral with right angles at A and B, prove that any line which is perpendicular to $\overleftrightarrow{AB}$ at a point between A and B must intersect $\overline{CD}$.

2. If $\triangle ABC$ is an arbitrary triangle, if D and E are the midpoints of $\overline{AB}$ and $\overline{AC}$, respectively, and if F and G are, respectively, the feet of the perpendiculars from B and C to $\overleftrightarrow{DE}$, prove that $BCGF$ is a Saccheri quadrilateral.

3. In Exercise 2, prove that $DE < \frac{1}{2}BC$.

4. Prove that if two of the angles of a quadrilateral are right angles, the sum of the measures of the other two angles cannot be 180.

5. If $ABCD$ and $A'B'C'D'$ are two Lambert quadrilaterals with congruent acute angles at A and A', prove Theorem 5 by proving each of the following assertions:

   (a) If $AB = A'B'$, then $ABCD \cong A'B'C'D'$.
   (b) If $BC = B'C'$, then $ABCD \cong A'B'C'D'$.

6. If $ABCD$ and $A'B'C'D'$ are two Lambert quadrilaterals with acute angles at A and A', prove Theorem 6 by proving each of the following assertions:

   (a) If $BC = B'C'$ and if $CD = C'D'$, then $ABCD \cong A'B'C'D'$.
   (b) If $AB = A'B'$ and if $BC = B'C'$, then $ABCD \cong A'B'C'D'$.
   (c) If $AB = A'B'$ and if $CD = C'D'$, then $ABCD \cong A'B'C'D'$.
   (d) If $AB = A'B'$ and if $AD = A'D'$, then $ABCD \cong A'B'C'D'$.

7. If $ABCD$ and $A'B'C'D'$ are two Saccheri quadrilaterals with congruent acute angles at C, D, C', and D', prove each of the following assertions:

   (a) If $AB = A'B'$, then $ABCD \cong A'B'C'D'$.
   (b) If $BC = B'C'$, then $ABCD \cong A'B'C'D'$.
   (c) If $CD = C'D'$, then $ABCD \cong A'B'C'D'$.

8. If $ABCD$ and $A'B'C'D'$ are two Saccheri quadrilaterals with acute angles at C and D and at C' and D', prove each of the following assertions:

   (a) If $AB = A'B'$ and if $BC = B'C'$, then $ABCD \cong A'B'C'D'$.
   (b) If $BC = B'C'$ and if $CD = C'D'$, then $ABCD \cong A'B'C'D'$.
   (c) If $AB = A'B'$ and if $CD = C'D'$, then $ABCD \cong A'B'C'D'$.

4.5   The Angle-sum Theorem.   We are now in a position to prove one of the most famous theorems of hyperbolic geometry, the so-called **angle-sum theorem.**   For convenience in doing this, we shall present the argument in two parts, first for right triangles and then for completely general triangles:

Lemma 1.   The sum of the measures of the angles of any right triangle is less than 180.

Proof.   Let $\triangle ABC$ be a right triangle with right angle at C.   Then on the opposite side of $\overleftrightarrow{AB}$ from C there is a ray, $\overrightarrow{AD}$, such that

$$m\angle BAD = m\angle ABC$$

Let M be the midpoint of $\overline{AB}$, and let $\overline{MP}$ be the perpendicular segment from M to $\overleftrightarrow{CB}$ (see Fig. 4.18).   Since $\overleftrightarrow{MP}$ and $\overleftrightarrow{AC}$ are both per-

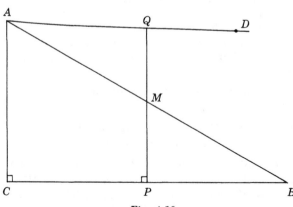

Fig. 4.18

pendicular to $\overleftrightarrow{CB}$, they cannot intersect.   Hence, by Pasch's theorem, $\overleftrightarrow{MP}$ must intersect $\overline{CB}$, that is, P lies between C and B.   Now on $\overrightarrow{AD}$ let Q be the point such that $AQ = BP$.   Then, since

$$\angle MAQ \cong \angle MBP, \qquad AM = BM, \qquad \text{and } AQ = BP$$

it follows that $\qquad\qquad \triangle MAQ \cong \triangle MBP$

and hence that $\qquad\qquad \angle AMQ \cong \angle BMP$

Therefore, by the converse of the vertical-angle theorem, $\overrightarrow{MQ}$ and $\overrightarrow{MP}$ must be collinear.   Also, since $\angle AQM \cong \angle BPM$, it follows that $\angle AQM$ is a right angle.   Thus A, C, P, and Q are the vertices of a Lambert quadrilateral with right angles at C, P, and Q.   Hence, by Theorem 4, Sec. 4.4, the fourth angle, $\angle CAQ$, must be acute.   But, since $\overrightarrow{AB}$ is between $\overrightarrow{AC}$ and $\overrightarrow{AD}$,

$$m\angle CAQ = m\angle CAB + m\angle BAQ = m\angle CAB + m\angle ABC$$

Thus $\qquad\qquad m\angle CAB + m\angle ABC < 90$

and hence $\qquad\quad m\angle ACB + m\angle CAB + m\angle ABC < 180$

as asserted.

Theorem 1.   The sum of the measures of the angles of any triangle is less than 180.

Proof.   Let $\triangle ABC$ be an arbitrary triangle.   If it is a right triangle, the assertion of the theorem is merely a restatement of the preceding lemma.   If it is not a right triangle, let $\overline{AB}$, say, be at least as long as either of the other two sides.   Then, by Theorem 5, Sec. 2.11, the foot of the perpendicular from C to $\overleftrightarrow{AB}$, say F, must lie

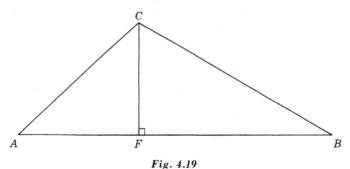

Fig. 4.19

between A and B (see Fig. 4.19).   Then, by the preceding lemma,

$$m\angle FAC + m\angle ACF + m\angle CFA < 180$$

or, since $\angle CFA$ is a right angle,

$$m\angle FAC + m\angle ACF < 90$$

Similarly, $$m\angle FCB + m\angle CBF < 90$$

and therefore, adding,

$$m\angle FAC + m\angle ACF + m\angle FCB + m\angle CBF < 180$$

Finally, since $\overrightarrow{CF}$ is between $\overrightarrow{CA}$ and $\overrightarrow{CB}$,

$$m\angle ACF + m\angle FCB = m\angle ACB$$

and therefore, substituting,

$$m\angle FAC + m\angle ACB + m\angle CBF < 180$$

as asserted.

As an immediate consequence of Theorem 1, we have the following result, special cases of which we have already encountered in Saccheri and Lambert quadrilaterals:

Corollary 1. The sum of the measures of the angles of any quadrilateral is less than 360.

Since Theorems 1 and 4, Sec. 2.9, were proved without the use of the euclidean parallel postulate, they hold equally well in hyperbolic and euclidean geometry. Using the preceding results, we can now prove an even stronger congruence result which has no counterpart in euclidean geometry:

Theorem 2. If there exists a correspondence between two triangles in which corresponding angles are congruent, then the triangles are congruent.

Proof. Let $\triangle ABC$ and $\triangle A'B'C'$ be two triangles such that

$$\angle A \cong \angle A', \qquad \angle B \cong \angle B', \qquad \text{and } \angle C \cong \angle C'$$

If $AB = A'B'$, it follows immediately from Theorem 1, Sec. 2.9, that the triangles are congruent. Hence let us suppose that $AB \neq A'B'$ or, specifically, $AB > A'B'$. Then between A and B there is a unique point, D, such that $AD = A'B'$. Furthermore, on $\overrightarrow{AC}$ there is a unique point, E, such that $AE = A'C'$. Now C cannot lie between A and E, that is, D and E cannot lie on opposite sides of $\overleftrightarrow{BC}$. In fact, if this were the case (Fig. 4.20a), then since

$$AD = A'B', \qquad AE = A'C', \qquad \text{and } \angle A \cong \angle A'$$

it follows that

$$\triangle ADE \cong \triangle A'B'C'$$

and hence $$\angle AED \cong \angle A'C'B' \cong \angle ACB$$

But this asserts that the measure of one of the exterior angles of $\triangle PCE$ is equal to the measure of one of the opposite interior angles, which is impossible.

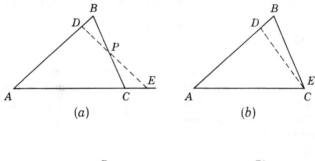

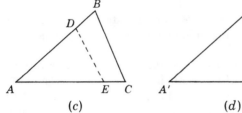

Fig. 4.20

Moreover, E cannot coincide with C, because if this were the case (Fig. 4.20b), then again

$$\triangle ADE \cong \triangle A'B'C'$$

and it would follow, as before, that

$$\angle AED \cong \angle A'C'B' \cong \angle ACB$$

But this contradicts the protractor postulate, since $\overrightarrow{EB}$ and $\overrightarrow{ED}$ are different rays.

Finally, E cannot lie between A and C (Fig. 4.20c), because again we would have

$$\triangle ADE \cong \triangle A'B'C'$$

Hence it would follow that

$$\angle ADE \cong \angle A'B'C' \cong \angle ABC$$

and
$$\angle AED \cong \angle A'C'B' \cong \angle ACB$$

Therefore, in the quadrilateral $DECB$ we would have

$$m\angle BDE + m\angle DEC + m\angle ECB + m\angle CBD$$
$$= (180 - m\angle ADE) + (180 - m\angle AED) + m\angle ACB + m\angle ABC$$
$$= 360$$

which is impossible, according to Corollary 1, Theorem 1.   Hence, no matter where E is located on $\overrightarrow{AC}$, we are led to a contradiction.   Thus the assumption that $AB \neq A'B'$ must be abandoned, and therefore

$$\triangle ABC \cong \triangle A'B'C'$$

as asserted.

Because of the exterior-angle theorem for triangles and asymptotic triangles, it is clear that if two lines have a common perpendicular, they can neither intersect nor be parallel, that is, they must be nonintersecting. Using the properties of Saccheri quadrilaterals and Corollary 1, Theorem 1, we can now prove the converse of this important result:

Theorem 3.   If two lines neither intersect nor are parallel, they have a unique common perpendicular.

Proof.   Let a and b be two lines which neither intersect nor are parallel, let A and A' be two points on a, and let B and B' be, respectively, the feet of the perpendiculars from A and A' to b (see Fig. 4.21). If $AB = A'B'$, then $ABB'A'$ is a Saccheri quadrilateral, and

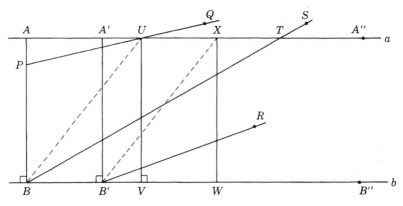

Fig. 4.21

by Theorem 1, the perpendicular bisector of $\overline{BB'}$ is also the perpendicular bisector of $\overline{AA'}$ and hence is the required common perpendicular. Suppose, therefore, that $AB \neq A'B'$, and for definiteness let $AB > A'B'$. Then between A and B there is a unique point, P, such that

$$BP = B'A'$$

and on the same side of $\overleftrightarrow{AB}$ as A' there is a unique ray, $\overrightarrow{PQ}$, such that

$$\angle BPQ \cong \angle B'A'A''$$

where A'' is a point on the ray opposite to $\overrightarrow{A'A}$. Now since a and b are nonintersecting lines, the ray, $\overrightarrow{B'R}$, which is parallel to $\overleftrightarrow{AA'}$ through B' must lie between $\overrightarrow{B'A'}$ and $\overrightarrow{B'B''}$, where B'' is a point on the ray opposite to $\overrightarrow{B'B}$. Furthermore, on the same side of $\overleftrightarrow{AB}$ as A' there is a unique ray, $\overrightarrow{BS}$, such that

$$\angle B'BS \cong \angle B''B'R$$

Then, from the exterior-angle theorem for asymptotic triangles, the ray which is parallel to $\overrightarrow{AA'}$ and $\overrightarrow{B'R}$ through B must lie between $\overrightarrow{BS}$ and $\overrightarrow{BB'}$. Hence $\overrightarrow{BS}$ must intersect $\overrightarrow{AA'}$, say at T. Now $\overrightarrow{B'R}$ is parallel to $\overrightarrow{A'A''}$. Therefore, since $\angle BPQ \cong \angle B'A'A''$, $\angle PBT \cong \angle A'B'R$, and $PB = A'B'$, it follows from Theorem 7, Sec. 4.3, applied to $\triangle(A'A'',B'R)$ and the union of $\overrightarrow{PQ}$, $\overrightarrow{BT}$, and $\overline{PB}$, that $\overrightarrow{PQ} \parallel \overrightarrow{BT}$. Hence $\overleftrightarrow{PQ}$, which intersects the side $\overline{AB}$ of $\triangle ABT$, cannot intersect $\overline{BT}$ and therefore must intersect $\overline{AT}$, say at U. Now let V be the foot of the perpendicular from U to b. Also, let W be the unique point on the ray opposite to $\overrightarrow{B'B}$ such that

$$BV = B'W$$

and let X be the unique point on the ray opposite to $\overrightarrow{A'A}$ such that

$$PU = A'X$$

Then, since $PB = A'B'$, $PU = A'X$, and $\angle BPU \cong \angle B'A'X$, it follows that

$$\triangle PBU \cong \triangle A'B'X$$

Therefore $\quad BU = B'X \quad$ and $\quad \angle UBP \cong \angle XB'A'$

Moreover, from the last congruence and from the fact that $\angle PBB''$ and $\angle A'B'B''$ are both right angles, it follows that

$$\angle UBV \cong \angle XB'W$$

Hence $$\triangle XB'W \cong \triangle UBV$$

and therefore

$$XW = UV \quad \text{and} \quad m\angle XWB' = m\angle UVB = 90$$

Thus, finally, $UVWX$ is a Saccheri quadrilateral. Hence $\overline{UX}$ and $\overline{VW}$, and therefore a and b, have a common perpendicular, as asserted. The uniqueness of the common perpendicular follows at once from the fact that if there were two common perpendiculars, their intersections with a and b would be the vertices of a quadrilateral whose angles had measures totaling 360, which is impossible, in view of Corollary 1, Theorem 1.

> **Corollary 1.** The perpendicular distance from a point on one of two nonintersecting lines to the other is a monotonically increasing function of the distance of the point from the common perpendicular of the two lines.

One of the famous results of euclidean geometry asserts that the perpendicular bisectors of the sides of a triangle are concurrent. This is not necessarily true in hyperbolic geometry, but with the results which we have at our disposal we can now determine what the corresponding properties are. The three possibilities are covered by the following theorems, the first of which is the familiar euclidean result, proved in the usual euclidean fashion:

> **Theorem 4.** If the perpendicular bisectors of two of the sides of a triangle intersect, the perpendicular bisector of the third side also passes through that intersection.

Proof. Let b and a be, respectively, the perpendicular bisectors of the sides $\overline{AC}$ and $\overline{BC}$ of $\triangle ABC$, and let a and b intersect in the point G (see Fig. 4.22). Let B' and A' be, respectively, the midpoints of $\overline{AC}$ and $\overline{BC}$. Then

$$\triangle BGA' \cong \triangle CGA'$$

since two sides and the included angle of one triangle are congruent to the corresponding parts of the second. Hence

$$BG = CG$$

Similarly, $$\triangle CGB' \cong \triangle AGB'$$

and therefore $$CG = AG$$

Thus $BG = AG$, since each is equal to CG. Now let C' be the midpoint of $\overline{AB}$. Then, because the sides of one are congruent to the corresponding sides of the other,

$$\triangle AC'G \cong \triangle BC'G$$
and therefore
$$\angle AC'G \cong \angle BC'G$$

Since these congruent angles form a linear pair, each must be a right

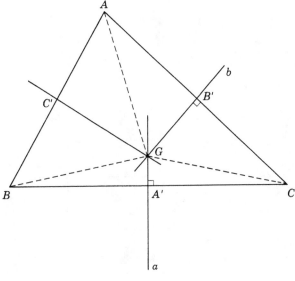

Fig. 4.22

angle. Hence $\overleftrightarrow{C'G}$ is perpendicular to $\overline{AB}$ at its midpoint, C', and thus the perpendicular bisector of $\overline{AB}$, namely $\overleftrightarrow{C'G}$, passes through G, as asserted.

Theorem 5. If the perpendicular bisectors of two of the sides of a triangle are perpendicular to a line, l, the perpendicular bisector of the third side is also perpendicular to l.

Proof. Let A' and B' be, respectively, the midpoints of the sides $\overline{BC}$ and $\overline{CA}$ of $\triangle ABC$; let a and b be the perpendicular bisectors of these sides; and let a and b be perpendicular to a line, l, at the points A_1 and B_1, respectively. Let A'', B'', and C'' be, respectively, the feet of the perpendiculars to l from A, B, and C (see Fig. 4.23). Now since

$$B'A = B'C \qquad \text{and, of course,} \qquad B'B_1 = B'B_1$$

it follows from Theorem 6, Sec. 4.4, that $AB'B_1A''$ and $CB'B_1C''$ are

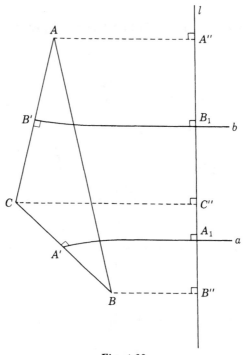

Fig. 4.23

congruent Lambert quadrilaterals. Therefore $AA'' = CC''$. Similarly, $BB'' = CC''$. Hence $AA'' = BB''$, and therefore $AA''B''B$ is a Saccheri quadrilateral. But, by Theorem 1, Sec. 4.4, the perpendicular bisector of $\overline{AB}$ must also be the perpendicular bisector of $\overline{A''B''}$. In other words, the perpendicular bisector of $\overline{AB}$ is also perpendicular to l, as asserted.

Theorem 6. If the perpendicular bisectors of two sides of a triangle are parallel, the perpendicular bisector of the third side is parallel to these bisectors in the same sense.

Proof. Let a, b, and c be, respectively, the perpendicular bisectors of $\overline{BC}$, $\overline{CA}$, and $\overline{AB}$ in $\triangle ABC$, and let a and b be parallel. Now c

cannot intersect either a or b, because if it did, then by Theorem 4, a, b, and c would all be concurrent, contrary to the hypothesis that $a\|b$. Likewise, if c is a nonintersector of a or b, then by Theorem 5, a, b, and c have a common perpendicular, contrary to the hypothesis that $a\|b$. Hence c must be parallel to a and b, as asserted.

It remains to show, however, that c is parallel to a and b in the same sense. To do this, let A_1, B_1, and C_1 be, respectively, the midpoints of $\overline{BC}$, $\overline{CA}$, and $\overline{AB}$. Now unless $\triangle ABC$ is equilateral, in which case it is obvious that the perpendicular bisectors are concurrent, at least one of the lines a, b, c must intersect a second side of $\triangle ABC$ in a point which is not a vertex. For definiteness, let us suppose that the perpendicular bisector of $\overline{BC}$ intersects $\overline{AC}$ at B_2, say. Since a, b, and c are all parallel, it is clear that B_1 and B_2 are distinct points. If B_2 is between A and B_1 (Fig. 4.24a), the perpendicular bisector, b, must intersect $\overline{BC}$ in a point, A_2, between A_1 and C. If B_2 is between B_1 and C, b can either intersect $\overline{AB}$, say in C_2 (Fig. 4.24b), or it can

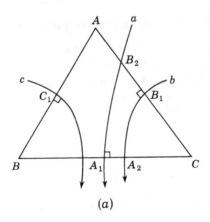

(a)

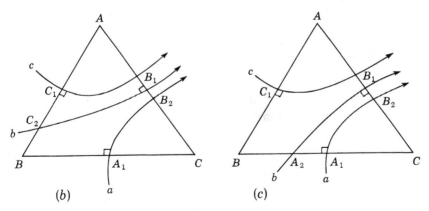

(b)   (c)

Fig. 4.24

intersect $\overline{BC}$ in a point, A_2, between B and A_1 (Fig. 4.24c). Now if two rays are parallel, they necessarily lie on the same side of the line determined by their endpoints. Moreover, the amplitude of parallelism is always less than 90. Hence, of the four possibilities for the parallelism of a and b in each case, we must have, respectively,

$$(a)\ \overleftrightarrow{B_2A_1}\|\overleftrightarrow{B_1A_2}, \qquad (b)\ \overleftrightarrow{A_1B_2}\|\overleftrightarrow{C_2B_1}, \qquad (c)\ \overleftrightarrow{A_1B_2}\|\overleftrightarrow{A_2B_1}$$

In case (a), c cannot be parallel to both $\overleftrightarrow{A_1B_2}$ and $\overleftrightarrow{A_2B_1}$, because since C_1 is on the same side of $\overleftrightarrow{B_1B_2}$ as A_1 and A_2, this would imply a ray on c parallel to both $\overleftrightarrow{A_1B_2}$ and $\overleftrightarrow{A_2B_1}$, which in turn would imply that $\overleftrightarrow{A_1B_2}\|\overleftrightarrow{A_2B_1}$, contrary to what we have just observed. Thus at least one of the relations

$$\overleftrightarrow{B_2A_1}\|c, \qquad \overleftrightarrow{B_1A_2}\|c$$

must hold. But from the transitive property of parallelism and the fact that $\overleftrightarrow{B_2A_1}\|\overleftrightarrow{B_1A_2}$, if either of these holds, so does the other, and hence a, b, and c are parallel in the same sense. A similar argument shows that in the other two cases, a, b, and c are also parallel in the same sense, as asserted.

EXERCISES

1. Prove Corollary 1, Theorem 3.
2. Prove Theorem 1 using the results of Exercise 2, Sec. 4.4.
3. Prove that the measure of any exterior angle of a triangle is greater than the sum of the measures of the opposite interior angles.
4. Prove that the sum of the measures of the angles of $\triangle BCD$ is greater than the sum of the measures of the angles of $\triangle(AB,CD)$.
5. If E is a point between A and C, prove that the sum of the measures of the angles of $\triangle BED$ is greater than the sum of the measures of the angles of $\triangle(AB,CD)$.
6. If A' is between A and B and if C' is between C and D, prove that the sum of the measures of the angles of $\triangle(AB,CD)$ is less than the sum of the measures of the angles of $\triangle(A'B,C'D)$.
7. If A' and C' are two points in the interior of $\triangle(AB,CD)$ and if B' and D' are two points such that $\overrightarrow{A'B'}\|\overrightarrow{AB}\|\overrightarrow{CD}\|\overrightarrow{C'D'}$, prove that the sum of the measures of the angles of $\triangle(A'B',C'D')$ is greater than the sum of the measures of the angles of $\triangle(AB,CD)$.
8. If A' and C' are interior points of the segment $\overline{AC}$ and if E is any interior point of $\triangle(AB,CD)$, prove that the sum of the measures of the angles

of $\triangle A'C'E$ is greater than the sum of the measures of the angles of $\triangle(AB,CD)$.

9. If E, F, and G are any three noncollinear points in the interior of

$$\triangle(AB,CD)$$

prove that the sum of the measures of the angles of $\triangle EFG$ is greater than the sum of the measures of the angles of $\triangle(AB,CD)$.

10. If D is any point in the interior of $\triangle ABC$, prove that the sum of the measures of the angles of $\triangle ABD$ is greater than the sum of the measures of the angles of $\triangle ABC$.

11. If D is a point in the interior of $\triangle ABC$ and if E and F are two points between B and C, prove that the sum of the measures of the angles of $\triangle DEF$ is greater than the sum of the measures of the angles of $\triangle ABC$.

12. In $\triangle ABC$ and $\triangle A'B'C'$, $m\angle B = m\angle C = m\angle B' = m\angle C'$. If

$$BC < B'C'$$

what relation exists between $m\angle A$ and $m\angle A'$?

13. If $\overleftrightarrow{BC} \parallel \overleftrightarrow{B'C'}$, if $\overleftrightarrow{BB'}$ and $\overleftrightarrow{CC'}$ intersect in a point A which is not between $\overleftrightarrow{BC}$ and $\overleftrightarrow{B'C'}$, and if $\triangle ABC$ is isosceles, is $\triangle AB'C'$ isosceles? Justify your conclusion.

14. Assuming the euclidean angle-sum theorem, prove the euclidean parallel postulate.

15. Complete the proof of Theorem 6 by giving in detail the argument in cases (b) and (c).

4.6 Further Properties of Parallel Lines.   The following important result has no counterpart in euclidean geometry:

> **Theorem 1.**   If the lines determined by two noncollinear rays intersect, there is a unique line which is parallel to both rays.

Proof.   It is clearly no specialization to assume that the given rays have the same endpoint (see Fig. 4.25). Hence, let us consider the noncollinear rays $\overrightarrow{OA}$ and $\overrightarrow{OB}$, on which, for convenience, we shall suppose that $OA = OB$. Let $\overrightarrow{AC}$ and $\overrightarrow{BD}$ be parallel, respectively, to $\overrightarrow{OB}$ and $\overrightarrow{OA}$; let A' and B' be points on the rays opposite to $\overrightarrow{AO}$ and $\overrightarrow{BO}$, respectively; and let $\overrightarrow{AU}$ and $\overrightarrow{BV}$ be the rays which bisect $\angle A'AC$ and $\angle DBB'$. Clearly, both $\overrightarrow{AC}$ and $\overrightarrow{BD}$ extend into the interior of $\angle A'OB'$. Hence, considering $\overleftrightarrow{AC}$ in relation to the asymptotic triangle $\triangle(OA',BD)$, it is evident that $\overrightarrow{AC}$ must intersect $\overrightarrow{BD}$, say in E. Now in $\triangle(OA',BD)$ and $\triangle(OB',AC)$ we have

$$OA = OB \qquad \text{and} \qquad m\angle AOB' = m\angle BOA'$$

Hence, from Theorem 4, Sec. 4.3,

(1) $$m\angle OAC = m\angle OBD$$

Therefore, the angles which form linear pairs with these angles have

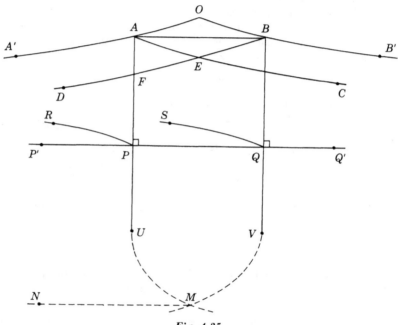

Fig. 4.25

the same measure:

$$m\angle A'AC = m\angle B'BD$$

and

(2) $$m\angle A'AU = m\angle UAC = m\angle DBV = m\angle VBB'$$

Now $\overleftrightarrow{AU}$ and $\overleftrightarrow{BV}$ are nonintersecting lines. To see this, suppose, first, that $\overrightarrow{AU}$ and $\overrightarrow{BV}$ intersect, say in M. Then, since $\triangle AOB$ is isosceles,

(3) $$m\angle OAB = m\angle OBA$$

From this, together with (1) and (2), it follows that

$$m\angle BAM = m\angle ABM$$

Hence $\triangle ABM$ is isosceles and therefore $AM = BM$. Now let $\overleftrightarrow{MN}$ be the parallel to $\overleftrightarrow{OA}$ and $\overleftrightarrow{BD}$ through M. Then in the asymptotic triangles $\triangle(MN,AA')$ and $\triangle(MN,BD)$ we have

$$AM = BM \quad \text{and} \quad m\angle MAA' = m\angle MBD$$

Hence, by Theorem 4, Sec. 4.3, $m\angle NMA = m\angle NMB$, which is a contradiction of the angle-construction theorem, since A and B lie on the same side of $\overleftrightarrow{MN}$, and $\overrightarrow{MA}$ and $\overrightarrow{MB}$ are distinct rays. Therefore $\overrightarrow{AU}$ and $\overrightarrow{BV}$ cannot intersect. In exactly the same way, we can show that the rays opposite to $\overrightarrow{AU}$ and $\overrightarrow{BV}$ cannot intersect. Hence, the lines $\overleftrightarrow{AU}$ and $\overleftrightarrow{BV}$ cannot intersect.

Likewise, neither of the relations $\overleftrightarrow{AU} \parallel \overleftrightarrow{BV}$, $\overleftrightarrow{UA} \parallel \overleftrightarrow{VB}$ can hold. For suppose the first of these holds. Since $\overrightarrow{AU}$ is between $\overrightarrow{AA'}$ and $\overrightarrow{AC}$, it must intersect $\overrightarrow{ED}$, say at F. Then in the asymptotic triangles $\triangle(AA',FD)$ and $\triangle(FU,BV)$, we have

$$m\angle AFD = m\angle BFU$$

and, by (2),

$$m\angle A'AF = m\angle FBV$$

Hence, by Theorem 6, Sec. 4.3, $AF = BF$. Therefore $\triangle AFB$ is isosceles, and we have

$$m\angle FAB = m\angle FBA$$

But

$$m\angle FBA = m\angle EAB$$

and therefore

$$m\angle FAB = m\angle EAB$$

But this also contradicts the angle-construction theorem, since F and E are on the same side of $\overleftrightarrow{AB}$ and $\overrightarrow{AF} \neq \overrightarrow{AE}$. Thus it is impossible that $\overleftrightarrow{AU} \parallel \overleftrightarrow{BV}$, and a similar argument shows that $\overleftrightarrow{UA} \parallel \overleftrightarrow{VB}$ is also impossible.

We have thus shown that $\overrightarrow{AU}$ and $\overrightarrow{BV}$ are nonintersecting lines, as asserted. Hence, by Theorem 3, Sec. 4.5, they must have a common perpendicular, and this, in fact, is the required common parallel to $\overleftrightarrow{OA}$ and $\overleftrightarrow{OB}$. To prove this, let $\overleftrightarrow{P'Q'}$ be the common perpendicular to $\overleftrightarrow{AU}$ and $\overleftrightarrow{BV}$ and let P and Q be the intersections of $\overleftrightarrow{P'Q'}$ with $\overleftrightarrow{AU}$ and $\overleftrightarrow{BV}$, respectively. Then, since $m\angle PAB = m\angle QBA$, it follows from Corollary 1, Theorem 2, Sec. 4.4, that $AP = BQ$. Now if $\overleftrightarrow{QP}$ is not parallel to $\overleftrightarrow{OA}$, let $\overleftrightarrow{PR}$ be the parallel to $\overleftrightarrow{OA}$ through P and let $\overleftrightarrow{QS}$ be the parallel to $\overleftrightarrow{OA}$ through Q. Then in the asymptotic triangles $\triangle(AA',PR)$ and $\triangle(BD,QS)$, we have

$$m\angle A'AP = m\angle DBQ \quad \text{and} \quad AP = BQ$$

Hence
$$m\angle RPA = m\angle SQB$$

and therefore
$$m\angle RPQ = m\angle SQQ'$$

where Q' is a point on the ray opposite to $\overrightarrow{QP'}$. But this is impossible, since, by Theorem 3, Sec. 4.3, the measure of an exterior angle ($\angle SQQ'$) of an asymptotic triangle $[\triangle(PR,QS)]$ must be greater than the measure of the opposite interior angle ($\angle RPQ$). Hence $\overleftrightarrow{QP} \parallel \overleftrightarrow{OA}$ and similarly $\overleftrightarrow{PQ} \parallel \overleftrightarrow{OB}$, as asserted.

To prove that there is only one line which is parallel to both $\overleftrightarrow{OA}$ and $\overleftrightarrow{OB}$, assume the contrary, and let l_1 and l_2 be two lines each parallel to both $\overleftrightarrow{OA}$ and $\overleftrightarrow{OB}$ and hence parallel to each other. Then by Corollary 4, Theorem 2, Sec. 4.2, the bisector of $\angle AOB$ is perpendicular to both l_1 and l_2. But l_1 and l_2, being parallel, cannot have a common perpendicular. Thus we must reject the possibility that there are two lines each parallel to both $\overleftrightarrow{OA}$ and $\overleftrightarrow{OB}$. Hence the common parallel to $\overleftrightarrow{OA}$ and $\overleftrightarrow{OB}$ is unique, as asserted.

> **Corollary 1.** There is a unique line parallel to any two noncollinear rays.

Using the last theorem, we can now prove the following important result:

> **Theorem 2.** Given any acute angle, there is a unique line which is perpendicular to a given one of its sides and parallel to the other.

Proof. Let $\angle AOB$ be an arbitrary angle, and let $\overrightarrow{OA}$ and $\overrightarrow{OB}$ be the sides to which the required line is to be respectively perpendicular and parallel (see Fig. 4.26). Let B' be a point on the opposite side of $\overleftrightarrow{OA}$ from B such that $\angle B'OA \cong \angle BOA$. By the angle-construction theorem, the ray $\overrightarrow{OB'}$ is, of course, unique. Now by the last theorem, there is a unique line which is parallel to both $\overleftrightarrow{OB}$ and $\overleftrightarrow{OB'}$, and by Corollary 4, Theorem 2, Sec. 4.2, this line is also perpendicular to $\overleftrightarrow{OA}$. That there can be only one such line follows immediately from the

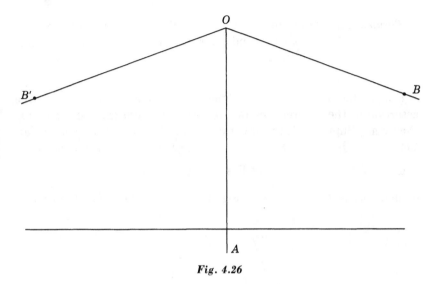

Fig. 4.26

observation that if there were two, they would both be parallel to $\overleftrightarrow{OB}$ and hence parallel to each other, yet they would also have a common perpendicular, which is impossible.

From Corollary 5, Theorem 2, Sec. 4.2, we inferred that for every distance, d, there is a unique amplitude of parallelism, α, which by Theorem 3, Sec. 4.3, is a monotonically decreasing function of d. The last theorem now justifies the converse assertion, namely, that for every number between 0 and 90 there is a unique distance, d, for which that number is the amplitude of parallelism. For given any number, α, such that $0 < \alpha < 90$, there is an angle, $\angle AOB$, such that $m\angle AOB = \alpha$. Then by the last theorem there is a unique line, l, meeting $\overrightarrow{OA}$ in D, say, which is perpendicular to $\overrightarrow{OA}$ and parallel to $\overrightarrow{OB}$. Then, clearly, the distance OD is the unique distance for which α is the amplitude of parallelism. Since the amplitude of parallelism, $\alpha = \pi(d)$, is a monotonically decreasing function of d for all d and since d is a monotonically decreasing function for all α between 0 and 90, it follows from a well-known result in analysis that the amplitude of parallelism is a continuous function of d with domain $0 < d < \infty$ and range $0 < \pi(d) < 90$.

The following theorem is hardly unexpected, but until we prove it, we cannot be sure whether it is one of the obvious results which is indeed true or one which, in spite of our intuition, is actually false in hyperbolic geometry:

Theorem 3.   On either of two intersecting lines there exist points whose perpendicular distance from the other line is arbitrarily large.

Proof.   Let p and q be two lines intersecting at O.   If $p \perp q$, the assertion of the theorem follows immediately from the point-plotting theorem.   Suppose, therefore, that p and q are not perpendicular. Let P_1 and Q_1 be points on p and q, respectively, such that $\angle Q_1 O P_1$ is acute, and let l be the unique line which is parallel to $\overleftrightarrow{OP_1}$ and perpendicular, say at Q_2, to $\overleftrightarrow{OQ_1}$ (see Fig. 4.27).   Now corresponding to any

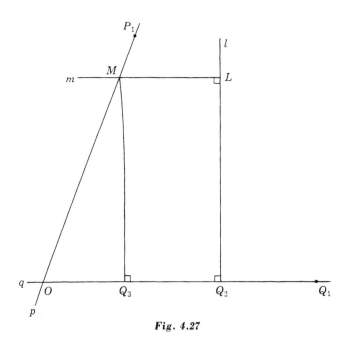

Fig. 4.27

positive number, r, however large, there is a unique point, L, on l and on the same side of q as P_1 such that

$$Q_2 L = r$$

Let m be the line which is perpendicular to l at L.   Since m and q have a common perpendicular, they cannot intersect.   Hence, by Theorem 2, Sec. 4.3, applied to the asymptotic triangle $\triangle(OP_1,Q_2L)$, m must intersect $\overrightarrow{OP_1}$, say at M.   Let Q_3 be the foot of the perpendicular from M to q.   Then MQ_3Q_2L is a Lambert quadrilateral with right angles at Q_3, Q_2, and L.   Hence $\angle Q_3ML$ must be acute, and therefore, by Theorem 3, Sec. 4.4,

$$MQ_3 > LQ_2 = r$$

as asserted.

From Corollary 1, Theorem 2, Sec. 4.2, we know that two parallel lines cannot have a common perpendicular.   Hence we cannot speak of *the* distance between two parallel lines as we do in euclidean geometry. However, the perpendicular distance from a point on one of two parallel lines to the other line is a reasonable measure of the "local" distance between the two lines, and we shall now investigate several of its properties.

Theorem 4.   If $\overleftrightarrow{AA'}$ is parallel to $\overleftrightarrow{BB'}$, then the perpendicular distance from A to $\overleftrightarrow{BB'}$ is greater than the perpendicular distance from A' to $\overleftrightarrow{BB'}$.

Proof.   Let $\overleftrightarrow{AA'} \parallel \overleftrightarrow{BB'}$, let F and F' be, respectively, the feet of the perpendiculars from A and A' to $\overleftrightarrow{BB'}$, and let A'' be a point distinct from A' on the ray opposite to $\overrightarrow{A'A}$ (see Fig. 4.28).   Then since $\angle F'A'A''$

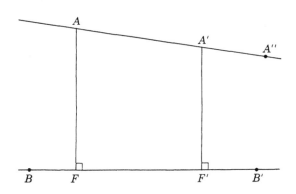

Fig. 4.28

is acute, it follows that $\angle F'A'A$ is obtuse.   Hence,

$$m\angle F'A'A > m\angle FAA'$$

and therefore, by Theorem 3, Sec. 4.4,

$$AF > A'F'$$

as asserted.

Theorem 4 is sometimes restated less precisely by saying that the distance between two parallel lines decreases monotonically in the direction of parallelism and increases monotonically in the opposite direction.   These observations are made more precise in the next two theorems.

Theorem 5.   There exist points on either of two parallel lines whose perpendicular distance from the other line is arbitrarily small.

Proof.   Let $a = \overleftrightarrow{AA'}$ and $b = \overleftrightarrow{BB'}$ be two lines such that $\overleftrightarrow{AA'} \| \overleftrightarrow{BB'}$, let F be the foot of the perpendicular from A to b, and let ε be an arbitrarily small positive number.   If $AF < \varepsilon$, there is nothing more to prove.   If $AF = \varepsilon$, then by the preceding theorem, the perpendicular distance from any point on the ray $\overrightarrow{AA'}$ to b is less than ε, and the assertion of the theorem is verified.   Therefore, let us suppose that $AF > \varepsilon$.   Then there is a unique point, P, between A and F such that $FP = \varepsilon$ (see Fig. 4.29).   At P there is a unique ray, $\overrightarrow{PP'}$,

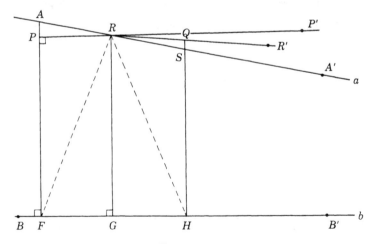

Fig. 4.29

perpendicular to $\overline{AF}$ on the same side of $\overleftrightarrow{AF}$ as A'. Moreover, from the extension of Pasch's theorem to asymptotic triangles, since $\overrightarrow{PP'}$ cannot intersect $\overrightarrow{FB'}$, it must intersect $\overrightarrow{AA'}$, say in R. Let G be the foot of the perpendicular from R to b. Then $PFGR$ is a Lambert quadrilateral with right angles at P, F, and G, and therefore $\angle GRP$ is acute. Furthermore, since $\overleftrightarrow{PP'}$ and $\overleftrightarrow{BB'}$ are nonintersecting lines, it follows that $m\angle GRP$ is greater than the amplitude of parallelism, that is, $m\angle GRA'$, for the distance GR. Hence the ray, $\overrightarrow{RR'}$, which lies on the opposite side of $\overleftrightarrow{RG}$ from P and forms with $\overrightarrow{RG}$ an angle congruent to $\angle GRP$, lies between $\overrightarrow{RP'}$ and $\overrightarrow{RA'}$. Now on b on the opposite side of G from F let H be the unique point such that $HG = FG$, and on $\overrightarrow{RR'}$ let Q be the unique point such that $QR = PR$. Now $\triangle HGR \cong \triangle FGR$ and so $HR = FR$. Hence $\triangle HRQ \cong \triangle FRP$, and therefore $HQ = FP = \varepsilon$ and $m\angle QHG = m\angle PFG = 90$. Finally, since Q and H are clearly on opposite sides of a, it follows that a must intersect $\overline{QH}$ in a point S. Thus S is a point on a such that its perpendicular distance, SH, from b is less than $QH = \varepsilon$, as required.

> **Theorem 6.** On either of two parallel lines there exist points whose perpendicular distance from the other line is arbitrarily large.

Proof. Let $a = \overleftrightarrow{AA'}$ and $b = \overleftrightarrow{BB'}$ be two lines such that $\overleftrightarrow{AA'} \| \overleftrightarrow{BB'}$, let F be the foot of the perpendicular from A to b, let R be a point on the opposite side of $\overleftrightarrow{AF}$ from A' such that $\overrightarrow{AR} \perp \overleftrightarrow{AF}$, and let r be an arbitrary positive number (see Fig. 4.30). Then, by Theorem 3, there exists a point, P, on the ray opposite to $\overrightarrow{AA'}$ such that the perpendicular distance, PR', from P to $\overleftrightarrow{AR}$ is greater than r. Let F' be the foot of the perpendicular from P to b. Since two lines cannot have more than one common perpendicular, F' cannot be collinear with P and R'. Now $\angle PR'F'$ is obtuse. Hence $m\angle PF'R' < m\angle PR'F$, and therefore, by Theorem 4, Sec. 2.11,

$$PF' > PR' = r$$

as required.

By an argument almost identical to the one we used to prove Theorem 6, the following result can be established:

> **Theorem 7.**   On either of two nonintersecting lines there exist points whose perpendicular distance from the other line is arbitrarily large.

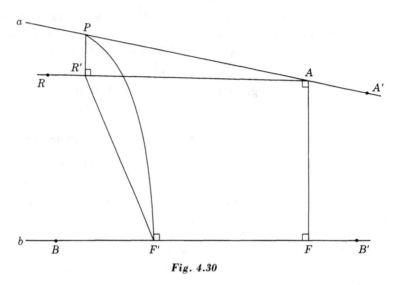

Fig. 4.30

EXERCISES

1.   In the proof of Theorem 1, show that the rays opposite to $\overrightarrow{AU}$ and $\overrightarrow{BV}$ cannot intersect.

2.   In the proof of Theorem 1, supply the details necessary to show that $\overrightarrow{UA}$ and $\overrightarrow{VB}$ cannot be parallel.

3.   In the proof of Theorem 1 show that the common perpendicular to $\overleftrightarrow{AU}$ and $\overleftrightarrow{BV}$ must lie on the opposite side of $\overleftrightarrow{AB}$ from O, as suggested by Fig. 4.25.

4.   Is the proof of Theorem 1 defective because Fig. 4.25 shows the assumed parallels $\overrightarrow{PR}$ and $\overrightarrow{QS}$ on the same side of $\overleftrightarrow{P'Q'}$ as O?

5.   Prove Corollary 1, Theorem 1.

6.   In the proof of Theorem 6, prove $\overrightarrow{PF'}$ always intersects $\overleftrightarrow{AR'}$ in a point between A and R', as shown in Fig. 4.30.

7.   Prove Theorem 7.

8.   Prove that on either of two intersecting lines there are points distinct from the intersection of the two lines whose perpendicular distance from the other line is arbitrarily small.

9.   Prove that the perpendicular distance from a point on one of two intersecting lines to the other is a monotonically increasing function of the distances of the point from the intersection of the two lines.

10. If P, Q, R, and S are four points such that $\overrightarrow{PS} \parallel \overrightarrow{QR}$ and $\overrightarrow{QP} \parallel \overrightarrow{RS}$, can $PS = QR$? Can $PQ = RS$?

4.7 Corresponding Points and Related Loci.

In addition to circles, which obviously exist in hyperbolic geometry as well as in euclidean geometry, there are two closely related loci of considerable importance in hyperbolic geometry. Their properties, some of which we shall investigate in this section, are based upon the concept of **corresponding points** on two lines:

Definition 1.   If $p = \overleftrightarrow{PP_1}$ and $q = \overleftrightarrow{QQ_1}$ are two lines and if P_1 and Q_1 are on the same side of $\overleftrightarrow{PQ}$, then P and Q are said to be corresponding points on p and q if and only if $\angle P_1PQ \cong \angle Q_1QP$.

From this definition it is clear that two points are not in and of themselves either corresponding or noncorresponding. It is only in relation to two lines, one through each of the points, that the property of being corresponding points is meaningful. Thus, in Fig. 4.31a, P and Q are

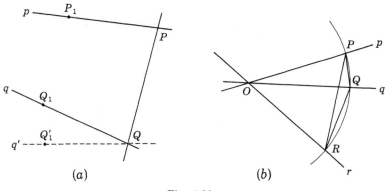

(a)   (b)

Fig. 4.31

corresponding points on the lines p and q because $\angle P_1PQ \cong \angle Q_1QP$. However, the same points, P and Q, are not corresponding points on the lines p and q', since $\angle P_1PQ$ is not congruent to $\angle Q_1'QP$.

If p and q intersect, say at O, then from the properties of isosceles triangles it follows that P and Q are corresponding points if and only if $OP = OQ$. Hence it is clear that (Fig. 4.31b)

1. On any line, q, through O there is a unique point, Q, which corresponds to a given point, P, on any other line, p, through O.
2. If p, q, and r are three lines through O and if P, Q, and R are points on p, q, and r such that P and Q are corresponding points on p and q, and Q and R are corresponding points on q and r, then P and R are corresponding points on p and r.
3. The locus of points which correspond to a given point, P, on a given line of the family of lines on O is a circle with center O and radius OP.

As the following theorems show, properties 1 and 2, with obvious modifications, also hold for corresponding points on the lines of a parallel family and for corresponding points on the lines of a family with a common perpendicular.

> ***Theorem 1.*** If two lines are nonintersecting, then to a given point on one of the lines there corresponds a unique point on the other.

Proof. Let P be an arbitrary point on one of two nonintersecting lines, p and q (see Fig. 4.32). By Theorem 3, Sec. 4.5, the lines p and q

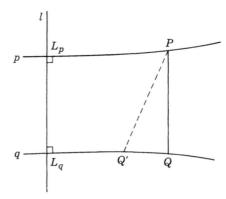

Fig. 4.32

must have a common perpendicular, l, meeting p and q at, say, L_p and L_q. If $P = L_p$, the required corresponding point is obviously and uniquely L_q. Let us assume therefore that P is a point on p distinct from L_p. Let Q be the unique point on q on the same side of $\overleftrightarrow{L_pL_q}$ as P such that $L_pP = L_qQ$. Then PL_pL_qQ is a Saccheri quadrilateral and therefore

$$\angle L_pPQ \cong \angle L_qQP$$

This proves that P and Q are corresponding points. To prove that Q is unique, assume the existence of a second point, Q', which also corresponds to P. Then, assuming for definiteness that $\overrightarrow{PQ'}$ is between $\overrightarrow{PL_p}$ and $\overrightarrow{PQ}$, we have

$$m\angle L_pPQ' < m\angle L_pPQ$$

and also, from the properties of corresponding points,

$$m\angle L_pPQ' = m\angle PQ'L_q$$
$$m\angle L_pPQ = m\angle PQL_q$$

Hence
$$m\angle PQ'L_q < m\angle PQL_q$$

which contradicts the exterior-angle theorem. Hence on q there can be but one point, Q, corresponding to P, as asserted.

Theorem 2. If p, q, and r are three lines with a common perpendicular, if a point, P, on p corresponds to a point, Q, on q, and if Q on q corresponds to a point, R, on r, then P on p corresponds to R on r.

Proof. Let p, q, and r be three lines with a common perpendicular, l, meeting p, q, and r in L_p, L_q, and L_r, say, and let P, Q, and R be three points such that P on p corresponds to Q on q and Q on q corresponds to R on r (see Fig. 4.33). If $P = L_p$, then $Q = L_q$, $R = L_r$, and the assertion of the theorem is obviously true. Let us assume, therefore, that $P \neq L_p$, and consequently $Q \neq L_q$ and $R \neq L_r$. Now since P

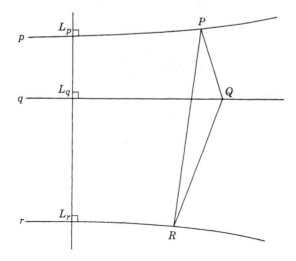

Fig. 4.33

and Q are corresponding points, it follows that

$$m\angle L_pPQ = m\angle L_qQP$$

Hence, by Corollary 1, Theorem 2, Sec. 4.4,

$$L_pP = L_qQ$$

Also, since Q and R are corresponding points,

$$m\angle L_qQR = m\angle L_rRQ$$

and consequently $\qquad L_qQ = L_rR$

Thus $\qquad\qquad\qquad L_pP = L_rR$

and therefore L_pPRL_r is a Saccheri quadrilateral.   Hence

$$m\angle L_pPR = m\angle L_rRP$$

which proves that P and R are corresponding points on p and r, as asserted.

Corollary 1.   If p, q, and r are three lines with a common perpendicular and if P on p corresponds to Q on q and Q on q corresponds to R on r, then P, Q, and R are noncollinear except when they are, respectively, the intersections of p, q, and r with their common perpendicular.

Theorem 3.   If $p = \overleftrightarrow{PP'}$ and $q = \overleftrightarrow{QQ'}$ are two lines such that $\overleftrightarrow{PP'} \| \overleftrightarrow{QQ'}$, there is one and only one point on q which corresponds to a given point on p.

Proof.   Let $\overleftrightarrow{PP'} \| \overleftrightarrow{QQ'}$, let $\overrightarrow{PS}$ and $\overrightarrow{QT}$ be the rays which bisect $\angle QPP'$ and $\angle PQQ'$, respectively, and let A be the point of intersection of $\overrightarrow{PS}$ and $\overrightarrow{QT}$ (see Fig. 4.34).   Let a be the line which passes through A and is parallel to $\overrightarrow{PP'}$ and $\overrightarrow{QQ'}$, let L be the foot of the perpendicular from P to a, and let R be the intersection of $\overrightarrow{PL}$ and $\overrightarrow{QQ'}$.   Then R is the point on q which corresponds to P on p.

To prove this, let B and C be the feet of the perpendiculars from L to p and q, and let D and E be the feet of the perpendiculars from A to p and q, respectively.   Clearly, because of the exterior-angle theorem, B must lie on $\overline{PD}$ and C must lie on $\overline{RE}$, as indicated in Fig. 4.34.   Now, by Exercise 10, Sec. 4.2, we have $AD = AE$.   Hence, by Theorem 7, Sec. 4.2,

$$LB = LC$$

Therefore $\qquad\qquad m\angle BLA = m\angle CLA$

since each is the amplitude of parallelism for the same distance.   Thus

$$\angle PLB \cong \angle RLC$$

since they are complements of congruent angles.   Hence

$$\triangle LPB \cong \triangle LRC$$

Finally, from this congruence, we conclude that

$$m\angle LPB = m\angle LRC$$

which proves that P and R are corresponding points on p and q, as asserted.

To prove that R is unique, suppose that there is a second point,

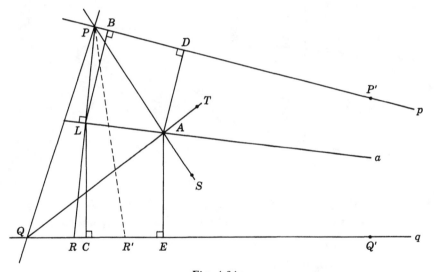

Fig. 4.34

R', on q which also corresponds to P, and let the notation be such that $\overrightarrow{PR'}$ is between $\overrightarrow{PP'}$ and $\overrightarrow{PR}$.   Then

$$m\angle RPP' > m\angle R'PP'$$

But, by hypothesis,

$$m\angle RPP' = m\angle PRQ'$$

and $\qquad\qquad m\angle R'PP' = m\angle PR'Q'$

Hence $\qquad\qquad m\angle PRQ' > m\angle PR'Q'$

which contradicts the exterior-angle theorem.   Thus there cannot be a second point, R', on q corresponding to P on p.

Theorem 4.   If p, q, and r are three lines which are parallel in the same sense, and if P on p corresponds to Q on q, and Q on q corresponds to R on r, then P, Q, and R cannot be collinear.

Proof.   Let $p = \overleftrightarrow{PP'}$, $q = \overleftrightarrow{QQ'}$, and $r = \overleftrightarrow{RR'}$ be three lines such that $\overleftrightarrow{PP'} \| \overleftrightarrow{QQ'} \| \overleftrightarrow{RR'}$, and let P, Q, and R be three points such that P on p corresponds to Q on q and Q on q corresponds to R on r (see Fig. 4.35).

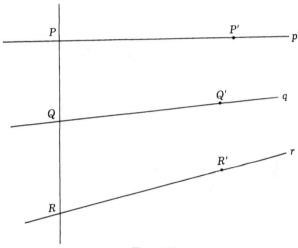

Fig. 4.35

To prove the assertion of the theorem, let us assume the contrary, and suppose that P, Q, and R are collinear.   Then, by hypothesis,

$$m\angle P'PQ = m\angle PQQ' \qquad \text{and} \qquad m\angle Q'QR = m\angle QRR'$$

Hence, since $\angle PQQ'$ and $\angle Q'QR$ form a linear pair, it follows that

$$m\angle P'PQ + m\angle QRR' = 180$$

But, since $\overleftrightarrow{PP'} \| \overleftrightarrow{RR'}$, this contradicts Corollary 2, Theorem 6, Sec. 4.2. Hence P, Q, and R cannot be collinear and the theorem is established.

Theorem 5.   If p, q, and r are three lines which are parallel in the same sense, and if P on p corresponds to Q on q and if Q on q corresponds to R on r, then P on p corresponds to R on r.

Proof. Let $p = \overleftrightarrow{PP'}$, $q = \overleftrightarrow{QQ'}$, and $r = \overleftrightarrow{RR'}$ be three lines such that $\overleftrightarrow{PP'} \| \overleftrightarrow{QQ'} \| \overleftrightarrow{RR'}$, and let P on p correspond to Q on q and Q on q correspond to R on r (see Fig. 4.36). By the preceding theorem, P,

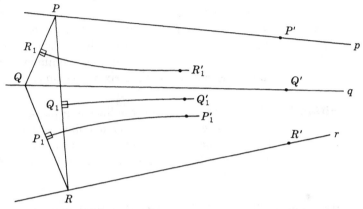

Fig. 4.36

Q, and R cannot be collinear. Let P_1, Q_1, and R_1 be the midpoints of $\overline{QR}$, $\overline{RP}$, and $\overline{PQ}$, respectively, and let $\overleftrightarrow{P_1P_1'}$, $\overleftrightarrow{Q_1Q_1'}$, and $\overleftrightarrow{R_1R_1'}$ be the parallels to $\overleftrightarrow{PP'}$, $\overleftrightarrow{QQ'}$, and $\overleftrightarrow{RR'}$ through P_1, Q_1, and R_1. Now since $R_1P = R_1Q$ and $m\angle R_1PP' = m\angle R_1QQ'$, it follows from Theorem 4, Sec. 4.3, that $m\angle PR_1R_1' = m\angle QR_1R_1'$. Hence, since these angles form a linear pair, each is a right angle, and $\overleftrightarrow{R_1R_1'}$ is in fact the perpendicular bisector of $\overline{PQ}$. Similarly, $\overleftrightarrow{P_1P_1'}$ is the perpendicular bisector of $\overline{QR}$. Now, by definition, $\overleftrightarrow{R_1R_1'} \| \overleftrightarrow{P_1P_1'}$. Moreover, by Theorem 5, Sec. 4.5, if the perpendicular bisectors of two sides of a triangle are parallel, the perpendicular bisector of the third side is parallel to these lines in the same sense. From this it follows that $\overleftrightarrow{Q_1Q_1'}$, being the unique parallel to the perpendicular bisectors $\overleftrightarrow{P_1P_1'}$ and $\overleftrightarrow{R_1R_1'}$ through the midpoint, Q_1, of $\overline{PR}$, must be the perpendicular bisector of $\overline{PR}$. But if this is the case, then

$$m\angle Q_1PP' = m\angle Q_1RR'$$

since each is the amplitude of parallelism for the distance $Q_1P = Q_1R$. But if this is the case, then P on p corresponds to R on r, as asserted.

As we pointed out at the beginning of this section, the locus of points on the lines of a concurrent family which correspond to a given point on one of the lines is a circle.   On the lines of a parallel family or the lines of a family with a common perpendicular, the locus is not a circle, but it is still of considerable interest.

Definition 2.   The locus of points on the lines of a family with a common perpendicular which correspond to a given point on one of the lines is called an equidistant curve.   The lines of the family are called axes of the equidistant curve, and their common perpendicular is called the directrix of the curve.

Definition 3.   The locus of points on the lines of a parallel family which correspond to a given point on one of the lines is called a horocycle or a limiting curve, and the lines of the parallel family are called axes of the horocycle.

From the proof of Theorem 2, it is clear that the name "equidistant curve" is well chosen, for the perpendicular distance from any point on the locus to the common perpendicular or directrix of the curve is a constant.   Also, from Corollary 1, Theorem 2, it follows that except in the trivial case when this distance is zero, an equidistant curve is not a straight line, contrary to the situation in euclidean geometry, where the locus is a line parallel to the common perpendicular.   Likewise, from Theorem 4, it follows that, without exception, a horocycle is never a straight line, contrary again to the situation in euclidean geometry, where the locus is a line perpendicular to each line of the parallel family.   Moreover, since no three points of a general equidistant curve or any horocycle can be collinear, it follows that no line can meet either an equidistant curve or a horocycle in more than two points.

One of the surprising properties of an equidistant curve is that it is everywhere concave to its directrix.   More precisely, we have the following theorem:

Theorem 6.   If P, Q, and R are three points on any equidistant curve and if the perpendiculars p, q, and r from P, Q, and R to the directrix of the curve are such that q is between p and r, then Q and the directrix of the curve lie on opposite sides of $\overleftrightarrow{PR}$.

The most striking property of horocycles is probably the fact that any two such curves are congruent:

Definition 4.   Two horocycles, one on the family of parallels $\{\overleftrightarrow{PP'}\}$, the other on the family $\{\overleftrightarrow{QQ'}\}$, are said to be congruent if for every set of points $\{P_i\}$ on one curve, there exists a corresponding set $\{Q_i\}$ on the other such that

$$\overline{P_iP_{i+1}} \cong \overline{Q_iQ_{i+1}}$$
$$\angle P_i'P_iP_{i+1} \cong \angle Q_i'Q_iQ_{i+1}$$
$$\angle P_{i+1}'P_{i+1}P_i \cong \angle Q_{i+1}'Q_{i+1}Q_i$$

Theorem 7.   Any two horocycles are congruent.

Proof.   Let P_1, P_2, P_3, . . . be an arbitrary set of points on the given horocycle on the parallel family $\{\overleftrightarrow{PP'}\}$, and let Q_1 be an arbitrary point on the horocycle on the parallel family $\{\overleftrightarrow{QQ'}\}$ (see Fig. 4.37).

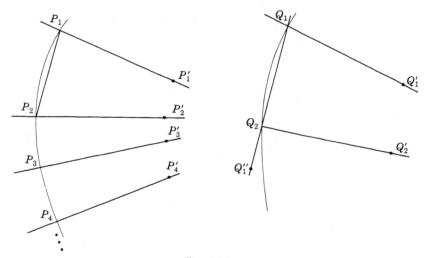

Fig. 4.37

Let $\overrightarrow{Q_1Q_1''}$ be either of the rays with the property that

$$\angle Q_1'Q_1Q_1'' \cong \angle P_1'P_1P_2$$

Let Q_2 be the unique point on $\overrightarrow{Q_1Q_1''}$ such that

$$Q_1Q_2 = P_1P_2$$

and let $\overrightarrow{Q_2Q_2'}$ be the unique parallel to $\overrightarrow{Q_1Q_1'}$ through Q_2.   Then by

Theorem 4, Sec. 4.3,
$$\angle Q_1 Q_2 Q_2' \cong \angle P_1 P_2 P_2'$$
But
$$\angle P_1 P_2 P_2' \cong \angle P_1' P_1 P_2$$

since P_1 and P_2 are corresponding points on the first horocycle. Moreover

$$\angle P_1' P_1 P_2 \cong \angle Q_1' Q_1 Q_2$$

from the definition of $\overrightarrow{Q_1 Q_1''}$. Hence, combining these relations, we have

$$\angle Q_1 Q_2 Q_2' \cong \angle Q_1' Q_1 Q_2$$

which proves that Q_1 and Q_2 are corresponding points on $\overleftrightarrow{Q_1 Q_1'}$ and $\overleftrightarrow{Q_2 Q_2'}$ and hence that Q_2 is on the second horocycle. Thus Q_2 is the point on the second horocycle which corresponds to P_2, as required. The continuation is now obvious, and the remaining points, Q_3, Q_4, . . . , on the second horocycle can be found in exactly the same fashion. We should note, however, that whereas in the first step of the proof $\overrightarrow{Q_1 Q_1''}$ could be chosen on either side of $\overleftrightarrow{Q_1 Q_1'}$, in subsequent steps the rays $\overrightarrow{Q_2 Q_2''}$, $\overrightarrow{Q_3 Q_3''}$, $\overrightarrow{Q_4 Q_4''}$, . . . are uniquely determined by the necessity of preserving among the Q's the same order which exists among the P's.

In Theorem 1, Sec. 2.16, we determined conditions under which a line and a circle would have two, one, or no points in common. However, the proof we gave there made use of the Pythagorean theorem, which does not hold in hyperbolic geometry. Hence, before we can use these important results here, we must deduce them from principles which are valid in hyperbolic geometry.

To do this, we shall use a fundamental property of the real numbers, known as the **Dedekind* Cut Property,** which is available to us because of the $1:1$ correspondence set up between the points of any line and the real numbers by the ruler postulate. For our purposes, the principle in question can be stated in the following form:

Dedekind's theorem.   If the nonnegative numbers are separated into two nonempty sets, S_1 and S_2, such that

(1) Every nonnegative number is either a member of S_1 or a member of S_2

(2) If x_1 is any member of S_1 and x_2 is any member of S_2, then $x_1 < x_2$

* Named for the great German mathematician J. W. R. Dedekind (1831–1916).

then there exists a number, d (which may be in either S_1 or S_2, depending upon how these sets are defined), such that every number less than d is in S_1 and every number greater than d is in S_2.

With this theorem available, we can now establish the following theorem without appeal to the Pythagorean theorem:[*]

Theorem 8. Let C be a circle with center P and radius r, let l be an arbitrary line in the plane of C, let O be the foot of the perpendicular from P to l, and let the perpendicular distance, PO, from P to l be p. Then

  (1) If $p > r$, every point of l lies in the exterior of C, and l and C have no points in common.
  (2) If $p = r$, the line l has only O in common with C, and l and C are tangent at O.
  (3) If $p < r$, the line l intersects the circle C in two points which are on opposite sides of and equidistant from O.

Proof. The first two assertions of the theorem follow immediately from Corollary 1, Theorem 4, Sec. 2.11. To prove the third, let O be the foot of the perpendicular to the given line, l, from the center, P, of the given circle; let $PO < r$; and let a coordinate system having O as origin be established on l. By a simple application of the ruler postulate and the triangle inequality, it is clear that on the ray $\overrightarrow{OA}$ consisting of the points of l with nonnegative coordinates, there exist points whose distance from P is less than r and also points whose distance from P is greater than r. Moreover, if L and L' are two points of $\overrightarrow{OA}$ such that $OL' > OL$, then $PL' > PL$, and conversely (see Fig. 4.38). In other words, for points on $\overrightarrow{OA}$, PL is a monotonically increasing function of the coordinate of L. Let us now divide the nonnegative numbers into two sets in the following way:

S_1: Those numbers which are the coordinates of points, L, such that $PL < r$

S_2: Those numbers which are the coordinates of points, L, such that $PL \geqq r$

---

[*] Since the ruler postulate, and hence all the properties of the real numbers, is available in both euclidean and hyperbolic geometry, the proof we are about to present is equally valid in both geometries and could have been given in Sec. 2.16.

Clearly, neither S_1 nor S_2 is empty. Moreover, every nonnegative number is either in S_1 or in S_2. Finally, from our earlier observation about the monotonic relation between PL and OL, it is evident that every number in S_1 is less than any number in S_2. Thus the hypotheses of Dedekind's theorem are fulfilled, and we may conclude that the

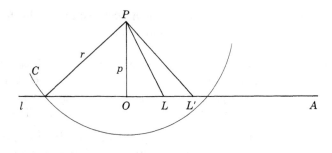

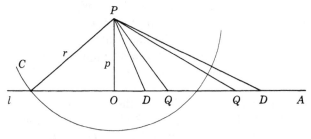

Fig. 4.38

"cut" we have established among the nonnegative numbers defines a number, d, such that every nonnegative number less than d is in S_1 and every nonnegative number greater than d is in S_2. We now assert that d is the coordinate of a point which is simultaneously on l and also on the circle, C.

In the first place, we observe that d cannot be in S_1. In fact, if it is, then from the definition of S_1

$$PD < r$$

Now consider the point Q, on l, whose coordinate is

(1) $$q = OQ = OD + (r - PD)$$

From the triangle inequality applied to $\triangle PDQ$ we have

$$PQ < PD + DQ = PD + (r - PD) = r$$

Thus q is a number which belongs to S_1. However from (1) it is clear that

$$q > d$$

Hence we have a contradiction since, by Dedekind's theorem, every number greater than the "cut" number, d, must belong to S_2. Thus d cannot be in S_1.

Similarly, if d is in S_2 it follows from the definition of S_2 that

$$PD \geqq r$$

If the equality sign holds, then obviously D is a point which is simultaneously on l and also on the circle, and our proof is essentially complete. Suppose therefore that D is a point such that

$$PD > r$$

In this case, consider the point Q, on l, whose coordinate is

(2) $$q = OQ = OD - (PD - r)$$

From the triangle inequality applied to $\triangle PQD$ we have

$$PQ + QD > PD \qquad \text{or} \qquad PQ + (PD - r) > PD$$

or, finally, $$PQ > r$$

Thus Q is a point whose coordinate, q, belongs to S_2. Moreover, since $PD > r$, it follows from (2) that $OQ < OD$, that is, that

$$q < d$$

But this is impossible because, according to Dedekind's theorem, every number less than d is in S_1, whereas we have just seen that q is in S_2. Thus the possibility that $PD > r$ must be rejected. Hence $PD = r$ and D is necessarily a point on the circle, as asserted. An identical argument, of course, establishes the existence of a second point of intersection, D', on the ray opposite to $\overrightarrow{OA}$.

Results comparable to those in Theorem 8 for the intersections of lines and equidistant curves and lines and horocycles will be found among the exercises.

EXERCISES

1. Investigate what changes, if any, are required in the proof of Theorem 1 if Q' is such that $\overrightarrow{PQ}$ is between $\overrightarrow{PL_p}$ and $\overrightarrow{PQ'}$ (Fig. 4.32).
2. Prove Corollary 1, Theorem 2.
3. Prove Theorem 6.
4. Is there a relation between the concavity of a horocycle and the direction of parallelism of the lines of the family on which the horocycle is defined? Prove your conjecture.

5. How might congruent equidistant curves be defined? Are every two equidistant curves congruent?

6. If P and Q are two points on a horocycle, prove that the perpendicular bisector of $\overline{PQ}$ is an axis of the curve.

7. Define a tangent to a horocycle. Prove that any line which contains a point of a horocycle and is perpendicular to the axis of the horocycle which passes through that point meets the horocycle in no other point.

8. Prove that if a line has one point in common with a horocycle and is neither an axis nor a tangent to the horocycle, it has a second point in common with the horocycle.

9. Define the interior and exterior of a horocycle. Prove that a line which is perpendicular to an axis of a horocycle at a point in the exterior of the horocycle has no point in common with the horocycle.

10. If a quadrilateral is inscribed in a horocycle, prove that the sum of the measures of one pair of opposite angles is equal to the sum of the measures of the other pair.

11. In $\triangle ABC$, if $\pi(b/2) + \pi(c/2) = m\angle A$, prove that A, B, and C lie on a horocycle. What can we conclude if $\pi(b/2) + \pi(c/2) > m\angle A$? if $\pi(b/2) + \pi(c/2) < m\angle A$?

12. If H and H' are two horocycles on the same parallel family and if (P,P'), (Q,Q'), (R,R'), . . . are the points in which the lines p, q, r, . . . of the parallel family intersect H and H', respectively, prove that

$$PP' = QQ' = RR' = \cdots$$

13. Prove that a line which is perpendicular to an axis of an equidistant curve at a point of the curve can meet the curve in no other point.

14. Prove that a line which is perpendicular to an axis of an equidistant curve at a point on the opposite side of the curve from its directrix cannot intersect the curve.

15. If P is a point on an equidistant curve, if F is the foot of the perpendicular from P to the directrix of the curve, and if Q is a point between P and F, prove that the line which is perpendicular to $\overline{PF}$ at Q intersects the curve in two points.

16. If C is an equidistant curve whose points are all at the distance k from the directrix, d, if P and Q are two points on the same side of d as C, if the distance from P to d is less than k, and if the distance from Q to d is greater than k, prove that $\overline{PQ}$ intersects C in one and only one point.

4.8 Defect and Area.

In Sec. 4.5 we established the remarkable result that the sum of the measures of the angles of a triangle is always less than 180. At that time, we did not determine a figure for the sum, and in fact, we could not have done so, for in hyperbolic geometry the sum of the measures of the angles of a triangle varies from triangle to triangle and is not a constant as it is in euclidean geometry. This is a fact of great importance in hyperbolic geometry, because the amount by which the sum of the measures of the angles of a triangle falls short of 180 has properties which make it a suitable measure of the area of triangular regions. In this section we shall undertake a brief investigation of this matter, beginning with the concept of **equivalent regions**.

Definition 1. If two polygonal regions are each the union of a finite number of triangular regions having no interior points in common and if there exists a correspondence between the triangular regions comprising one polygonal region and the triangular regions comprising the other polygonal region such that corresponding triangles are congruent, the two polygonal regions are said to be equivalent.

Lemma 1. The regions determined by two congruent polygons are equivalent.

Theorem 1. Two polygonal regions each equivalent to a third polygonal region are equivalent to each other.

Proof. Let P_1 and P_2 be two polygonal regions, each equivalent to a third polygonal region, R. This means that we have two partitions of the region R into triangular regions. Clearly, the segments which effect one of the partitions, say the second, divide the triangular regions of the first partition into subregions, some still triangular and others polygonal. For instance, if we consider the polygonal regions P_1, P_2, and R shown in Fig. 4.39, the segments defining the second partition of R divide the triangular region, 2, into two triangular regions, 2_1 and 2_3, and one quadrilateral region, 2_2. Now each of the polygonal regions formed when the two partitions of R are superposed can be subdivided into triangular regions by the introduction of additional segments. In our illustrative figure, this requires only one new segment, namely, a segment to decompose the quadrilateral region 2_2 into two triangular regions, $2_2'$ and $2_2''$. In this way, a new partition of R consisting exclusively of triangular regions is obtained. Moreover, since $\triangle_1$ in the first partition of R is, by hypothesis, congruent to $\triangle_1$ in P_1, it follows by Lemma 1 that $\triangle_1$ in P_1 can be partitioned into four triangular regions, respectively congruent to regions $1_1, 1_2, 1_3, 1_4$, as indicated. Similarly, $\triangle_2$ in P_1 can be partitioned into regions respectively congruent to regions $2_1, 2_2', 2_2'', 2_3$; and $\triangle_3$, in P_1, can be partitioned into regions respectively congruent to regions 3_1 and 3_2.

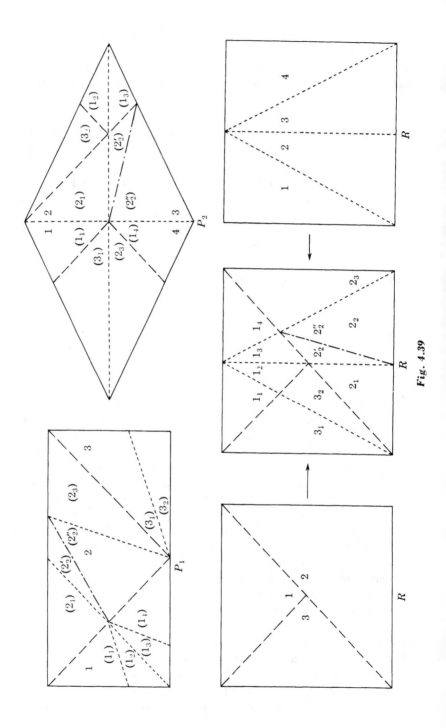

Fig. 4.39

In the same manner, the triangular regions in P_2 can be partitioned into subregions congruent to the appropriate regions in the refined subdivision of R. For instance, $\triangle_1$ in P_2 can be partitioned into regions congruent to regions 1_1 and 3_1; $\triangle_2$ in P_2 can be partitioned into regions congruent to 1_2, 2_1, and 3_2; $\triangle_3$ in P_2 can be partitioned into regions congruent to regions 1_3, $2'_2$, and $2''_2$; and $\triangle_4$ in P_2 can be partitioned into triangular regions congruent to regions 1_4 and 2_3. Thus, in general, the two polygonal regions can be partitioned into the same number of triangular regions each congruent to the same triangular region in the refined, or superposed, partition of R. Therefore P_1 and P_2 are themselves partitioned into triangular regions which are congruent in pairs and hence are equivalent, as asserted.

Definition 2. The amount by which the sum of the measures of the angles of a triangle falls short of 180 is called the defect of the triangle.

The importance of the concept of the defect of a triangle lies in the fact that triangles with the same defect are equivalent. To prove this, we find it convenient to consider first the special case of two triangles with a side of one congruent to a side of the other:

Theorem 2. Two triangles with the same defect and a side of one congruent to a side of the other are equivalent.

Proof. Let $\triangle ABC$ and $\triangle A'B'C'$ be two triangles having the same defect, and let $BC = B'C'$. Let us first consider $\triangle ABC$. Let D and E be the midpoints of $\overline{AB}$ and $\overline{AC}$, and let F, G, and H be, respectively, the feet of the perpendiculars from A, B, and C to $\overleftrightarrow{DE}$ (see Fig. 4.40).

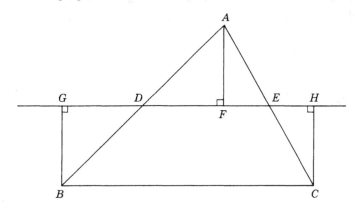

Fig. 4.40

Assuming that F lies between D and E, it is clear that

$$\triangle AFE \cong \triangle CHE$$
Hence $\qquad\qquad AF = CH$
Likewise $\qquad\quad \triangle AFD \cong \triangle BGD$
and so $\qquad\qquad AF = BG$
Therefore $\qquad\quad CH = BG$

and thus $CHGB$ is a Saccheri quadrilateral. Hence $\angle GBC$ and $\angle HCB$ are congruent acute angles. Moreover, because of the congruence of the pairs of triangles we have just noted, it is evident that $\triangle ABC$ is equivalent to the quadrilateral $CHGB$. Furthermore,

$$\begin{aligned}
m\angle GBC + m\angle HCB &= (m\angle GBD + m\angle ABC) + (m\angle HCE + m\angle ACB)\\
&= (m\angle FAD + m\angle ABC) + (m\angle FAE + m\angle ACB)\\
&= (m\angle FAD + m\angle FAE) + m\angle ABC + m\angle ACB\\
&= m\angle BAC + m\angle ABC + m\angle ACB\\
&= \text{angle sum for } \triangle ABC
\end{aligned}$$

Finally, since $m\angle GBC = m\angle HCB$, it follows that

$$m\angle GBC = \tfrac{1}{2}(\text{angle sum of } \triangle ABC)$$

In essentially the same fashion, the same result is obtained if $\triangle ABC$ is such that F is not between D and E.

Now let us apply the preceding argument to $\triangle A'B'C'$. It, too, is equivalent to a Saccheri quadrilateral, $C'H'G'B'$, in which the side $\overline{B'C'}$ is congruent to the corresponding side, $\overline{BC}$, of the Saccheri quadrilateral associated with $\triangle ABC$. Moreover, the measure of the congruent acute angles, $\angle G'B'C'$ and $\angle H'C'B'$, of the quadrilateral $C'H'G'B'$ is equal to one-half the angle sum of $\triangle A'B'C'$. But since $\triangle ABC$ and $\triangle A'B'C'$ have the same defect, they obviously must have the same angle sum. Hence the acute angles of the two Saccheri quadrilaterals have the same measure. Therefore, by Exercise 7, Sec. 4.4, the two quadrilaterals are congruent and hence equivalent. From the transitive property of equivalence guaranteed by Theorem 1, it thus follows that $\triangle ABC$ is equivalent to $\triangle A'B'C'$, as asserted.

> **Corollary 1.** The locus of vertices of triangles having the same base and the same defect is an equidistant curve whose directrix is the line determined by the midpoints of the two sides other than the common base.

Proof. That every point of the equidistant curve is the vertex of a triangle having the given defect or angle sum follows immediately because in each case the triangle has an angle sum equal to twice the measure of one of the acute angles of a fixed Saccheri quadrilateral (the quadrilateral $CHGB$ in the proof of the last theorem). To prove that every point which is the vertex of a triangle with the given base and the given defect is a point of the locus, let us assume the contrary and suppose that $\triangle ABC$ has the given base, $\overline{BC}$, and the given defect but that its third vertex, A, does not lie on the equidistant curve. Let F be the foot of the perpendicular from A to the directrix of the equidistant curve and G be the intersection of $\overleftrightarrow{AF}$ and the equidistant curve (see Fig. 4.41). Then by the first part of our proof, $\triangle GBC$ has

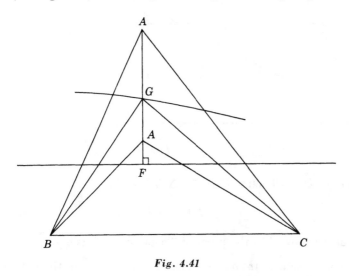

Fig. 4.41

the same angle sum as $\triangle ABC$. But by an easy application of the exterior-angle theorem (see Exercise 10, Sec. 4.5), it follows that the angle sum of $\triangle ABC$ is less than the angle sum of $\triangle GBC$ if G is between A and F and is greater than the angle sum of $\triangle GBC$ if G is a point distinct from A which is not between A and F. This contradiction overthrows the possibility that A is not a point of the equidistant curve, and the corollary is established.

Theorem 3. Two triangles with the same defect and a side of one greater than a side of the other are equivalent.

Proof. Let $\triangle ABC$ and $\triangle A'B'C'$ be two triangles with the same defect, and let $AC < A'C'$. Let D and E be the midpoints of $\overline{AB}$ and $\overline{AC}$, respectively, and let F be the foot of the perpendicular from C to $\overleftrightarrow{DE}$ (see Fig. 4.42). Now $CF \leqq CE = \tfrac{1}{2}AC < \tfrac{1}{2}A'C'$. Hence, by Theorem 8, Sec. 4.7, there is a point, G, on the ray opposite to $\overrightarrow{FE}$

such that $CG = \frac{1}{2}A'C'$.   Also, there is a point, H, on the ray opposite
to $\overrightarrow{GC}$ such that $GH = GC$.   Let K be the foot of the perpendicular

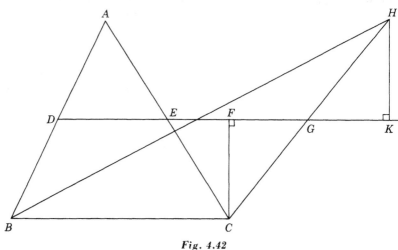

Fig. 4.42

from H to $\overleftrightarrow{DE}$.   Then $\triangle HKG \cong \triangle CFG$ and so

$$HK = CF$$

But, as we showed in the proof of Theorem 2, CF is equal to the perpendicular distance from A to $\overleftrightarrow{DE}$.   Hence H lies on the equidistant
curve which passes through A and has $\overleftrightarrow{DE}$ as directrix.   Therefore,
by the corollary to the last theorem, $\triangle HBC$ has the same defect as
$\triangle ABC$ and, by Theorem 2, is equivalent to it.   Now, by hypothesis,
$\triangle A'B'C'$ has the same defect as $\triangle ABC$ and hence as $\triangle HBC$.   Moreover, $\triangle A'B'C'$ and $\triangle HBC$ are such that $A'C' = HC$.   Therefore, by
Theorem 2, $\triangle A'B'C'$ and $\triangle HBC$ are equivalent.   Finally, from the
transitive property of equivalence, it follows that $\triangle A'B'C'$ and $\triangle ABC$
are equivalent, since each is equivalent to $\triangle HBC$.   Thus the theorem
is established.

Since any two triangles with the same defect are covered by either
Theorem 2 or Theorem 3 (or possibly by both), it is clear that they
can be combined into the following single assertion:

Theorem 4.   Two triangles with the same defect are equivalent.

The following theorems can now be proved without great difficulty:

Theorem 5.   If the region R bounded by a triangle, $\triangle$, is the union of a finite number of triangular regions, $R_1, R_2, \ldots, R_k$, with boundaries $\triangle_1, \triangle_2, \ldots, \triangle_k$, no two of the regions having any interior points in common, then the defect of $\triangle$ is equal to the sum of the defects of $\triangle_1, \triangle_2, \ldots, \triangle_k$.

Theorem 6.   If the region R bounded by a triangle, $\triangle$, is equivalent to the union of a finite number of triangular regions, $R_1, R_2, \ldots, R_k$, with boundaries $\triangle_1, \triangle_2, \ldots, \triangle_k$, no two of the regions having an interior point in common, then the defect of $\triangle$ is equal to the sum of the defects of $\triangle_1, \triangle_2, \ldots, \triangle_k$.

Theorem 7.   Two equivalent triangles have the same defect.

Theorem 8.   Two triangles are equivalent if and only if they have the same defect.

In view of the preceding theorems, it is natural and convenient to define the area of a triangle to be proportional to the defect of the triangle. From this point of view, the choice of a particular proportionality constant, k^2, amounts essentially to the choice of a particular unit of area. Using this definition for the area of a triangle, Theorem 8 can be restated in the following form:

Theorem 9.   Two triangles have the same area if and only if they are equivalent.

The idea of defect can obviously be extended from triangles to polygons in general and used to define the area of general polygonal regions. However, we shall leave the details of this extension as exercises.

EXERCISES

1.   If D and E are two points in the interior of $\triangle ABC$, prove that the sum of the measures of the angles of $\triangle ADE$ is greater than the sum of the measures of the angles of $\triangle ABC$.
2.   If A', B', and C' are, respectively, interior points of the sides $\overline{BC}$, $\overline{CA}$, and $\overline{AB}$ of $\triangle ABC$, prove that the angle sum of $\triangle ABC$ is less than the angle sum of $\triangle A'B'C'$.
3.   If D, E, and F are interior points of $\triangle ABC$, prove that the angle sum of $\triangle ABC$ is less than the angle sum of $\triangle DEF$.

4. Supply the details required to carry through the proof of Theorem 2 if F is not between D and E (Fig. 4.40).
5. If $\triangle ABC$ is an arbitrary triangle, prove Theorem 5 in the special case in which one vertex of each component triangle is A and its other vertices are points of $\overline{BC}$.
6. If $\triangle ABC$ is an arbitrary triangle, prove Theorem 5 in the special case in which the vertices of the component triangles all lie on $\overline{AB}$ and $\overline{AC}$.
7. Using the results of Exercises 5 and 6, prove Theorem 5 in the general case.
8. Prove Theorem 6.
9. Prove Theorem 7.
10. Prove Theorem 8.

4.9 Numerical Relations in Right Triangles. Many of the more advanced results in hyperbolic geometry depend upon certain numerical relations between the measures of the parts of right triangles. In this section we shall investigate some of the simpler results of this sort.

Let $\triangle ABC$ be an arbitrary right triangle, with right angle at C, and let a, b, and c be, respectively, the lengths of the sides opposite the vertices A, B, and C. The measures of the angles at A and B we shall denote, respectively, by α and β, the distances for which these are the amplitudes of parallelism we shall denote by a' and b', and the amplitude of parallelism for the distances a and b we shall denote by α' and β'. Thus

$$m\angle A = \alpha = \pi(a') \quad \text{or} \quad a' = \pi^{-1}(\alpha)$$
$$a = \pi^{-1}(\alpha') \quad \text{or} \quad \pi(a) = \alpha'$$

$$m\angle B = \beta = \pi(b') \quad \text{or} \quad b' = \pi^{-1}(\beta)$$
$$b = \pi^{-1}(\beta') \quad \text{or} \quad \pi(b) = \beta'$$

Now let B' be the point on the ray opposite to $\overrightarrow{AB}$ such that $AB' = a'$, let C' be a point on the ray opposite to $\overrightarrow{AC}$, and let D and E be points on the opposite side of $\overleftrightarrow{AB}$ from C such that $\overleftrightarrow{B'D} \perp \overleftrightarrow{BA}$ and $\overleftrightarrow{BE} \| \overleftrightarrow{B'D}$ (see Fig. 4.43). Then

$$m\angle B'BE = \pi(a' + c)$$

Moreover, since $m\angle B'AC' = m\angle BAC = \alpha = \pi(a')$

it follows that $\qquad\qquad \overleftrightarrow{AC'} \| \overleftrightarrow{B'D}$

and hence that $\qquad \overleftrightarrow{BE} \| \overleftrightarrow{AC'} \quad$ that is, $\overleftrightarrow{BE} \| \overleftrightarrow{CC'}$

Therefore $$m\angle CBE = \pi(a)$$

Finally, since $\overrightarrow{BA}$ is between $\overrightarrow{BC}$ and $\overrightarrow{BE}$, we have

$$\beta + m\angle B'BE = \pi(a)$$

or

$$(1a) \qquad \beta + \pi(c + a') = \pi(a)$$

Similarly,

$$(1b) \qquad \alpha + \pi(c + b') = \pi(b)$$

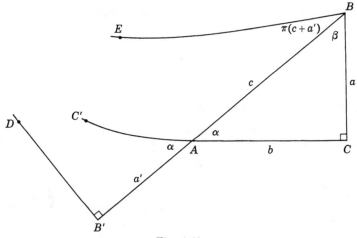

Fig. 4.43

Now suppose that $a' < c$, and let D be the point on $\overline{AB}$ such that $AD = a'$ (see Fig. 4.44).   Also, let B' and D' be points on the same side of $\overleftrightarrow{AB}$ as C such that $\overleftrightarrow{DD'} \perp \overleftrightarrow{AB}$ and $\overleftrightarrow{BB'} \| \overleftrightarrow{AC}$.   Then $m\angle CBB' = \pi(a)$. Moreover, since $\alpha = \pi(a')$, it follows that $\overleftrightarrow{AC} \| \overleftrightarrow{DD'}$ and hence that

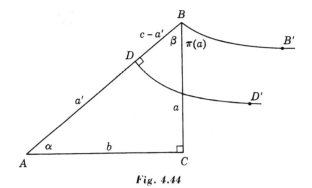

Fig. 4.44

$\overleftrightarrow{BB'} \| \overleftrightarrow{DD'}$.   Thus $m\angle DBB' = \pi(c - a')$.   Then, since $\overrightarrow{BC}$ is between $\overrightarrow{BA}$ and $\overrightarrow{BB'}$, we have

$$\beta + m\angle CBB' = m\angle DBB'$$

or

(2a) $$\beta + \pi(a) = \pi(c - a')$$

Similarly, of course, we have

(2b) $$\alpha + \pi(b) = \pi(c - b')$$

Furthermore, from (1a) and (2a), by adding and subtracting, we have

(3a) $$\pi(a) = \frac{\pi(c - a') + \pi(c + a')}{2}$$

(3b) $$\beta = \frac{\pi(c - a') - \pi(c + a')}{2}$$

and similarly, from (1b) and (2b),

(4a) $$\pi(b) = \frac{\pi(c - b') + \pi(c + b')}{2}$$

(4b) $$\alpha = \frac{\pi(c - b') - \pi(c + b')}{2}$$

If $a' > c$, let D be the point on the ray opposite to $\overrightarrow{BA}$ such that $AD = a'$, and as before, let B' and D' be points on the same side of

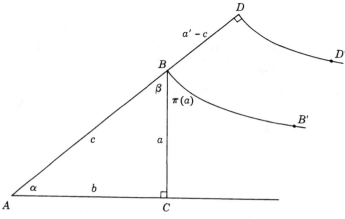

Fig. 4.45

$\overleftrightarrow{AB}$ as C such that $\overleftrightarrow{DD'} \perp \overleftrightarrow{AB}$ and $\overleftrightarrow{BB'} \| \overleftrightarrow{AC}$ (see Fig. 4.45).   Then

$$m\angle CBB' = \pi(a)$$

Moreover, since $\alpha = \pi(a')$, it follows that

$$\overleftrightarrow{AC}\,\|\,\overleftrightarrow{DD'} \qquad \text{and hence that} \qquad \overleftrightarrow{BB'}\,\|\,\overleftrightarrow{DD'}$$

Therefore

$$m\angle B'BD = \pi(a' - c)$$

Then, since $\overrightarrow{BB'}$ is between $\overrightarrow{BC}$ and $\overrightarrow{BD}$, we have

$$180 - \beta - m\angle CBB' = m\angle B'BD$$

or

(5a) $$180 - \beta - \pi(a) = \pi(a' - c)$$

and, similarly,

(5b) $$180 - \alpha - \pi(b) = \pi(b' - c)$$

It is interesting to note that (5a) and (5b) are identical with (2a) and (2b), respectively, if we define

(6) $$\pi(-x) = 180 - \pi(x)$$

For this and similar reasons, we shall in fact accept (6) as the definition of $\pi(x)$ for negative arguments.

Since the hypotenuse, $\overline{BA}$, of any right triangle, $\triangle ABC$, obviously intersects the side $\overline{CA}$, it is clear that β must be less than the amplitude

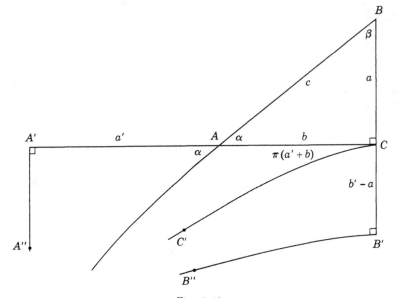

Fig. 4.46

of parallelism for the distance a, that is,

$$\beta = \pi(b') < \pi(a)$$

Hence, from the monotonic character of $\pi(x)$, it follows that

$$b' > a$$

Now let B' be the point on the ray opposite to $\overrightarrow{CB}$ such that $BB' = b'$, and let A' be the point on the ray opposite to $\overrightarrow{AC}$ such that $AA' = a'$. Also, let A'' be a point on the same side of $\overleftrightarrow{AC}$ as B' such that $\overleftrightarrow{A'A''} \perp \overleftrightarrow{AC}$, let B'' be a point on the same side of $\overleftrightarrow{BC}$ as A' such that $\overleftrightarrow{B'B''} \perp \overleftrightarrow{BC}$, and let $\overleftrightarrow{CC'} \| \overleftrightarrow{B'B''}$ (see Fig. 4.46). Now $\overleftrightarrow{BA} \| \overleftrightarrow{B'B''}$, since $\beta = \pi(b')$, and $\overleftrightarrow{A'A''} \| \overleftrightarrow{BA}$, since $\alpha = \pi(a')$. Therefore $\overleftrightarrow{A'A''} \| \overleftrightarrow{B'B''}$. Also $\overleftrightarrow{CC'} \| \overleftrightarrow{A'A''}$, since each is parallel to $\overleftrightarrow{B'B''}$. Therefore, $m\angle A'CC' = \pi(a' + b)$ and $m\angle C'CB' = \pi(b' - a)$. Hence, since $\overrightarrow{CC'}$ is between $\overrightarrow{CA}$ and $\overrightarrow{CB'}$,

(7a) $$\pi(a' + b) + \pi(b' - a) = 90$$

and similarly,

(7b) $$\pi(b' + a) + \pi(a' - b) = 90$$

EXERCISES

1. Derive formula (1b).
2. Derive formula (2b).
3. Derive formulas (4a) and (4b).
4. Derive formula (5b).
5. Derive formula (7b).
6. Assuming, as can be proved,* that

$$\pi(x) = \cos^{-1}(\tanh x)$$

solve the following right triangles: (a) $c = 1$, $m\angle A = 30$; (b) $c = 1$, $m\angle A = 45$; (c) $c = 1$, $m\angle A = 60$; (d) $a = 1$, $m\angle B = 45$; (e) $b = 1$, $m\angle A = 30$.

4.10 Numerical Relations in Lambert Quadrilaterals. Let $DPQR$ be an arbitrary Lambert quadrilateral with acute angle at D.

---

* A derivation of this formula in one particular model of the hyperbolic plane will be found in Sec. 5.4.

Let $m\angle D = \delta$, and let the lengths of the sides be as shown in Fig. 4.47. Let $\overleftrightarrow{DD'}$ and $\overleftrightarrow{RR'}$ be the parallels to $\overleftrightarrow{PQ}$ through D and R, respectively. Let S be the point on the ray opposite to $\overrightarrow{RD}$ such that

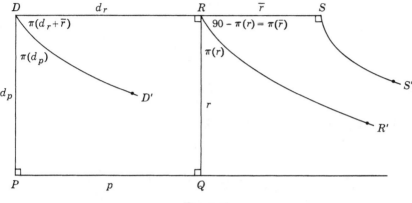

Fig. 4.47

$RS = \bar{r}$, where $\bar{r}$ is the distance complementary to r, and let S' be a point on the same side of $\overleftrightarrow{DR}$ as R' such that $\overleftrightarrow{SS'} \perp RS$. Now $\overleftrightarrow{RR'} \| \overleftrightarrow{SS'}$ since $m\angle R'RS = 90 - \pi(r) = \pi(\bar{r})$ is the amplitude of parallelism for the distance $RS = \bar{r}$. Moreover, $\overleftrightarrow{DD'} \| \overleftrightarrow{RR'}$, since each is parallel to $\overleftrightarrow{PQ}$. Hence

$$\overrightarrow{PQ} \| \overleftrightarrow{DD'} \| \overleftrightarrow{RR'} \| \overleftrightarrow{SS'}$$

Therefore

(1a) $$\delta = \pi(d_p) + \pi(d_r + \bar{r})$$

and, similarly,

(1b) $$\delta = \pi(d_r) + \pi(d_p + \bar{p})$$

Now suppose that $d_r > \bar{r}$, and let S be the point of $\overline{DR}$ such that $RS = \bar{r}$ (see Fig. 4.48). Let $\overleftrightarrow{DD'}$ and $\overleftrightarrow{RR'}$ be the parallels to $\overleftrightarrow{QP}$ through D and R, respectively, and let S' be a point on the same side of $\overleftrightarrow{DR}$ as D' and R' such that $\overleftrightarrow{SS'} \perp DR$. Then $\overleftrightarrow{RR'} \| \overleftrightarrow{SS'}$, since $m\angle SRR' = 90 - \pi(r) = \pi(\bar{r})$, and thus

$$\overrightarrow{QP} \| \overleftrightarrow{DD'} \| \overleftrightarrow{RR'} \| \overleftrightarrow{SS'}$$

Therefore, $m\angle D'DP = \pi(d_p)$ and $m\angle D'DS = \pi(d_r - \bar{r})$. Hence, since $m\angle D'DS = m\angle D'DP + m\angle PDS$, we have

(2a) $\pi(d_r - \bar{r}) = \pi(d_p) + \delta$

and, similarly,

(2b) $\pi(d_p - \bar{p}) = \pi(d_r) + \delta$

Combining (1a) and (2a), we obtain

(3a) $\delta = \dfrac{\pi(d_r - \bar{r}) + \pi(d_r + \bar{r})}{2}$

(3b) $\pi(d_p) = \dfrac{\pi(d_r - \bar{r}) - \pi(d_r + \bar{r})}{2}$

Similarly, from (1b) and (2b), we have

(4a) $\delta = \dfrac{\pi(d_p - \bar{p}) + \pi(d_p + \bar{p})}{2}$

(4b) $\pi(d_r) = \dfrac{\pi(d_p - \bar{p}) - \pi(d_p + \bar{p})}{2}$

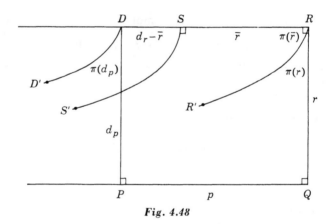

Fig. 4.48

If $d_r < \bar{r}$, let S be the unique point on the ray opposite to $\overrightarrow{DR}$ such that $RS = \bar{r}$, let $\overleftrightarrow{DD'} \| \overleftrightarrow{QP}$, $\overleftrightarrow{RR'} \| \overleftrightarrow{QP}$, and let S' be a point on the same side of $\overleftrightarrow{DR}$ as R' such that $\overleftrightarrow{SS'} \perp \overleftrightarrow{DR}$ (see Fig. 4.49).   Then $\overleftrightarrow{SS'} \| \overleftrightarrow{RR'}$, since $m\angle SRR' = 90 - \pi(r) = \pi(\bar{r})$, and thus

$$\overleftrightarrow{QP} \| \overleftrightarrow{DD'} \| \overleftrightarrow{RR'} \| \overleftrightarrow{SS'}$$

Therefore, since $m\angle SDD' + m\angle D'DP + \delta = 180$, we have

(5a) $\pi(\bar{r} - d_r) = 180 - \pi(d_p) - \delta$

and, similarly,

(5b) $\pi(\bar{p} - d_p) = 180 - \pi(d_r) - \delta$

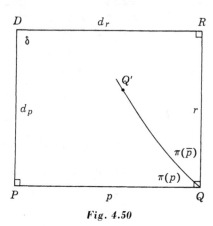

Fig. 4.49

Again it is interesting to note that (5a) and (5b) are equivalent to (2a) and (2b), respectively, under the convention that

$$\pi(-x) = 180 - \pi(x)$$

Now in the Lambert quadrilateral $DPQR$, let $\overset{\longleftrightarrow}{QQ'}$ be the parallel

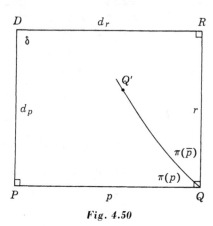

Fig. 4.50

to $\overset{\longleftrightarrow}{PD}$ through Q (see Fig. 4.50). It follows at once that $\overrightarrow{QQ'}$ must intersect $\overline{DR}$. Hence

$$m\angle RQQ' = 90 - \pi(p) = \pi(\bar{p}) < \pi(r)$$

and, therefore, from the monotonic character of $\pi(x)$,

$$\bar{p} > r$$

Similarly, $\bar{r} > p$

Now let d be the distance for which the amplitude of parallelism is δ, i.e., let d be the distance such that $\delta = \pi(d)$, let D' be the point on the ray opposite to $\overrightarrow{DP}$ such that $DD' = d$, and let S be the point on the ray opposite to $\overrightarrow{PQ}$ such that $QS = \bar{r}$ (see Fig. 4.51). Let $\overleftrightarrow{PP'}$

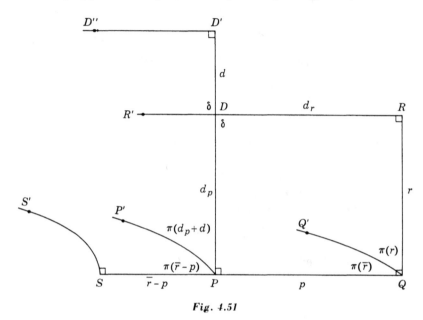

Fig. 4.51

and $\overleftrightarrow{QQ'}$ be the parallels to $\overleftrightarrow{RD}$ through P and Q, respectively, and let S' and D'' be points on the same side of $\overleftrightarrow{DP}$ as S such that $\overleftrightarrow{SS'} \perp PQ$ and $\overleftrightarrow{D'D''} \perp PD$. Then $\overleftrightarrow{QQ'} \| \overleftrightarrow{SS'}$, since $m\angle SQQ' = \pi(\bar{r})$, and therefore $\overleftrightarrow{SS'} \| \overleftrightarrow{RD}$, since each is parallel to $\overleftrightarrow{QQ'}$. Also $\overleftrightarrow{D'D''} \| \overleftrightarrow{RD}$, since $\delta = \pi(d)$, and therefore $\overleftrightarrow{SS'} \| \overleftrightarrow{D'D''} \| \overleftrightarrow{PP'}$, since each is parallel to $\overleftrightarrow{RD}$. Hence

(6a)
$$\pi(d_p + d) + \pi(\bar{r} - p) = 90$$

and, similarly,

(6b)
$$\pi(d_r + d) + \pi(\bar{p} - r) = 90$$

In hyperbolic as well as in euclidean geometry, it is obvious that a right triangle is uniquely determined when the length of the hypotenuse, c, and the measure of one of the acute angles, say β, are given. Likewise it is clear that a trirectangular quadrilateral is uniquely determined when d_r and r, or d_p and p, are given. It follows, therefore, that if the pairs of quantities $\{d_r, r\}$ and $\{c, \beta\}$ are related, then the triangle is uniquely determined when the quadrilateral is given, and vice versa.

The most fruitful way to do this is to set

(7) $$c = d_r$$

and

(8a) $$\beta = 90 - \pi(r)$$

which is equivalent to

(8b) $$b' = \bar{r} \quad \text{or} \quad \overline{b'} = r$$

Now if we compare (4a), Sec. 4.9, and (3a) under the relations (7) and (8b), we have

$$\pi(b) = \frac{\pi(c - b') + \pi(c + b')}{2} = \frac{\pi(d_r - \bar{r}) + \pi(d_r + \bar{r})}{2} = \delta$$

Thus, under the relations (7) and (8b),

(9) $$\delta = \pi(b)$$

or, since $\delta = \pi(d)$,

(9a) $$b = d$$

Similarly, if we compare (4b), Sec. 4.9, and (3b) under the same relations, we have

$$\alpha = \frac{\pi(c - b') - \pi(c + b')}{2} = \frac{\pi(d_r - \bar{r}) - \pi(d_r + \bar{r})}{2} = \pi(d_p)$$

Hence

(10a) $$\alpha = \pi(d_p)$$

or, since $\alpha = \pi(a')$,

(10b) $$d_p = a'$$

Finally, from (7a), Sec. 4.9, and (6a), we have, respectively,

$$\pi(b' - a) = 90 - \pi(a' + b)$$
$$\pi(\bar{r} - p) = 90 - \pi(d_p + d)$$

But we have just seen that $a' = d_p$ and $b = d$. Hence, since $b' = \bar{r}$,

$$\pi(\bar{r} - a) = \pi(\bar{r} - p)$$

and therefore

(11) $$a = p$$

The full correspondence between the right triangle and the related Lambert quadrilateral can be shown graphically as follows: In the quadrilateral $DPQR$, let the ray which lies between $\overrightarrow{QP}$ and $\overrightarrow{QR}$ and which forms with $\overrightarrow{QP}$ an angle congruent to $\angle ABC$ intersect $\overline{PD}$ at

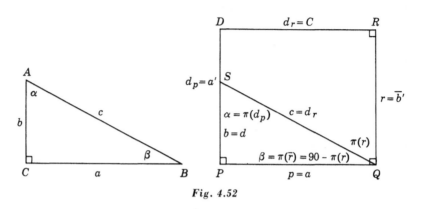

Fig. 4.52

S (see Fig. 4.52). From (11), it follows that the length of the side $\overline{PQ}$ is $p = a$. Hence

$$\triangle SQP \cong \triangle ABC$$

Therefore

$$SQ = c = d_r$$
$$SP = b = d$$
$$m\angle PSQ = \alpha = \pi(d_p)$$
$$DP = d_p = a'$$

Furthermore, since $m\angle SQR = \pi(r)$, it follows that $\overleftrightarrow{QS} \| \overleftrightarrow{RD}$.

By means of the correspondence we have just established, it is possible to describe how the parallel from a given point to a given line in a given direction can actually be constructed: Let l be a given line and D a given point not on l. Let P be the foot of the perpendicular from D to l, let Q be an arbitrary point on l distinct from P, and let R be the foot of the perpendicular from D to the line which is perpendicular to l at Q (see Fig. 4.53). This establishes a Lambert quadrilateral, $DPQR$, in which $DR > PQ$. Now, by Theorem 8, Sec. 4.7, the circle with center at Q and radius DR will intersect $\overline{DP}$ in a point, S. Then $\triangle SQP$ and the quadrilateral $DPQR$ are related as we described above, and therefore

$$m\angle PSQ = \pi(d_p)$$

Hence the required parallel is the line determined by the ray which forms with $\overrightarrow{DP}$ an angle of measure $\alpha = \pi(d_p)$ on the desired side of $\overleftrightarrow{DP}$.

The construction we have just described for the parallel from a given point to a given line in a given direction is especially important because it permits us to carry out various other constructions. In

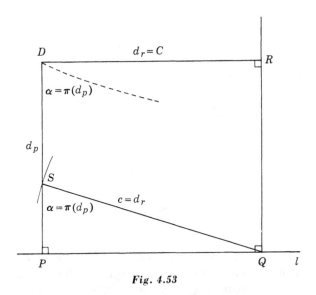

Fig. 4.53

fact, our proof of the existence of a common parallel to two noncollinear rays (Theorem 1, Sec. 4.6), our proof of the existence of a line perpendicular to one side of an acute angle and parallel to the other (Theorem 2, Sec. 4.6), and our proof of the existence of a common perpendicular to two nonintersecting lines (Theorem 3, Sec. 4.5) were actually descriptions of how to construct the lines in question, provided that a parallel from a given point to a given line in a given direction could be constructed. Since we now know how to do this, the other constructions can also be carried out.

EXERCISES

1. Derive formula (1*b*).   2. Derive formula (2*b*).
3. Derive formulas (4*a*) and (4*b*).   4. Prove that $\bar{r} > p$.
5. Derive formula (6*b*).

5

A EUCLIDEAN
MODEL OF THE
HYPERBOLIC PLANE

5.1 The Description of the Model. When one has made even
the modest excursion into hyperbolic geometry that we made in the
last chapter and observed the striking differences between it and
euclidean geometry, two questions naturally come to mind. The
first is fundamentally mathematical: "Is hyperbolic geometry consistent?" The second, which implicitly assumes that the first is to
be answered in the affirmative, is fundamentally physical: "Is it possible that hyperbolic geometry is the geometry of the world around us?"

One obvious way to answer the second question would seem to be
by measurement. But a moment's reflection should convince us that
measurement can decide the issue only if the quantities we propose
to measure are large enough to be detected, and they may not be!
The sum of the measures of the angles of a triangle may actually be
less than 180, yet the defect may be less than the inevitable error
in our most refined observations. Two parallel lines may not be
equidistant, yet for the portions of the lines accessible to us, the varia-
tion in the perpendicular distance from one to the other may be too
small for our most delicate procedures to detect.* Since the defect
of a triangle increases as the dimensions of the triangle increase and

---

* It is said that Gauss once attempted to decide whether the geometry of the
physical world is euclidean or hyperbolic by measuring the angles of a triangle
whose vertices were carefully surveyed points on three widely separated mountain
peaks. Had the sum of the angle measures been sufficiently different from 180,
the experiment would have been conclusive. However, to the accuracy with which
the measurements could be made, the sum turned out to be 180, which proved
not that in this respect the geometry of the universe is euclidean, but merely that
if it is not, the defect of terrestrial triangles is less than a very small amount.

since the distance between parallel lines becomes arbitrarily large in the direction opposite to the direction of parallelism, it is conceivable that as man moves out into space, he may find cosmic configurations whose deviations from euclidean characteristics are large enough to be measured. If so, the question will be decided; but if not, the matter will remain forever in doubt.

The question of what the geometry of the physical world really is may well be meaningless, but the question of the consistency of hyperbolic geometry is not. Because of the inevitable inaccuracy of all measurements, a physical model demonstrating the absolute consistency of hyperbolic geometry is hardly to be expected. However, mathematical models exist which prove the relative consistency of hyperbolic geometry, and in this chapter we shall investigate one such system. The model we shall consider will be a portion of the euclidean plane in which, after suitable definitions of distance and angle measures have been given, the postulates of hyperbolic geometry can all be verified. When we have accomplished this, we shall have shown that if euclidean geometry is consistent, then hyperbolic geometry is also consistent, since any inconsistency among the postulates of hyperbolic geometry or their consequences would imply an inconsistency within a subsystem of the euclidean plane. Or to put it differently, strange as the properties of hyperbolic geometry may seem, if hyperbolic geometry is inconsistent, then so is euclidean geometry. And, of course, with the relative consistency of hyperbolic geometry established, the impossibility of proving the euclidean parallel postulate from the other euclidean postulates is established.

In constructing our model of hyperbolic geometry we shall work in the coordinatized euclidean plane, that is, the cartesian plane, and we shall assume without proof the results of elementary analytic geometry. We must begin, of course, by describing the objects which comprise our particular system; that is, we must define what we shall mean by points, lines, and the incidence relation between points and lines. Then we must define parallel lines, the distance between two points, and the measurement of angles. Finally, after the necessary definitions have been given, we must verify that all the postulates of hyperbolic geometry are satisfied in the model, H_2, which we have created.

Definition 1.   By the system H_2 we shall mean the set of all points of the euclidean plane, E_2, which lie in the interior of the circle, C, whose equation is

$$f = -x^2 - y^2 + 1 = 0$$

The lines of H_2 are the chords of C. In H_2, a point,

P, lies on a line, λ, if and only if *P* is a point of the chord which defines λ. Two lines of H_2 are said to be intersectors if the lines which they determine in E_2 intersect in a point of H_2. Two lines of H_2 are said to be parallel if the lines which they determine in E_2 intersect in a point of *C*. Two lines of H_2 are said to be nonintersectors if the lines which they determine in E_2 do not intersect in a point of H_2 or *C*, that is, if they are either parallel in E_2 or intersect in a point in the exterior of *C*.

The circle *C* we shall refer to as the **metric gauge conic,** and the expression *f* we shall call the **first metric gauge function.** The possible relations between lines of H_2 are suggested in Fig. 5.1. Obviously, the incidence postulates for points and lines, Postulates 1 and 2,

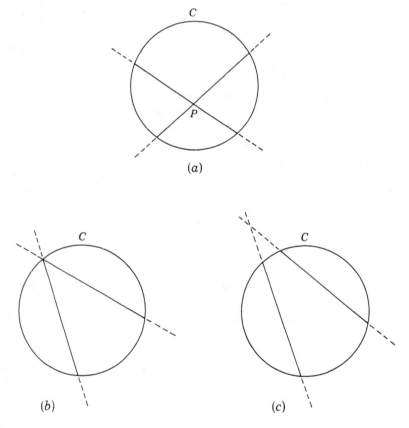

Fig. 5.1. (*a*) Intersecting lines in H_2; (*b*) parallel lines in H_2; (*c*) nonintersecting lines in H_2.

Sec. 2.3, are satisfied in H_2; and it appears plausible that the hyperbolic parallel postulate, Postulate 17_H, Sec. 4.2, is also satisfied.

EXERCISES

1. Explain carefully why Postulates 1 and 2, Sec. 2.3, are satisfied in H_2.
2. Explain why it is plausible that the hyperbolic parallel postulate is satisfied in H_2. Why isn't it certain at this stage?

5.2   The Measurement of Distance.   Before we can describe how distances are to be measured in H_2, there are a few preliminary ideas which we must develop.   A person familiar with projective geometry will recognize these as very special cases of projective properties, but we shall neither assume nor explore these relations.

Let us begin by considering, in E_2, an arbitrary point $P_1:(x_1,y_1)$ in the exterior of the circle C.   Obviously, two tangents can be drawn from P_1 to C, and from familiar results of elementary geometry, the points of contact of these tangents are the intersections of C and the circle whose center is the midpoint, $Q:(x_1/2, y_1/2)$, of the segment $\overline{OP}$ and whose radius is the distance OQ (Fig. 5.2).   Thus the coordinates

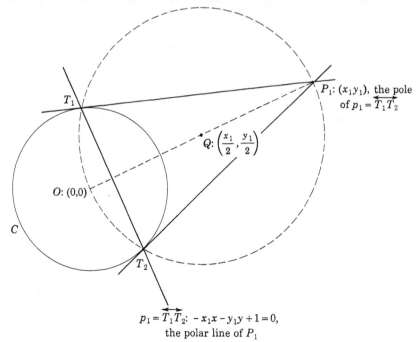

$$p_1 = \overleftrightarrow{T_1 T_2}:\ -x_1 x - y_1 y + 1 = 0,$$
the polar line of P_1

Fig. 5.2

of the two points of contact, T_1 and T_2, satisfy simultaneously the equation

$$\left(x - \frac{x_1}{2}\right)^2 + \left(y - \frac{y_1}{2}\right)^2 = \left(\frac{x_1}{2}\right)^2 + \left(\frac{y_1}{2}\right)^2$$

or
$$x^2 - x_1 x + y^2 - y_1 y = 0$$

and the equation of C, namely,

$$-x^2 - y^2 + 1 = 0$$

Adding these two equations, we find that the coordinates of the points of contact satisfy the linear equation

(1) $$\qquad\qquad -x_1 x - y_1 y + 1 = 0$$

In other words, (1) is the equation of the line, $\overleftrightarrow{T_1 T_2}$, determined by the points of contact of the tangents to C from P_1. This line we shall henceforth refer to as the **polar line** of the point P_1.

Conversely, any line, p_1, which intersects C in two points but does not pass through the center of the circle C has an equation of the form

(2) $$\qquad\qquad l_1 x + m_1 y + n_1 = 0 \qquad n_1 \neq 0$$

in which, since the perpendicular distance from the origin, O, to the line, p_1, is less than 1, we have

$$\left| \frac{n_1}{\sqrt{l_1^2 + m_1^2}} \right| < 1$$

or

(3) $$\qquad\qquad -l_1^2 - m_1^2 + n_1^2 < 0$$

When we compare Eq. (2) with Eq. (1), it is clear that p_1 is the polar line of the point

$$P_1 : \left(-\frac{l_1}{n_1}, -\frac{m_1}{n_1} \right)$$

Moreover, using the euclidean distance formula, we have

$$OP_1 = \sqrt{\frac{l_1^2}{n_1^2} + \frac{m_1^2}{n_1^2}}$$

and by (3) this is greater than 1. Hence P_1 lies in the exterior of C. The point P_1 we shall call the **pole** of the line p_1.

The function $-x_1 x + y_1 y + 1$ which appears on the left in Eq. (1) is obviously closely related to the function

$$f = -x^2 - y^2 + 1$$

which defines the metric gauge conic, C, and our definition of distance in H_2 will be based upon these two functions. For convenience, we shall denote by $f(P_i,P_i)$, or sometimes just by f_{ii}, the result of substituting the coordinates of a point $P_i:(x_i,y_i)$ into f. Similarly, we shall denote by $f(P_i,P_j)$, or sometimes just by f_{ij}, the result of substituting the coordinates of a point $P_j:(x_j,y_j)$ into the function defining the polar line of the point, $P_i:(x_i,y_i)$. Thus

$$f(P_i,P_i) = f_{ii} = -x_i^2 - y_i^2 + 1$$
$$f(P_i,P_j) = f_{ij} = -x_ix_j - y_iy_j + 1$$

Clearly, the vanishing of $f(P_i,P_j)$ expresses the fact that the point P_j lies on the polar line of the point P_i. Moreover, since $f(P_i,P_j)$ is symmetric in the coordinates of P_i and P_j, the relation $f(P_i,P_j) = 0$ is also the condition that P_i lies on the polar line of P_j. Hence we have the following interesting result:

Theorem 1. If P_j lies on the polar line of P_i, then P_i lies on the polar line of P_j.

In the work ahead of us, we shall have to make frequent use of certain inequalities involving the quantities $f(P_i,P_i)$ and $f(P_i,P_j)$. Specifically, we shall need the following results:

Theorem 2. The quantity $f(P_i,P_i)$ is positive, zero, or negative according as P_i lies in the interior of C, on C, or in the exterior of C.

Proof. Using polar coordinates in E_2, we can write

$$x_i = r_i \cos \theta_i$$
$$y_i = r_i \sin \theta_i.$$

Hence $f(P_i,P_i) = -r_i^2 \cos^2 \theta_i - r_i^2 \sin^2 \theta_i + 1 = 1 - r_i^2$

If P_i lies in the interior of C, that is, if P_i is a point of H_2, then r_i is less than 1 and $f(P_i,P_i)$ is clearly positive. If P_i is a point of C, then r_i is equal to 1 and $f(P_i,P_i)$ is equal to zero. Finally, if P_i lies in the exterior of C, then r_i is greater than 1 and $f(P_i,P_i)$ is negative, as asserted.

Theorem 3. If P_i and P_j are both points of H_2, then $f(P_i,P_j)$ is positive.

Proof. Again using polar coordinates, we have

$$f(P_i,P_j) = -(r_i \cos \theta_i)(r_j \cos \theta_j) - (r_i \sin \theta_i)(r_j \sin \theta_j) + 1$$
$$= 1 - r_ir_j \cos (\theta_i - \theta_j)$$

If P_i and P_j are both points of H_2, then both r_i and r_j are less than 1 and $f(P_i,P_j)$ is clearly positive, as asserted. On the other hand, if P_i and P_j are not both points of H_2, one or both of the r's may be arbitrarily large, and $f(P_i,P_j)$ may be positive, negative, or zero.

Theorem 4. If P_i and P_j are both points of H_2, then

$$f(P_i,P_i)f(P_j,P_j) - f^2(P_i,P_j) \leqq 0$$

the equality sign holding if and only if $P_i = P_j$.

Proof. Using results obtained in the proofs of the last two theorems, we have

$$
\begin{aligned}
f(P_i,P_i)f(P_j,P_j) - f^2(P_i,P_j) &= (1 - r_i^2)(1 - r_j^2) \\
&\quad - [1 - r_i r_j \cos(\theta_i - \theta_j)]^2 \\
&= -r_i^2 - r_j^2 + r_i^2 r_j^2 + 2r_i r_j \cos(\theta_i - \theta_j) \\
&\quad - r_i^2 r_j^2 \cos^2(\theta_i - \theta_j) \\
&= -(r_i - r_j)^2 - 2r_i r_j [1 - \cos(\theta_i - \theta_j)] \\
&\quad + r_i^2 r_j^2 \sin^2(\theta_i - \theta_j) \\
&= -(r_i - r_j)^2 - 4r_i r_j \sin^2 \frac{\theta_i - \theta_j}{2} \\
&\quad + 4r_i^2 r_j^2 \sin^2 \frac{\theta_i - \theta_j}{2} \cos^2 \frac{\theta_i - \theta_j}{2} \\
&= -(r_i - r_j)^2 - 4r_i r_j \sin^2 \left(\frac{\theta_i - \theta_j}{2} \right) \\
&\quad \left(1 - r_i r_j \cos^2 \frac{\theta_i - \theta_j}{2} \right)
\end{aligned}
$$

If both P_i and P_j are points of H_2, then r_i and r_j are each less than 1, and the last expression is surely equal to or less than zero. Moreover, it can equal zero if and only if $r_i = r_j$ and $\theta_i = \theta_j$, that is, if and only if $P_i = P_j$, as asserted.

From Theorem 4 it follows that if P_i and P_j are both points of H_2, then

$$\frac{f^2(P_i,P_j)}{f(P_i,P_i)f(P_j,P_j)} \geqq 1$$

or, since $f(P_i,P_i)$, $f(P_j,P_j)$, and $f(P_i,P_j)$ are all positive for points in H_2,

$$\frac{f(P_i,P_j)}{\sqrt{f(P_i,P_i)f(P_j,P_j)}} \geqq 1$$

From the last inequality it follows that there is always a unique nonnegative number, ϕ, such that

$$\cosh \phi = \frac{f(P_i,P_j)}{\sqrt{f(P_i,P_i)f(P_j,P_j)}}$$

and upon this fact we base our definition of distance in H_2:

Definition 1. If $P_1:(x_1,y_1)$ and $P_2:(x_2,y_2)$ are both points of H_2, the distance between P_1 and P_2 is given by the formula

$$(P_1P_2)_k = k \cosh^{-1} \frac{f(P_1,P_2)}{\sqrt{f(P_1,P_1)f(P_2,P_2)}}$$

At first glance this may seem a strange and unnatural definition for a quantity as intuitively clear as distance. However, just as we have insisted that both points and lines are any objects which have the properties the postulates ascribe to points and lines, so any correspondence between pairs of points in H_2 and the nonnegative numbers is a valid definition of distance if it has the appropriate properties. And the correspondence established by Definition 1 does indeed have all the properties required of a distance function, as we shall now verify.

In the first place, given any two points, A and A', Definition 1 establishes a correspondence which associates a unique nonnegative number with every pair of points in H_2 in such a way that the number 1 is assigned to the pair (A,A'). In fact, with A and A' given, the quantities $f(A,A)$, $f(A',A')$, and $f(A,A')$ are all determined, and hence the requirement that

$$(AA')_k = k \cosh^{-1} \frac{f(A,A')}{\sqrt{f(A,A)f(A',A')}} = 1$$

determines k. With k fixed, the distance between all other points of H_2 is then determined. Moreover, since $\cosh^{-1} u = 0$ if and only if $u = 1$, and since, according to Theorem 4,

$$\frac{f(P_1,P_2)}{\sqrt{f(P_1,P_1)f(P_2,P_2)}} = 1$$

if and only if $P_1 = P_2$, it follows that the number assigned to a pair of points by Definition 1 is zero if and only if the points are the same. Thus the first distance postulate, Postulate 8, Sec. 2.4, is satisfied. Incidentally, since $f(P_2,P_2) = 0$ if P_2 is a point of the metric gauge conic, C, and since $\cosh^{-1} u$ becomes infinite as u becomes infinite and conversely, it is clear that the distance between any point, P_1, of H_2 and any point, P_2, of the metric gauge conic, C, is infinite. The conic, C, can thus be described as the "conic at infinity" for the system H_2, and in the inclusive system E_2 it is now meaningful to say that in H_2 "parallel lines meet at infinity."

To verify Postulate 9, Sec. 2.4, namely,

$$m_\alpha(P_1,P_2) = m_\alpha(B,B')m_\beta(P_1,P_2)$$

let $\alpha = (A,A')$ and $\beta = (B,B')$ be two unit pairs, and let k_α and k_β be, respectively, the proportionality constants required for measurements in terms of these unit pairs. Then

$$m_\alpha(P_1,P_2) = k_\alpha \cosh^{-1} \frac{f(P_1,P_2)}{\sqrt{f(P_1,P_1)f(P_2,P_2)}}$$

$$m_\beta(P_1,P_2) = k_\beta \cosh^{-1} \frac{f(P_1,P_2)}{\sqrt{f(P_1,P_1)f(P_2,P_2)}}$$

$$m_\alpha(B,B') = k_\alpha \cosh^{-1} \frac{f(B,B')}{\sqrt{f(B,B)f(B',B')}}$$

$$m_\beta(B,B') = k_\beta \cosh^{-1} \frac{f(B,B')}{\sqrt{f(B,B)f(B',B')}} = 1$$

and therefore

$$m_\alpha(P_1,P_2) = k_\alpha \cosh^{-1} \frac{f(P_1,P_2)}{\sqrt{f(P_1,P_1)f(P_2,P_2)}}$$

$$= k_\alpha \cosh^{-1} \frac{f(P_1,P_2)}{\sqrt{f(P_1,P_1)f(P_2,P_2)}} \cdot k_\beta \cosh^{-1} \frac{f(B,B')}{\sqrt{f(B,B)f(B',B')}}$$

$$= k_\alpha \cosh^{-1} \frac{f(B,B')}{\sqrt{f(B,B)f(B',B')}} \cdot k_\beta \cosh^{-1} \frac{f(P_1,P_2)}{\sqrt{f(P_1,P_1)f(P_2,P_2)}}$$

$$= m_\alpha(B,B')m_\beta(P_1,P_2)$$

as required.

Furthermore, given any unit pair, α, that is, given a value of k, say k_α, and any point P_0 on an arbitrary line, λ, there are exactly two points, P_1 and P_1', on λ such that

$$m_\alpha(P_0,P_1) = m_\alpha(P_0,P_1') = 1$$

In fact, we have the following more general theorem:

Theorem 5. If P_0 is an arbitrary point of H_2 and if d is an arbitrary positive number, then on every line through P_0 there are exactly two points, P_d and P_d', one on each side of P_0, such that

$$m_\alpha(P_0,P_d) = m_\alpha(P_0,P_d') = d$$

Proof. Let λ be an arbitrary line through an arbitrary point $P_0:(x_0,y_0)$ of H_2, and let d be an arbitrary positive number. Then in E_2 the equation of the locus of points, $P:(x,y)$, whose hyperbolic distance from P_0 is d, is

$$\frac{f(P_0,P)}{\sqrt{f(P_0,P_0)f(P,P)}} = \cosh\frac{d}{k_\alpha} = \delta$$

say, or

(4) $$f^2(P_0,P) - \delta^2 f(P_0,P_0)f(P,P) = 0$$

From the definition of $f(P_0,P)$ and $f(P,P)$, it is clear that this is the equation of a conic, Γ. Moreover, according to Theorem 2, $f(P,P)$ is zero if P is a point of C and negative if P is a point in the exterior of C. Hence, since $f(P_0,P_0)$ is positive because P_0 is a point of H_2, it follows that if the coordinates of any point on C or in the exterior of C are substituted into the left member of Eq. (4), the resulting number is positive. Hence Γ can contain no point on C or in its exterior. Therefore, Γ must lie entirely within C and hence must be an ellipse. Furthermore, if the coordinates of P_0 are substituted into the left member of Eq. (4), we obtain

$$f^2(P_0,P_0) - \delta^2 f^2(P_0,P_0)$$

which is negative, since $\delta^2 = \cosh^2(d/k_\alpha) > 1$. Thus, since

$$f^2(P_0,P) - \delta^2 f^2(P_0,P_0)f(P,P)$$

is positive for points outside C and hence outside Γ, but negative for P_0, the point P_0 must lie in the interior of Γ. Thus any line through P_0 must intersect the ellipse Γ in two points on opposite sides of P_0. Hence, on any line through P_0 there are exactly two points, one on each side of P_0, whose hyperbolic distance from P_0 is any given positive number, d. By taking d to be 1, it is clear that on any line through P_0 there are exactly two points at a unit distance from P_0, which verifies Postulate 10, Sec. 2.4.

Now consider the "hyperbolic circles"

$$\Gamma_1: f^2(P_0,P) - \delta_1^2 f(P_0,P_0)f(P,P) = 0$$
$$\Gamma_2: f^2(P_0,P) - \delta_2^2 f(P_0,P_0)f(P,P) = 0$$

corresponding to two hyperbolic distances, d_1 and d_2, such that $d_1 < d_2$ and therefore $\delta_1^2 < \delta_2^2$. Since the second equation can be written $[f^2(P_0,P) - \delta_1^2 f(P_0,P_0)f(P,P)] - (\delta_2^2 - \delta_1^2)f(P_0,P_0)f(P,P) = 0$, it is clear that for any point, P, on Γ_1, the left member of the equation of Γ_2 is negative. Hence Γ_1 lies entirely within Γ_2, which means that of two points of λ on the same side of P_0, the one whose hyperbolic distance from P_0 is the greater is also the one whose euclidean distance from P_0 is the greater, and conversely. Thus we have established the following result:

Corollary 1. If P_0 and P_1 are particular points on a line, λ, of H_2 and if P is a general point of λ on the same side of P_0 as P_1, then the hyperbolic distance from P_0 to P is a monotonically increasing function of the cartesian distance from P_0 to P, and conversely.

EXAMPLE 1

If the measurement of distance in H_2 is to be such that the distance between $A: (-\frac{1}{2}, -\frac{1}{2})$ and $A': (-\frac{1}{4}, -\frac{1}{4})$ is 1, what is the locus of points whose hyperbolic distance from $(\frac{3}{4},0)$ is 1? 2?

Our first step, of course, must be to determine k_α. Now for A and A' we have

$$f(A,A) = \tfrac{1}{2}, \qquad f(A,A') = \tfrac{3}{4}, \qquad f(A',A') = \tfrac{7}{8}$$

Hence

$$1 = k_\alpha \cosh^{-1} \frac{\frac{3}{4}}{\sqrt{\frac{7}{16}}} = k_\alpha \cosh^{-1} 1.134, \qquad \text{whence } k_\alpha = \frac{1}{0.512}$$

The equation of the locus corresponding to $d = 1$ is therefore

$$(-\tfrac{3}{4}x + 1)^2 - (\cosh^2 0.512)(\tfrac{7}{16})(-x^2 - y^2 + 1) = 0$$

or
$$(x - 0.67)^2 + 0.85y^2 = (0.27)^2$$

The equation of the locus corresponding to $d = 2$ is

$$(-\tfrac{3}{4}x + 1)^2 - (\cosh^2 1.024)(\tfrac{7}{16})(-x^2 - y^2 + 1) = 0$$

or
$$(x - 0.45)^2 + 0.65y^2 = (0.50)^2$$

Plots of these two "hyperbolic circles" are shown in Fig. 5.3.

Now let P_0 be an arbitrary point on a line, λ, and let P_1 be either of the two points on λ such that $P_0P_1 = 1$, in terms of some chosen unit pair. Then to every point, Q, which in E_2 lies on the same side of P_0 as P_1 there corresponds a unique positive number, namely, its hyperbolic distance from P_0. And conversely, by Theorem 5, to every positive number there corresponds a unique point, Q, on the same side of P_0 as P_1, namely, the point whose hyperbolic distance from P_0 is that number. Similarly, a correspondence can be set up between the negative numbers and the points on the opposite side of P_0 from P_1 by assigning to each point the negative of its distance from P_0. The assignment of the number 0 to P_0 completes the correspondence between the points of λ and the set of all real numbers. Moreover, the $1:1$ correspondence thus established has the property that if q_1 and q_2 are the coordinates of any points, Q_1 and Q_2, on $\overleftrightarrow{P_0P_1}$,

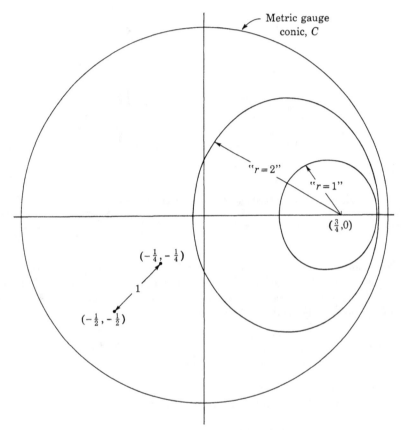

Fig. 5.3. Two concentric circles in H_2.

then the length of the segment $\overline{Q_1Q_2}$ is

$$Q_1Q_2 = |q_1 - q_2|$$

To prove this, let us suppose first that Q_1 and Q_2 are two points whose coordinates on λ are of opposite sign. Then

$$|q_1 - q_2| = |d_1 + d_2| = \left| k \cosh^{-1} \frac{f(P_0,Q_1)}{\sqrt{f(P_0,P_0)f(Q_1,Q_1)}} \right. $$
$$\left. + k \cosh^{-1} \frac{f(P_0,Q_2)}{\sqrt{f(P_0,P_0)f(Q_2,Q_2)}} \right|$$

If, for convenience, we set

$$f(P_0,P_0) = f_{00}, \quad f(Q_1,Q_1) = f_{11}, \quad f(Q_2,Q_2) = f_{22}$$
$$f(P_0,Q_1) = f_{01}, \quad f(P_0,Q_2) = f_{02}, \quad f(Q_1,Q_2) = f_{12}$$

and then let

$$u = \cosh^{-1} \frac{f_{01}}{\sqrt{f_{00}f_{11}}} \quad \text{and} \quad v = \cosh^{-1} \frac{f_{02}}{\sqrt{f_{00}f_{22}}}$$

we have $|q_1 - q_2| = |k(u + v)|$ where

$$\cosh u = \frac{f_{01}}{\sqrt{f_{00}f_{11}}} \qquad \sinh u = \sqrt{\frac{f_{01}^2 - f_{00}f_{11}}{f_{00}f_{11}}}$$

$$\cosh v = \frac{f_{02}}{\sqrt{f_{00}f_{22}}} \qquad \sinh v = \sqrt{\frac{f_{02}^2 - f_{00}f_{22}}{f_{00}f_{22}}}$$

Hence

$$\cosh (u + v) = \cosh u \cosh v + \sinh u \sinh v$$

(5)
$$= \frac{f_{01}f_{02} + \sqrt{f_{01}^2 - f_{00}f_{11}} \sqrt{f_{02}^2 - f_{00}f_{22}}}{f_{00} \sqrt{f_{11}f_{22}}}$$

Now since $P_0:(x_0,y_0)$, $Q_1:(x_1 y_1)$, and $Q_2:(x_2,y_2)$ are collinear points, it follows from the results of elementary analytic geometry that the coordinates of P_0 can be expressed as linear combinations of the coordinates of Q_1 and Q_2 of the form

$$x_0 = \frac{\mu_1 x_1 + \mu_2 x_2}{\mu_1 + \mu_2}, \qquad y_0 = \frac{\mu_1 y_1 + \mu_2 y_2}{\mu_1 + \mu_2}$$

μ_1 and μ_2 being of like sign if P_0 lies between Q_1 and Q_2 and of opposite sign if P_0 is not between Q_1 and Q_2. Hence

$$f_{00} = -\left(\frac{\mu_1 x_1 + \mu_2 x_2}{\mu_1 + \mu_2}\right)^2 - \left(\frac{\mu_1 y_1 + \mu_2 y_2}{\mu_1 + \mu_2}\right)^2 + 1$$
$$= \frac{\mu_1^2 f_{11} + 2\mu_1\mu_2 f_{12} + \mu_2^2 f_{22}}{(\mu_1 + \mu_2)^2}$$

$$f_{01} = -\left(\frac{\mu_1 x_1 + \mu_2 x_2}{\mu_1 + \mu_2}\right) x_1 - \left(\frac{\mu_1 y_1 + \mu_2 y_2}{\mu_1 + \mu_2}\right) y_1 + 1$$
$$= \frac{\mu_1 f_{11} + \mu_2 f_{12}}{\mu_1 + \mu_2}$$

$$f_{02} = -\left(\frac{\mu_1 x_1 + \mu_2 x_2}{\mu_1 + \mu_2}\right) x_2 - \left(\frac{\mu_1 y_1 + \mu_2 y_2}{\mu_1 + \mu_2}\right) y_2 + 1$$
$$= \frac{\mu_1 f_{12} + \mu_2 f_{22}}{\mu_1 + \mu_2}$$

Therefore, substituting into the radicals in the numerator of (5), we have

$$\sqrt{f_{01}^2 - f_{00}f_{11}}$$
$$= \sqrt{\frac{\mu_1^2 f_{11}^2 + 2\mu_1\mu_2 f_{11}f_{12} + \mu_2^2 f_{12}^2 - (\mu_1^2 f_{11} + 2\mu_1\mu_2 f_{12} + \mu_2^2 f_{22})f_{11}}{(\mu_1 + \mu_2)^2}}$$
$$= \sqrt{\frac{\mu_2^2(f_{12}^2 - f_{11}f_{22})}{(\mu_1 + \mu_2)^2}}$$

and, similarly,

$$\sqrt{f_{02}^2 - f_{00}f_{22}} = \sqrt{\frac{\mu_1^2(f_{12}^2 - f_{11}f_{22})}{(\mu_1 + \mu_2)^2}}$$

Hence Eq. (5) becomes

$$\cosh(u + v) = \frac{(\mu_1 f_{11} + \mu_2 f_{12})(\mu_1 f_{12} + \mu_2 f_{22}) + \sqrt{\mu_1^2 \mu_2^2 (f_{12}^2 - f_{11}f_{22})^2}}{(\mu_1 + \mu_2)^2 f_{00} \sqrt{f_{11}f_{22}}}$$

Now by Theorem 4, the quantity $f_{12}^2 - f_{11}f_{22}$ is positive. Moreover, since we are presently considering the case in which Q_1 and Q_2 are on opposite sides of P_0, it follows that μ_1 and μ_2 have the same sign and therefore $\mu_1\mu_2$ is positive. Hence, simplifying the radical in the numerator of the last expression, we have

$$\frac{\mu_1^2 f_{11}f_{12} + \mu_1\mu_2(f_{11}f_{22} + f_{12}^2) + \mu_2^2 f_{12}f_{22} + \mu_1\mu_2(f_{12}^2 - f_{11}f_{22})}{(\mu_1 + \mu_2)^2 f_{00} \sqrt{f_{11}f_{22}}}$$
$$= \frac{f_{12}(\mu_1^2 f_{11} + 2\mu_1\mu_2 f_{12} + \mu_2^2 f_{22})}{\sqrt{f_{11}f_{22}}\,(\mu_1 + \mu_2)^2 f_{00}}$$
$$= \frac{f_{12}}{\sqrt{f_{11}f_{22}}}$$

since $\mu_1^2 f_{11} + 2\mu_1\mu_2 f_{12} + \mu_2^2 f_{22} = f_{00}(\mu_1 + \mu_2)^2$. Thus

$$\cosh(u + v) = \frac{f_{12}}{\sqrt{f_{11}f_{22}}} \qquad \text{or} \qquad u + v = \cosh^{-1}\frac{f_{12}}{\sqrt{f_{11}f_{22}}}$$

and therefore

$$|q_1 - q_2| = |k(u + v)| = \left| k \cosh^{-1}\frac{f_{12}}{\sqrt{f_{11}f_{22}}} \right| = Q_1 Q_2$$

If Q_1 and Q_2 have coordinates which are not of opposite sign, we have $|q_1 - q_2| = |d_1 - d_2|$. However, proceeding just as before, we reach the same conclusion, the change of sign connecting d_1 and d_2 being compensated for by the fact that now, since P_0 is not between Q_1 and Q_2, the product $\mu_1\mu_2$ is negative, and therefore $\sqrt{\mu_1^2\mu_2^2} = -\mu_1\mu_2$ and not $\mu_1\mu_2$, as before.   Thus the ruler postulate, Postulate 11, Sec. 2.4, is verified, and Definition 1 does indeed provide a valid distance function.

We can now define segments and rays in H_2 and prove the same theorems about them that we did in E_2.   Moreover, from Corollary 1, Theorem 5, it is clear that every segment in H_2 is also a segment in the euclidean sense, and every euclidean segment contained in H_2 is also a segment in the hyperbolic sense.   Hence order relations for collinear points in H_2 are in every case identical with the order relations of the same points regarded as points of E_2.   If l is the euclidean line determined by an arbitrary hyperbolic line, λ, the two halfplanes of H_2 which have λ as edge can be defined as the intersections of the interior of C and the respective halfplanes of E_2 which have l as edge. With this definition, it follows immediately that the plane-separation postulate, Postulate 12, Sec. 2.5, is also satisfied in H_2.

EXERCISES

1. Show that a circle in H_2 is a circle in E_2 if and only if its center is the origin.

2. If $(AB)_H$ and $(AB)_E$ denote, respectively, the hyperbolic and euclidean lengths of the segment $\overline{AB}$, under what conditions, if any, does the relation

    $$(P_0P_1)_H < (P_0P_2)_H$$

    imply that $(P_0P_1)_E < (P_0P_2)_E$

3. What is the value of k if the hyperbolic length of the segment whose endpoints are $(0,0)$ and $(0,\frac{1}{4})$ is 1 ?

In the following exercises, assume that $k = 1$.

4. What is the hyperbolic distance between

    (a) $(0,0)$ and $(0,\frac{1}{2})$ (b) $(0,0)$ and $(\frac{1}{2},\frac{1}{2})$
    (c) $(0,\frac{1}{2})$ and $(\frac{1}{2},0)$ (d) $(-\frac{1}{2},0)$ and $(\frac{1}{2},0)$

5. What are the cartesian coordinates of the two points on the line $x = 0$ whose hyperbolic distance from $(0,\frac{1}{2})$ is 1 ?

6. What are the cartesian coordinates of the points on the positive x and y axes whose hyperbolic distances from the origin are 1, 2, 3, 4, and 5 ?

7. What are the cartesian coordinates of the two points on the x axis whose hyperbolic distance from $(0,\frac{1}{2})$ is 2 ?

8. Find and graph the cartesian equation of the locus of points whose hyperbolic distances from $(0,\frac{1}{4})$ and $(\frac{1}{2},0)$ are equal.   Verify that the pole of the line determined by the given points is a point of the locus.

9. An equilateral triangle in H_2 is to have $(-\frac{1}{2},0)$ and $(\frac{1}{2},0)$ as two of its vertices. What are the cartesian coordinates of the third vertex?

10. What are the cartesian coordinates of the midpoint of the hyperbolic segment whose endpoints are the origin and the point $(a,0)$?

5.3 The Measurement of Angles. Before we can describe how angle measures are to be assigned in H_2, we must first introduce the notion of the coordinates of a line and then a second metric gauge function.

In elementary analytic geometry it is customary to identify points by pairs of coordinates and lines by single equations. This asymmetry usually passes unnoticed, and in any event causes little or no inconvenience, even though it can easily be eliminated. In fact, the line defined by the equation

$$(1) \qquad lx + my + n = 0 \qquad l,m \text{ not both zero}$$

is completely determined by the three coefficients (l,m,n), which can thus be regarded as coordinates of the line. However, unlike the coordinates of a point, which are uniquely determined when the point is given, the coordinates of a line are not fixed when the line is given. In fact, since

$$klx + kmy + kn = 0$$

defines the same line as does Eq. (1), provided that $k \neq 0$, it is clear that if (l,m,n) are coordinates of a line, then for $k \neq 0$, (kl,km,kn) can also serve as coordinates of the line.

The coordinates of a line can be determined equally well from an equation of the line or from the coordinates of two points on the line. In fact, since the equation of the line determined by the two points $P_1:(x_1,y_1)$ and $P_2:(x_2,y_2)$ can be written

$$\begin{vmatrix} x & y & 1 \\ x_1 & y_1 & 1 \\ x_2 & y_2 & 1 \end{vmatrix} = (y_1 - y_2)x + (x_2 - x_1)y + (x_1y_2 - x_2y_1) = 0$$

it follows that the coordinates of the line can be taken to be

$$(l,m,n) = [(y_1 - y_2),(x_2 - x_1),(x_1y_2 - x_2y_1)]$$

More specifically, in order to introduce the idea of direction on a line and to establish coordinates for rays, we adopt the following definition:

Definition 1. If $P_1:(x_1,y_1)$ and $P_2:(x_2,y_2)$ are any two points of E_2, the coordinates of the line $\overleftrightarrow{P_1P_2}$ or of the ray $\overrightarrow{P_1P_2}$ are

$$(l,m,n) = [(y_1 - y_2),(x_2 - x_1),(x_1y_2 - x_2y_1)]$$

or any positive multiple of these numbers. The coordinates of the oppositely directed line, $\overleftrightarrow{P_2P_1}$, or of the oppositely directed ray, $\overrightarrow{P_2P_1}$, are any negative multiples of these numbers.

It is convenient to note that the coordinates of $\overleftrightarrow{P_1P_2}$ or of $\overrightarrow{P_1P_2}$ can be read from the matrix

$$\begin{Vmatrix} x_1 & y_1 & 1 \\ x_2 & y_2 & 1 \end{Vmatrix}$$

whose rows are, respectively, the coordinates of P_1 and P_2, by deleting the first, second, and third columns in turn and taking the resulting 2×2 determinants with signs alternately plus and minus. Similarly, the coordinates of $\overleftrightarrow{P_2P_1}$ or of $\overrightarrow{P_2P_1}$ can be read from the matrix

$$\begin{Vmatrix} x_2 & y_2 & 1 \\ x_1 & y_1 & 1 \end{Vmatrix}$$

The following result is also worthy of explicit mention, although we shall leave its proof as an exercise:

Theorem 1. If $\overrightarrow{VA_1}:(l_1,m_1,n_1)$ and $\overrightarrow{VA_2}:(l_2,m_2,n_2)$ are two non-collinear rays, then the coordinates (l_3,m_3,n_3) of any ray, $\overrightarrow{VA_3}$, between $\overrightarrow{VA_1}$ and $\overrightarrow{VA_2}$ are of the form

$$l_3 = \mu_1 l_1 + \mu_2 l_2$$
$$m_3 = \mu_1 m_1 + \mu_2 m_2$$
$$n_3 = \mu_1 n_1 + \mu_2 n_2$$

where μ_1 and μ_2 are positive.

We now observe that the perpendicular distance from the origin to any tangent to the metric gauge conic,

$$C: f = -x^2 - y^2 + 1 = 0$$

is 1, and conversely. Hence, recalling the normal form of the equation of a line, it is clear that the coefficients in the equation

$$lx + my + n = 0$$

of an arbitrary tangent to C must satisfy the relation

$$\left| \frac{n}{\sqrt{l^2 + m^2}} \right| = 1$$

or
$$F = l^2 + m^2 - n^2 = 0$$

This, then, is the equation satisfied by the coordinates of any tangent to the metric gauge conic. The collection of these tangents, that is, the set of lines whose coordinates satisfy the equation $F = 0$, we shall call the **metric gauge envelope**, and the function F we shall call the **second metric gauge function**. The result of substituting the coordinates of a particular line, $\lambda_i:(l_i,m_i,n_i)$, into the function F we shall denote by $F(\lambda_i,\lambda_i)$, or occasionally by F_{ii}. Since the perpendicular distance from the origin to the line $\lambda: lx + my + n = 0$ is less than 1 if λ intersects C in two points and is greater than 1 if λ does not intersect C, the function F has the properties described by the following theorem:

> **Theorem 2.** If $\lambda: lx + my + n = 0$ is an arbitrary line, then $F(\lambda,\lambda)$ is greater than 0, equal to 0, or less than 0 according as the intersection of λ and the metric gauge conic consists of two points, one point, or no points.

EXAMPLE 1

Sketch the envelope of the lines whose coordinates satisfy the equation

$$l^2 = 4mn$$

If $m = 0$, we must also have $l = 0$, and there is no line corresponding to these values. If $m \neq 0$, we may without restriction assume it to be 1. Then we can construct a table of values very much as though we were plotting a locus of points:

l	m	n	Equation
0	1	0	$y = 0$
$\pm\frac{1}{2}$	1	$\frac{1}{16}$	$\pm\frac{1}{2}x + y + \frac{1}{16} = 0$
± 1	1	$\frac{1}{4}$	$\pm\, x + y + \frac{1}{4} = 0$
± 2	1	1	$\pm 2x + y + 1 = 0$
± 3	1	$\frac{9}{4}$	$\pm 3x + y + \frac{9}{4} = 0$

A plot of these lines is shown in Fig. 5.4. It is not difficult to show that the envelope of this example consists of all the lines tangent to the parabola $y = x^2$.

If $\lambda_1:(l_1,m_1,n_1)$ is a given line, then associated with the equation of the metric gauge envelope

$$F = l^2 + m^2 - n^2 = 0$$

is the equation

(2) $$l_1 l + m_1 m - n_1 n = 0$$

This is obviously analogous to the equation of the polar line of a point;

and if λ_1 is a line which intersects the metric gauge conic, Eq. (2) also expresses an important property of poles and polars. For, as we observed in Sec. 5.2, if $n_1 \neq 0$, the pole of the line

$$\lambda_1 \colon l_1 x + m_1 y + n_1 = 0$$

is the point
$$P_1 \colon \left(-\frac{l_1}{n_1}, \ -\frac{m_1}{n_1} \right)$$

Furthermore, if $lx + my + n = 0$ is the equation of any line, λ, whose

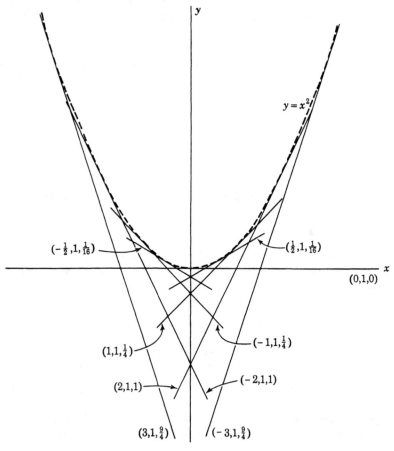

Fig. 5.4. The envelope $l^2 = 4mn$.

coordinates, (l,m,n), satisfy Eq. (2), then λ passes through P_1, since

$$l\left(-\frac{l_1}{n_1} \right) + m\left(-\frac{m_1}{n_1} \right) + n = 0$$

if (l,m,n) satisfy Eq. (2). On the other hand, if $n_1 = 0$ so that the given line, λ_1, passes through the origin, Eq. (2) becomes simply

$$l_1l + m_1m = 0 \qquad \text{or} \qquad \frac{l}{m} = -\frac{m_1}{l_1}$$

Hence, any line, λ, whose coordinates satisfy Eq. (2) when $n_1 = 0$ is perpendicular to λ_1. Summarizing these observations, we have the following theorem:

> **Theorem 3.**   If $\lambda_1:(l_1,m_1,n_1)$ is a given line, then any line $\lambda:(l,m,n)$ whose coordinates satisfy the equation
>
> $$l_1l + m_1m - n_1n = 0$$
>
> passes through the pole of λ_1 if $n_1 \neq 0$ and is perpendicular to λ_1 if $n_1 = 0$, and conversely.

The result of substituting the coordinates of a line $\lambda_j:(l_j,m_j,n_j)$ into the expression $l_il + m_im - n_in$ we shall denote by $F(\lambda_i,\lambda_j)$, or simply by F_{ij}.

Since $F(\lambda_i,\lambda_j)$ is symmetric in the coordinates of λ_i and λ_j, the vanishing of $F(\lambda_i,\lambda_j)$ is simultaneously the condition that λ_j passes through the pole of λ_i and the condition that λ_i passes through the pole of λ_j. Thus we have the following analog of Theorem 1, Sec. 5.2:

> **Theorem 4.**   If λ_j passes through the pole of λ_i, then λ_i passes through the pole of λ_j.

Since reversing the signs of the coordinates of either λ_i or λ_j will change the sign of $F(\lambda_i,\lambda_j)$, it is clear that $F(\lambda_i,\lambda_j)$ may be either positive, negative, or zero. To explore this matter more fully, let $\overrightarrow{A_0A_1}:(l_1,m_1,n_1)$ be an arbitrary ray in H_2, and let A_2 be the pole of the line $\lambda_1: \overleftrightarrow{A_0A_1}$. Then if the coordinates of $\lambda_2: \overrightarrow{A_0A_2}$ are (l_2,m_2,n_2), it follows from Theorem 3 that

$$F(\lambda_1,\lambda_2) = 0$$

Now by Theorem 1, any ray, $\overrightarrow{A_0A_3}$, between $\overrightarrow{A_0A_1}$ and $\overrightarrow{A_0A_2}$ has coordinates of the form

$$l_3 = \mu_1l_1 + \mu_2l_2$$
$$m_3 = \mu_1m_1 + \mu_2m_2$$
$$n_3 = \mu_1n_1 + \mu_2n_2 \qquad \mu_1, \mu_2 > 0$$

Hence, if $\lambda_3 = \overleftrightarrow{A_0A_3}$,

$$
\begin{aligned}
F(\lambda_1,\lambda_3) &= l_1(\mu_1l_1 + \mu_2l_2) + m_1(\mu_1m_1 + \mu_2m_2) - n_1(\mu_1n_1 + \mu_2n_2) \\
&= \mu_1(l_1^2 + m_1^2 - n_1^2) + \mu_2(l_1l_2 + m_1m_2 - n_1n_2) \\
&= \mu_1F(\lambda_1,\lambda_1) + \mu_2F(\lambda_1,\lambda_2) \\
&= \mu_1F(\lambda_1,\lambda_1) > 0
\end{aligned}
$$

Similarly, the coordinates of any ray, $\overrightarrow{A_0A_3}$, between $\overrightarrow{A_0A_1}$ and the ray opposite to $\overrightarrow{A_0A_2}$ are of the form

$$
\begin{aligned}
l_3 &= \mu_1l_1 + \mu_2(-l_2) \\
m_3 &= \mu_1m_1 + \mu_2(-m_2) \\
n_3 &= \mu_1n_1 + \mu_2(-n_2) \qquad \mu_1, \mu_2 > 0
\end{aligned}
$$

and hence

$$
F(\lambda_1,\lambda_3) = \mu_1F(\lambda_1,\lambda_1) - \mu_2F(\lambda_1,\lambda_2) = \mu_1F(\lambda_1,\lambda_1) > 0
$$

For an arbitrary ray, $\overrightarrow{A_0A_3}$, between $\overrightarrow{A_0A_2}$ and the ray opposite to $\overrightarrow{A_0A_1}$, we have the coordinates

$$
\begin{aligned}
l_3 &= \mu_1(-l_1) + \mu_2l_2 \\
m_3 &= \mu_1(-m_1) + \mu_2m_2 \\
n_3 &= \mu_1(-n_1) + \mu_2n_2 \qquad \mu_1, \mu_2 > 0
\end{aligned}
$$

and hence

$$
F(\lambda_1,\lambda_3) = -\mu_1F(\lambda_1,\lambda_1) + \mu_2F(\lambda_1,\lambda_2) = -\mu_1F(\lambda_1,\lambda_1) < 0
$$

Finally, if $\overrightarrow{A_0A_3}$ is an arbitrary ray between the rays opposite to $\overrightarrow{A_0A_1}$ and $\overrightarrow{A_0A_2}$, its coordinates are

$$
\begin{aligned}
l_3 &= \mu_1(-l_1) + \mu_2(-l_2) \\
m_3 &= \mu_1(-m_1) + \mu_2(-m_2) \\
n_3 &= \mu_1(-n_1) + \mu_2(-n_2) \qquad \mu_1, \mu_2 > 0
\end{aligned}
$$

and $F(\lambda_1,\lambda_3) = -\mu_1F(\lambda_1,\lambda_1) - \mu_2F(\lambda_1,\lambda_2) = -\mu_1F(\lambda_1,\lambda_1) < 0$

The signs of $F(\lambda_1,\lambda)$ for lines, λ, passing through A_0 and directed into the various sectors formed by λ_1 and the line determined by A_0 and the pole of λ_1 are shown in Fig. 5.5. The resemblance between the distribution of the signs of $F(\lambda_1,\lambda)$ in the four sectors shown and the distribution of the signs of $\cos\theta$ in the four quadrants of the cartesian plane should be noted. Its significance will become apparent when we have defined angle measures in H_2.

The measurement of angles in H_2 is based upon the following theorem:

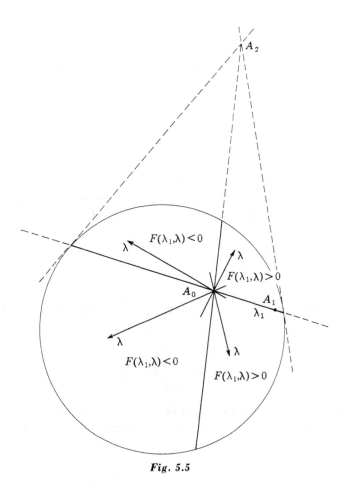

Fig. 5.5

Theorem 5.   If $P_1:(x_1,y_1)$, $P_2:(x_2,y_2)$, and $P_3:(x_3,y_3)$ are three non-collinear points of H_2 and the coordinates of $\overleftrightarrow{P_3P_1}$ and $\overleftrightarrow{P_3P_2}$ are (l_1,m_1,n_1) and (l_2,m_2,n_2), respectively, then

$$F_{11}F_{22} - F_{12}^2 > 0$$

Proof.   Since both $\overleftrightarrow{P_3P_1}$ and $\overleftrightarrow{P_3P_2}$ obviously pass through P_3, it follows that the coordinates of P_3 must satisfy the equations of $\overleftrightarrow{P_3P_1}$ and $\overleftrightarrow{P_3P_2}$. That is,

$$l_1x_3 + m_1y_3 + n_1 = 0$$
$$l_2x_3 + m_2y_3 + n_2 = 0$$

Hence $x_3 = \dfrac{m_1n_2 - m_2n_1}{l_1m_2 - l_2m_1}$ and $y_3 = \dfrac{n_1l_2 - n_2l_1}{l_1m_2 - l_2m_1}$

Moreover, since P_3 is a point of H_2, it follows that

$$x_3^2 + y_3^2 < 1$$

or

(3) $(m_1n_2 - m_2n_1)^2 + (n_1l_2 - n_2l_1)^2 - (l_1m_2 - l_2m_1)^2 < 0$

Now it is easy to verify that

$$
\begin{aligned}
F_{11}F_{22} - F_{12}^2 &= (l_1^2 + m_1^2 - n_1^2)(l_2^2 + m_2^2 - n_2^2) \\
&\qquad - (l_1l_2 + m_1m_2 - n_1n_2)^2 \\
&= -(m_1n_2 - m_2n_1)^2 - (n_1l_2 - n_2l_1)^2 + (l_1m_2 - l_2m_1)^2
\end{aligned}
$$

Hence, using (3), it is clear that

$$F_{11}F_{22} - F_{12}^2 > 0$$

as asserted.

From Theorem 5 it is evident that for intersecting lines in H_2

$$\frac{F_{12}^2}{F_{11}F_{22}} < 1$$

or $$-1 < \frac{F_{12}}{\sqrt{F_{11}F_{22}}} < 1$$

Hence $F_{12}/\sqrt{F_{11}F_{22}}$ is the cosine of some angle between 0 and π, and it becomes possible to introduce angle measures into H_2 via the following definition:

Definition 2.   If $\overrightarrow{VA}:(l_1,m_1,n_1)$ and $\overrightarrow{VB}:(l_2,m_2,n_2)$ are two non-collinear rays of H_2, then

$$m_k \angle AVB = k \cos^{-1} \frac{F_{12}}{\sqrt{F_{11}F_{22}}}$$

It remains now to verify that angle measures as thus defined have the properties asserted by the angle-measurement postulates, Postulates 13, 14, and 15, Sec. 2.6.

In the first place, if we let $\overrightarrow{VA}:(l_1,m_1,n_1)$ be an arbitrary ray in the edge of any halfplane in H_2, and $\overrightarrow{VX}:(l_2,m_2,n_2)$ be an arbitrary ray extending from V into the given halfplane, then, by Definition 2,

$$m_k \angle AVX = k \cos^{-1} \frac{F_{12}}{\sqrt{F_{11}F_{22}}}$$

and as soon as k is determined, a unique positive number is assigned to $\angle AVX$. This number we shall also assign to $\overrightarrow{VX}$. To determine k, we note that if we let $\overrightarrow{VX}$ be the ray opposite to $\overrightarrow{VA}$, so that

then
$$(l_2, m_2, n_2) = (-l_1, -m_1, -n_1)$$
$$F_{12} = -F_{11} = -F_{22}$$

and
$$k \cos^{-1} \frac{F_{12}}{\sqrt{F_{11}F_{22}}} = k \cos^{-1}(-1) = k\pi$$

If we wish the ray opposite to $\overrightarrow{VA}$ to correspond to the number R, according to the last formula, this requires that $k\pi = R$, or

$$k = \frac{R}{\pi}$$

Furthermore, if we let $\overrightarrow{VX} = \overrightarrow{VA}$, so that

then
$$(l_2, m_2, n_2) = (l_1, m_1, n_1)$$
$$F_{12} = F_{11} = F_{12}$$

and
$$k \cos^{-1} \frac{F_{12}}{\sqrt{F_{11}F_{22}}} = k \cos^{-1}(1) = 0$$

Hence by taking $k = R/\pi$ we can assign to any angle, $\angle AVX$, the unique positive number

$$m_R \angle AVX = k_R \cos^{-1} \frac{F_{12}}{\sqrt{F_{11}F_{22}}} < R \qquad k_R = \frac{R}{\pi}$$

Moreover, the correspondence

$$\overrightarrow{VX} \sim k_R \cos^{-1} \frac{F_{12}}{\sqrt{F_{11}F_{22}}}$$

assigns a unique positive number to the ray $\overrightarrow{VX}$ in such a way that $\overrightarrow{VA}$ corresponds to 0, the ray opposite to $\overrightarrow{VA}$ corresponds to R, and the number assigned to any ray extending from V into the given half-plane is between 0 and R. Thus, we have verified Postulate 13 and a portion of Postulate 15.

To check the conversion relation of Postulate 14, namely,

$$m_R \angle AVB = \frac{R}{S} m_S \angle AVB$$

we merely observe that

$$m_R \angle AVB = \frac{R}{\pi} \cos^{-1} \frac{F_{12}}{\sqrt{F_{11}F_{22}}}$$

$$= \frac{R}{S} \left(\frac{S}{\pi} \cos^{-1} \frac{F_{12}}{\sqrt{F_{11}F_{22}}} \right)$$

$$= \frac{R}{S} m_S \angle AVB$$

To check that there is a $1:1$ correspondence between the real numbers between 0 and R and the rays extending from V into a given halfplane containing V in its edge, we must not only show that to every ray there corresponds a unique positive number, which we have just done, but also that to every number between 0 and R there corresponds a unique ray extending from V into the given halfplane.

To do this, let $\overrightarrow{VA_1}$ be the ray to which the number 0 is assigned, let P be the pole of the line $\overleftrightarrow{VA_1}$, let $\overrightarrow{VA_2}$ be the ray which lies on $\overleftrightarrow{VP}$ and extends from V on the given side of $\overleftrightarrow{VA_1}$, and let ρ be any number between 0 and R. Then if the coordinates of $\overrightarrow{VA_1}$ and $\overrightarrow{VA_2}$ are (l_1, m_1, n_1) and (l_2, m_2, n_2), respectively, it follows from Theorem 3 that

$$F_{12} = 0$$

Now any ray, $\overrightarrow{VA_3}$, extending from V into the given halfplane has coordinates of the form

$$(4) \qquad \begin{aligned} l_3 &= \mu_1 l_1 + \mu_2 l_2 \\ m_3 &= \mu_1 m_1 + \mu_2 m_2 \\ n_3 &= \mu_1 n_1 + \mu_2 n_2 \end{aligned}$$

where μ_1 and μ_2 are both positive if $\overrightarrow{VA_3}$ is between $\overrightarrow{VA_1}$ and $\overrightarrow{VA_2}$, and μ_1 is negative and μ_2 positive if $\overrightarrow{VA_3}$ is between $\overrightarrow{VA_2}$ and the ray opposite to $\overrightarrow{VA_1}$. We must now show that there is a unique ray, $\overrightarrow{VA_3}$, that is, a unique pair of coordinates (μ_1, μ_2), such that

$$(5) \qquad k \cos^{-1} \frac{F_{13}}{\sqrt{F_{11}F_{33}}} = \frac{R}{\pi} \cos^{-1} \frac{F_{13}}{\sqrt{F_{11}F_{33}}} = \rho$$

Using (4), it is easy to verify that

$$\begin{aligned} F_{13} &= \mu_1 F_{11} + \mu_2 F_{12} = \mu_1 F_{11} \\ F_{33} &= \mu_1^2 F_{11} + 2\mu_1\mu_2 F_{12} + \mu_2^2 F_{22} = \mu_1^2 F_{11} + \mu_2^2 F_{22} \end{aligned}$$

Hence (5) becomes

$$\frac{R}{\pi} \cos^{-1} \frac{\mu_1 F_{11}}{\sqrt{F_{11}(\mu_1^2 F_{11} + \mu_2^2 F_{22})}} = \rho$$

or

$$\frac{\mu_1^2 F_{11}^2}{F_{11}(\mu_1^2 F_{11} + \mu_2^2 F_{22})} = \cos^2 \frac{\rho \pi}{R} = \alpha^2$$

say, or

$$\mu_1^2(1 - \alpha^2)F_{11} - \alpha^2 \mu_2^2 F_{22} = 0$$

Since α^2 is obviously less than 1, and since, by Theorem 2, F_{11} and F_{22} are both positive, it follows that the last equation has the real solutions

$$\frac{\mu_1}{\mu_2} = \pm \sqrt{\frac{\alpha^2 F_{22}}{(1 - \alpha^2)F_{11}}}$$

The positive square root identifies a ray between $\overrightarrow{VA_1}$ and $\overrightarrow{VA_2}$ corresponding to a value $\rho = \rho_1$ between 0 and $R/2$. The negative square root identifies a ray between $\overrightarrow{VA_2}$ and the ray opposite to $\overrightarrow{VA_1}$ corresponding to the supplementary value $\rho = \rho_2 = R - \rho_1$. Depending on whether the value of ρ is between 0 and $R/2$ or between $R/2$ and R, one or the other of these rays is the unique ray, $\overrightarrow{VA_3}$, corresponding to the given value of ρ. Thus the 1:1 character of the correspondence between the numbers in the interval $(0,R)$ and the rays issuing from V into the given halfplane is established.

It remains now to check the assertion of the protractor postulate, Postulate 15, Sec. 2.6, that if p_1 and p_2 are the coordinates of two rays, $\overrightarrow{VP_1}$ and $\overrightarrow{VP_2}$, extending from V into a given halfplane containing V in its edge, then

$$m_R \angle P_1 V P_2 = |p_1 - p_2|$$

To do this, we proceed very much as we did in verifying the ruler postulate in Sec. 5.2. Let (l_1, m_1, n_1) be the coordinates of the ray $\overrightarrow{VP_1}$ and let (l_2, m_2, n_2) be the coordinates of the ray $\overrightarrow{VP_2}$. Then if the coordinates of the ray, $\overrightarrow{VP_0}$, to which the number 0 is assigned be (l_0, m_0, n_0), we have

$$l_0 = \mu_1 l_1 + \mu_2 l_2$$
$$m_0 = \mu_1 m_1 + \mu_2 m_2$$
$$n_0 = \mu_1 n_1 + \mu_2 n_2$$

where, since $\overrightarrow{VP_0}$ is neither between $\overrightarrow{VP_1}$ and $\overrightarrow{VP_2}$ nor between the rays opposite to these rays, μ_1 and μ_2 are of opposite sign. It is now easy to verify that

$$F_{00} = \mu_1^2 F_{11} + 2\mu_1 \mu_2 F_{12} + \mu_2^2 F_{22}$$
$$F_{01} = \mu_1 F_{11} + \mu_2 F_{12}$$
$$F_{02} = \mu_1 F_{12} + \mu_2 F_{22}$$

Then $\quad |p_1 - p_2| = \left| \dfrac{R}{\pi} \cos^{-1} \dfrac{F_{01}}{\sqrt{F_{00}F_{11}}} - \dfrac{R}{\pi} \cos^{-1} \dfrac{F_{02}}{\sqrt{F_{00}F_{22}}} \right|$

$$= \frac{R}{\pi} |u - v|$$

say, where

$$\cos u = \frac{F_{01}}{\sqrt{F_{00}F_{11}}} \qquad \sin u = \sqrt{\frac{F_{00}F_{11} - F_{01}^2}{F_{00}F_{11}}}$$

$$\cos v = \frac{F_{02}}{\sqrt{F_{00}F_{22}}} \qquad \sin v = \sqrt{\frac{F_{00}F_{22} - F_{02}^2}{F_{00}F_{22}}}$$

Now, using the expressions noted above, we have

$$F_{00}F_{11} - F_{01}^2 = (\mu_1^2 F_{11} + 2\mu_1\mu_2 F_{12} + \mu_2^2 F_{22})F_{11} - (\mu_1 F_{11} + \mu_2 F_{12})^2$$
$$= \mu_2^2 (F_{11}F_{22} - F_{12}^2)$$

and, similarly, $\quad F_{00}F_{22} - F_{02}^2 = \mu_1^2 (F_{11}F_{22} - F_{12}^2)$

Hence

$\cos (u - v) = \cos u \cos v + \sin u \sin v$

$$= \frac{F_{01}F_{02} - \sqrt{(F_{00}F_{11} - F_{01}^2)(F_{00}F_{22} - F_{02}^2)}}{F_{00}\sqrt{F_{11}F_{22}}}$$

$$= \frac{(\mu_1 F_{11} + \mu_2 F_{12})(\mu_1 F_{12} + \mu_2 F_{22}) - \sqrt{\mu_1^2 \mu_2^2 (F_{11}F_{22} - F_{12}^2)^2}}{F_{00}\sqrt{F_{11}F_{22}}}$$

Simplifying, remembering that $\mu_1\mu_2 < 0$ and $(F_{11}F_{22} - F_{12}^2) > 0$, this becomes

$$\frac{(\mu_1^2 F_{11} + 2\mu_1\mu_2 F_{12} + \mu_2^2 F_{22})F_{12}}{F_{00}\sqrt{F_{11}F_{22}}}$$

or finally, since $F_{00} = \mu_1^2 F_{11} + 2\mu_1\mu_2 F_{12} + \mu_2^2 F_{22}$,

$$\frac{F_{12}}{\sqrt{F_{11}F_{22}}}$$

Therefore $\quad \cos(u - v) = \dfrac{F_{12}}{\sqrt{F_{11}F_{22}}} \quad$ or $\quad |u - v| = \cos^{-1} \dfrac{F_{12}}{\sqrt{F_{11}F_{22}}}$

and hence $\quad |p_1 - p_2| = \dfrac{R}{\pi} |u - v| = \dfrac{R}{\pi} \cos^{-1} \dfrac{F_{12}}{\sqrt{F_{11}F_{22}}} = m_R \angle P_1 V P_2$

This completes the verification that the postulates of angle measurement are satisfied by Definition 2.

EXERCISES

1. What is the pole of
   (a) The line whose coordinates are $(3,2,1)$?
   (b) The line whose equation is $x - 2y = 1$?
   (c) The line determined by the points $(-\frac{1}{2},0)$ and $(\frac{1}{2},\frac{3}{4})$?

2. Prove that the lines $\lambda_1: (l_1,m_1,n_1)$, $\lambda_2: (l_2,m_2,n_2)$, and $\lambda_3: (l_3,m_3,n_3)$ are concurrent if and only if

$$\begin{vmatrix} l_1 & m_1 & n_1 \\ l_2 & m_2 & n_2 \\ l_3 & m_3 & n_3 \end{vmatrix} = 0$$

3. Graph the envelopes whose equations are

   (a) $l = m$ (b) $l = n$ (c) $n = 0$
   (d) $4lm = n^2$ (e) $4l^2 + m^2 = n^2$ (f) $l^3 = mn^2$

4. Complete the verification begun in Exercise 2, Sec. 5.1, that the hyperbolic parallel postulate, Postulate 17_H, is satisfied in H_2.

In Exercises 5 to 10, assume that $k = 180/\pi$.

5. What are the hyperbolic measures of the angles determined by the lines $x + y = 1$ and $-x + 2y = 1$?

6. If $A = (0,0)$, $B = (\frac{1}{2},0)$, $C = (0,\frac{1}{2})$, and $D = (-\frac{1}{2},-\frac{1}{2})$, what is (a) $m\angle ABC$; (b) $m\angle ABD$; (c) $m\angle ACD$; (d) $m\angle BCD$?

7. In Exercise 6, prove that $m\angle BAC + m\angle CAD + m\angle DAB = 360$.

8. (a) What are the cartesian equations of the lines which bisect in H_2 the angles determined by the lines $x = \frac{1}{2}$ and $y = 0$? (b) What are the cartesian equations of the lines which pass through $(\frac{1}{2},\frac{1}{2})$ and determine angles of hyperbolic measure 60 with the line $x + y = 1$?

9. (a) What is the hyperbolic distance from the origin to the line whose cartesian equation is $x + y = 1$? (b) What is the hyperbolic distance from $(\frac{1}{2},0)$ to the line whose cartesian equation is $x + y = 1$?

10. If the area of the triangular region whose vertices are $(0,0)$, $(\frac{1}{4},0)$, and $(0,\frac{1}{4})$ is 1, what is
    (a) The area of the triangular region whose vertices are $(0,0)$, $(\frac{1}{2},0)$, and $(0,\frac{1}{2})$?
    (b) The area of the triangular region determined by the lines $x = 0$, $2x + 2y = 1$, $2x - 2y = 1$?
    (c) The area of the polygonal region whose vertices are $(0,0)$, $(\frac{1}{2},0)$, $(\frac{1}{2},\frac{1}{2})$, and $(0,\frac{1}{2})$?

11. Prove Theorem 1.

5.4 Triangles in H_2.

We have now introduced the measurement of lengths and angles into H_2 in such a way that the appropriate postulates are satisfied. Our final obligation is to verify that the congruence postulate, Postulate 16, Sec. 2.9, is satisfied in H_2. Before we can do this, however, we must derive formulas analogous to the law of cosines and the law of sines for triangles in H_2, and this, in turn, requires certain additional identities connecting the f_{ij}'s and the F_{ij}'s.

Let $A:(x_1,y_1)$, $B:(x_2,y_2)$, and $C:(x_3,y_3)$ be three noncollinear points

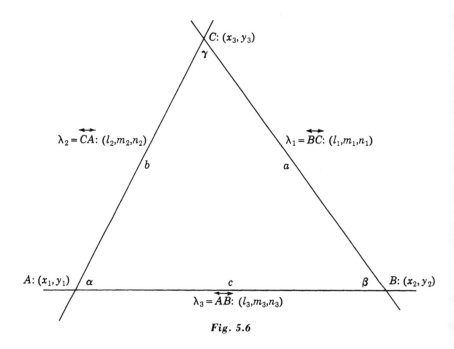

Fig. 5.6

of H_2, and let the lines determined by these points, cyclically directed, be (see Fig. 5.6)

$$\lambda_1 = \overleftrightarrow{BC}:(l_1,m_1,n_1)$$
$$\lambda_2 = \overleftrightarrow{CA}:(l_2,m_2,n_2)$$
$$\lambda_3 = \overleftrightarrow{AB}:(l_3,m_3,n_3)$$

As we have already seen, the coordinates of the λ's can be expressed in terms of the coordinates of A, B, and C, and we have

$$\lambda_1:(y_2 - y_3),\ (x_3 - x_2),\ (x_2y_3 - x_3y_2)$$
$$\lambda_2:(y_3 - y_1),\ (x_1 - x_3),\ (x_3y_1 - x_1y_3)$$
$$\lambda_3:(y_1 - y_2),\ (x_2 - x_1),\ (x_1y_2 - x_2y_1)$$

Conversely, the coordinates of A, B, and C can be expressed in terms of the coordinates of the λ's. For instance, since A lies on both λ_2 and λ_3, its coordinates must satisfy the equation of each of these lines; hence

$$l_2 x_1 + m_2 y_1 + n_2 = 0$$
$$l_3 x_1 + m_3 y_1 + n_3 = 0$$

whence $\qquad x_1 = \dfrac{m_2 n_3 - m_3 n_2}{l_2 m_3 - l_3 m_2} \qquad$ and $\qquad y_1 = \dfrac{n_2 l_3 - n_3 l_2}{l_2 m_3 - l_3 m_2}$

Similarly,

$$x_2 = \frac{m_3 n_1 - m_1 n_3}{l_3 m_1 - l_1 m_3}, \qquad y_2 = \frac{n_3 l_1 - n_1 l_3}{l_3 m_1 - l_1 m_3}$$

$$x_3 = \frac{m_1 n_2 - m_2 n_1}{l_1 m_2 - l_2 m_1}, \qquad y_3 = \frac{n_1 l_2 - n_2 l_1}{l_1 m_2 - l_2 m_1}$$

It is easy to check and important to note that

$$l_2 m_3 - l_3 m_2 = l_3 m_1 - l_1 m_3 = l_1 m_2 - l_2 m_1 = \begin{vmatrix} x_1 & y_1 & 1 \\ x_2 & y_2 & 1 \\ x_3 & y_3 & 1 \end{vmatrix} = \Delta$$

say, where, since A, B, and C are noncollinear, $\Delta \neq 0$.

Now if (u_1, v_1, w_1), (u_2, v_2, w_2), (u_3, v_3, w_3) are any three ordered triples, it is only a matter of elementary algebra to verify that

(1) $\quad (-u_i u_j - v_i v_j + w_i w_j)^2 - (-u_i^2 - v_i^2 + w_i^2)(-u_j^2 - v_j^2 + w_j^2)$
$$= (v_i w_j - v_j w_i)^2 + (w_i u_j - w_j u_i)^2 - (u_i v_j - u_j v_i)^2 \qquad i \neq j$$

(2) $\quad (-u_i u_j - v_i v_j + w_i w_j)(-u_k^2 - v_k^2 + w_k^2)$
$$- (-u_i u_k - v_i v_k + w_i w_k)(-u_j u_k - v_j v_k + w_j w_k)$$
$$= (v_j w_k - v_k w_j)(v_k w_i - v_i w_k)$$
$$+ (w_j u_k - w_k u_j)(w_k u_i - w_i u_k)$$
$$- (u_j v_k - u_k v_j)(u_k v_i - u_i v_k)$$
$$i \neq j \neq k$$

If we apply (1) and (2) to the triples $(x_1, y_1, 1)$, $(x_2, y_2, 1)$, $(x_3, y_3, 1)$, these equations become, respectively,

(3) $\qquad f_{ij}^2 - f_{ii} f_{jj} = l_k^2 + m_k^2 - n_k^2 = F_{kk}$
(4) $\qquad f_{ij} f_{kk} - f_{ik} f_{jk} = l_i l_j + m_i m_j - n_i n_j = F_{ij}$

Similarly, if we apply (1) and (2) to the triples (l_1, m_1, n_1), (l_2, m_2, n_2), (l_3, m_3, n_3), we obtain

(5) $\qquad F_{ij}^2 - F_{ii} F_{jj} = (x_k \Delta)^2 + (y_k \Delta)^2 - \Delta^2 = -f_{kk} \Delta^2$
(6) $\quad F_{ij} F_{kk} - F_{ik} F_{jk} = (x_i \Delta)(x_j \Delta) + (y_i \Delta)(y_j \Delta) - \Delta^2 = -f_{ij} \Delta^2$

Now taking $k = 1$, for convenience, we have for the length, a, of the side $\overline{BC}$ of $\triangle ABC$

(7) $$\cosh a = \frac{f_{23}}{\sqrt{f_{22}f_{33}}}$$

$$\sinh a = \sqrt{\cosh^2 a - 1} = \sqrt{\frac{f_{23}^2 - f_{22}f_{33}}{f_{22}f_{33}}}$$

(8) $$= \sqrt{\frac{F_{11}}{f_{22}f_{33}}}$$

by (3). Similar expressions, of course, hold for the lengths, b and c, of the other two sides.

For the measure, α, of $\angle A$ we have, taking $R = \pi$ for convenience,

(9) $$\cos \alpha = -\frac{F_{23}}{\sqrt{F_{22}F_{33}}}$$

the minus sign entering because we must reverse the signs of the coordinates (l_2, m_2, n_2) of the line $\overleftrightarrow{CA}$ in order to obtain the coordinates of the ray $\overrightarrow{AC}$ which, with $\overrightarrow{AB}$, determines $\angle A$. Hence

$$\sin \alpha = \sqrt{1 - \cos^2 \alpha} = \sqrt{\frac{F_{22}F_{33} - F_{23}^2}{F_{22}F_{33}}}$$

(10) $$= \sqrt{\frac{\Delta^2 f_{11}}{F_{22}F_{33}}}$$

by (5). Similar expressions can, of course, be obtained for the functions of β and γ by permuting the subscripts.

With the preceding formulas available, we can now prove the following important theorems:

Theorem 1.   In any triangle in H_2

$$\cosh a = \cosh b \cosh c - \sinh b \sinh c \cos \alpha$$

Proof. Evaluating the right-hand side of the given equation, we have

$$\frac{f_{13}}{\sqrt{f_{11}f_{33}}} \frac{f_{12}}{\sqrt{f_{11}f_{22}}} - \sqrt{\frac{F_{22}}{f_{11}f_{33}}} \sqrt{\frac{F_{33}}{f_{11}f_{22}}} \frac{-F_{23}}{\sqrt{F_{22}F_{33}}}$$

$$= \frac{f_{12}f_{13} + F_{23}}{f_{11}\sqrt{f_{22}f_{33}}}$$

$$= \frac{f_{12}f_{13} + (f_{23}f_{11} - f_{12}f_{13})}{f_{11}\sqrt{f_{22}f_{33}}} \qquad \text{by (4)}$$

$$= \frac{f_{23}}{\sqrt{f_{22}f_{33}}}$$

$$= \cosh a$$

as asserted.

By replacing $\cosh a$, $\cosh b$, $\sinh b$, $\cosh c$, and $\sinh c$ by the first two terms of their Maclaurin expansions, it is easy to show that the formula of Theorem 1 approaches the euclidean law of cosines

$$a^2 = b^2 + c^2 - 2bc \cos \alpha$$

as the dimensions of the triangle approach 0.

Theorem 2. In any triangle in H_2

$$\cos \alpha = -\cos \beta \cos \gamma + \sin \beta \sin \gamma \cosh a$$

Proof. Evaluating the right-hand side of the given equation, we have

$$-\left(\frac{-F_{13}}{\sqrt{F_{11}F_{33}}}\right)\left(\frac{-F_{12}}{\sqrt{F_{11}F_{22}}}\right) + \sqrt{\frac{f_{22}\Delta^2}{F_{11}F_{33}}} \sqrt{\frac{f_{33}\Delta^2}{F_{11}F_{22}}} \frac{f_{23}}{\sqrt{f_{22}f_{33}}}$$

$$= \frac{-F_{12}F_{13} + f_{23}\Delta^2}{F_{11}\sqrt{F_{22}F_{33}}}$$

$$= \frac{-F_{12}F_{13} + (F_{13}F_{12} - F_{11}F_{23})}{F_{11}\sqrt{F_{22}F_{33}}} \qquad \text{by (6)}$$

$$= \frac{-F_{23}}{\sqrt{F_{22}F_{33}}}$$

$$= \cos \alpha$$

as asserted.

Theorem 3. In any triangle in H_2

$$\frac{\sin \alpha}{\sinh a} = \frac{\sin \beta}{\sinh b} = \frac{\sin \gamma}{\sinh c} = \sqrt{\frac{\Delta^2 f_{11}f_{22}f_{33}}{F_{11}F_{22}F_{33}}}$$

Proof. By direct substitution, each fraction turns out immediately to have the asserted value.

Theorem 4. If there exists a one-to-one correspondence between two triangles or between a triangle and itself in which two sides and the angle determined by these sides in one triangle are congruent to the corresponding parts of the other triangle, then the correspondence is a congruence and the triangles are congruent.

Proof. Let $A \leftrightarrow A'$, $B \leftrightarrow B'$, and $C \leftrightarrow C'$ be a correspondence between $\triangle ABC$ and $\triangle A'B'C'$ such that

$$\overline{AB} \cong \overline{A'B'}, \quad \angle A \cong \angle A', \quad \overline{AC} \cong \overline{A'C'}$$

that is,

$$c = c', \quad \alpha = \alpha', \quad b = b'$$

Then, using Theorem 1, we have

$$\cosh a = \cosh b \cosh c - \sinh b \sinh c \cos \alpha$$
$$= \cosh b' \cosh c' - \sinh b' \sinh c' \cos \alpha'$$
$$= \cosh a'$$

Hence $a = a'$, that is, $\overline{BC} \cong \overline{B'C'}$. Moreover, using a permuted version of the formula of Theorem 1, we have

$$\cos \beta = \frac{\cosh b - \cosh a \cosh c}{\sinh a \sinh c}$$
$$= \frac{\cosh b' - \cosh a' \cosh c'}{\sinh a' \sinh c'}$$
$$= \cos \beta'$$

Hence $\beta = \beta'$, that is, $\angle B \cong \angle B'$. Similarly, we can show that $\angle C \cong \angle C'$, which completes the proof of the theorem.

Theorem 4 is, of course, precisely the congruence postulate, Postulate 16, Sec. 2.9. Now that it has been verified, our proof of the relative consistency of hyperbolic geometry is complete. However, before we conclude our discussion, we shall investigate the interpretation in H_2 of several other features of hyperbolic geometry.

One of the striking properties of hyperbolic geometry which we encountered in Chap. 4 was the fact that if corresponding angles of two triangles are congruent, the triangles are congruent. Since this is a consequence of postulates which we have now verified in H_2, it is obvious that this must be a theorem in H_2 for which we could give exactly the same proof we used in Chap. 4. However, a much shorter proof is available in H_2:

Theorem 5. If there exists a correspondence between two triangles in which corresponding angles are congruent, then the triangles are congruent.

Proof. Let $A \leftrightarrow A'$, $B \leftrightarrow B'$, $C \leftrightarrow C'$ be a correspondence between $\triangle ABC$ and $\triangle A'B'C'$ in which $\alpha = \alpha'$, $\beta = \beta'$, $\gamma = \gamma'$. Then, using the formula of Theorem 2, we have

$$\cosh a = \frac{\cos \alpha + \cos \beta \cos \gamma}{\sin \beta \sin \gamma}$$
$$= \frac{\cos \alpha' + \cos \beta' \cos \gamma'}{\sin \beta' \sin \gamma'}$$
$$= \cosh a'$$

Hence $a = a'$, that is, $\overline{BC} \cong \overline{B'C'}$. Similarly, of course, $\overline{AC} \cong \overline{A'C'}$ and $\overline{AB} \cong \overline{A'B'}$, which proves that the triangles are congruent.

Another interesting feature of hyperbolic geometry is the angle-sum theorem, for which a very neat proof is available in H_2:

Theorem 6.   If $k = 1$, the sum of the measures of the angles of any triangle in H_2 is less than π.

Proof.   Let $\triangle ABC$ be an arbitrary triangle in H_2 and, for convenience, let the notation be chosen so that

$$\alpha \geqq \beta \geqq \gamma$$

Then, since the measure of any angle in H_2 is between 0 and π when $k = 1$, we have $-\pi < \alpha - \beta - \gamma < \pi$, or

(11)
$$-\frac{\pi}{2} < \frac{\alpha - \beta - \gamma}{2} < \frac{\pi}{2}$$

Now, using the formula of Theorem 2, together with the expansion of $\cos (\beta + \gamma)$, we have

$$\cos \alpha = -\cos \beta \cos \gamma + \sin \beta \sin \gamma \cosh a$$
$$\cos (\beta + \gamma) = \cos \beta \cos \gamma - \sin \beta \sin \gamma$$

Then, adding,

$$\cos \alpha + \cos (\beta + \gamma) = \sin \beta \sin \gamma (\cosh a - 1)$$

and, converting the sum on the left to a product,

$$2 \cos \frac{\alpha + \beta + \gamma}{2} \cos \frac{\alpha - \beta - \gamma}{2} = \sin \beta \sin \gamma (\cosh a - 1)$$

Now each factor on the **right** of the last equation is positive. Hence the factors on the left must be of the same sign. But from (11) it follows that

$$\cos \frac{\alpha - \beta - \gamma}{2} > 0$$

Hence
$$\cos \frac{\alpha + \beta + \gamma}{2} > 0$$

which implies that
$$\frac{\alpha + \beta + \gamma}{2} < \frac{\pi}{2}$$

or
$$\alpha + \beta + \gamma < \pi$$

as asserted. This conclusion is based, of course, on the assumption that $k = 1$, that is, that $R = \pi$. For arbitrary R, the conclusion would be that the sum of the measures of the angles of **any** triangle in H_2 is less than R. However, regardless of the unit used in measuring angles, the central fact is the same, namely, that the sum of the measures of the angles of any triangle in H_2 is less than the sum of the measures of two right angles.

As a final observation, we shall apply the formula of Theorem 2 to a general right-angled asymptotic triangle and obtain an expression for the amplitude of parallelism, $\alpha = \pi(c)$, in H_2. To do this, let $C:(x_3,y_3)$ be a point of the metric gauge conic, and let $A:(x_1,y_1)$ and $B:(x_2,y_2)$ be points of H_2 such that $\angle ABC$ is a right angle (see Fig. 5.7). Then $\overrightarrow{AC}\|\overrightarrow{BC}$, $\triangle(AC,BC)$ is a right-angled asymptotic triangle, and $\alpha = m\angle A$ is the amplitude of parallelism for the distance $AB = c$.

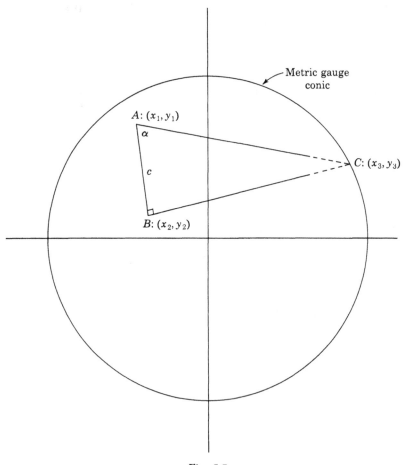

Fig. 5.7

Now C, though a well-defined point of E_2, is not a point of H_2. However, the proof of the identity of Theorem 2 is valid as long as the quantities involved are meaningful, that is, as long as formulas (7),

(8), (9), and (10) are meaningful. Hence, although we cannot apply to $\triangle(AC,BC)$ a version of this identity involving either of the infinite sides, it is correct to write

(12) $$\cos \gamma = -\cos \alpha \cos \beta + \sin \alpha \sin \beta \cosh c$$

provided that $\cos \gamma$ can be determined. Now, according to Eq. (9),

$$\cos \gamma = \frac{-F_{12}}{\sqrt{F_{11}F_{22}}}$$

Also, $f_{33} = 0$, since $C:(x_3,y_3)$ is a point on the metric gauge conic, $f = 0$. Therefore, by (3) and (4),

$$\begin{aligned}
F_{12} &= f_{12}f_{33} - f_{13}f_{23} = -f_{13}f_{23} \\
F_{11} &= f_{23}^2 - f_{22}f_{33} \quad = f_{23}^2 \\
F_{22} &= f_{13}^2 - f_{11}f_{33} \quad = f_{13}^2
\end{aligned}$$

and hence $$\cos \gamma = \frac{-(-f_{13}f_{23})}{\sqrt{f_{23}^2 f_{13}^2}} = 1$$

Returning now to Eq. (12), and remembering that $\beta = \dfrac{\pi}{2}$, we have

$$1 = \sin \alpha \cosh c$$

or, since $\alpha = \pi(c)$,

(13) $$\sin \pi(c) = \text{sech } c$$

A more conventional relation can be obtained by squaring both sides of (13) and using obvious identities:

$$\sin^2 \pi(c) = \text{sech}^2 c$$
$$1 - \cos^2 \pi(c) = 1 - \tanh^2 c$$
(14) $$\cos \pi(c) = \tanh c$$

EXERCISES

1. Using the appropriate formulas,
   (a) Prove that the triangle inequality holds in H_2.
   (b) Prove that Theorem 1, Sec. 2.9 (the angle-side-angle congruence theorem), holds in H_2.
   (c) Prove that Theorem 4, Sec. 2.9 (the side-side-side congruence theorem), holds in H_2.
2. (a) Verify that the formula of Theorem 1 reduces to the euclidean law of cosines as the dimensions of the triangle approach zero.
   (b) Verify that the formula of Theorem 3 reduces to the euclidean law of sines as the dimensions of the triangle approach zero.
3. Does a euclidean relation result from the formula of Theorem 2 as the dimensions of the triangle approach zero?

4. Using the formulas of Theorems 1, 2, and 3, show that if $\triangle ABC$ is a right triangle with right angle at A, then

(a) $\cosh a = \cosh b \cosh c$ (b) $\cos \beta \cos \gamma = \sin \beta \sin \gamma \cosh a$
(c) $\cos \beta = \sin \gamma \cosh b$ (d) $\cos \gamma = \sin \beta \cosh c$
(e) $\sinh b = \sin \beta \sinh a$ (f) $\sinh c = \sin \gamma \sinh a$

To what euclidean relation does each of these reduce as the dimensions of $\triangle ABC$ approach zero?

5. If $\triangle ABC$ is a right triangle with right angle at A, establish the following relations:

(a) $\sinh c = \tanh b \cot \beta$ (b) $\sinh b = \tanh c \cot \gamma$
(c) $\cos \beta = \tanh b \coth a$ (d) $\cos \gamma = \tanh c \coth a$
(e) $\cosh a = \cot \beta \cot \gamma$

To what euclidean relation does each of these reduce as the dimensions of $\triangle ABC$ approach zero?

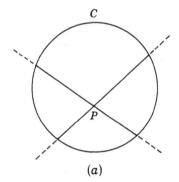

(a)

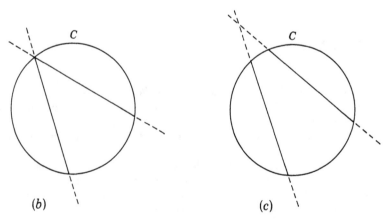

(b)   (c)

Fig. 5.1. (a) Intersecting lines in H_2; (b) parallel lines in H_2; (c) nonintersecting lines in H_2.

5.5 Summary. The work of this chapter has served two purposes. In the first place, it has established the relative consistency of hyperbolic geometry and demonstrated that the euclidean parallel postulate cannot be derived from the other euclidean postulates. Second, it has provided us with a model in which many of the most striking features of hyperbolic geometry are vividly apparent and which therefore can serve us as a ready reminder of these properties.

We have already noted how the very definition of the system H_2 gives us an immediate interpretation of intersecting, parallel, and nonintersecting lines (Fig. 5.1). We have also observed that a line, λ_2, is perpendicular to a second line, λ_1, if and only if $F_{12} = 0$, that is, if and only if λ_2 passes through the pole of λ_1. Hence the lines which are perpendicular to a given line, λ_1, are the lines of the pencil whose vertex is the pole, P_1, of λ_1 (Fig. 5.8). The existence and uniqueness of the line which passes through a given point and is perpendicular to a given line is thus obvious.

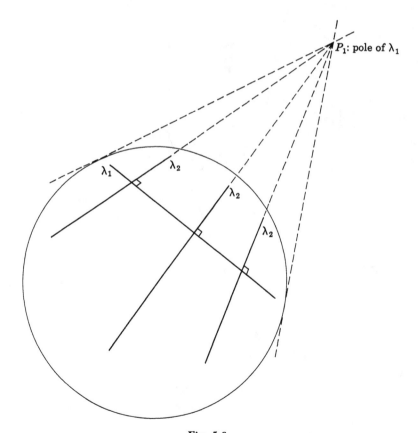

Fig. 5.8

Another striking property of hyperbolic geometry is the existence
of lines which are perpendicular to one of two intersecting lines and
parallel to the other. In H_2, this too is obvious. For if λ_1 and λ_2
are two intersecting lines, if P_1 is the pole of λ_1, and if Q_2 and Q_2' are
the intersections of λ_2 and the metric gauge conic, then both $\overleftrightarrow{P_1Q_2}$
and $\overleftrightarrow{P_1Q_2'}$ are perpendicular to λ_1 and parallel to λ_2 (Fig. 5.9).

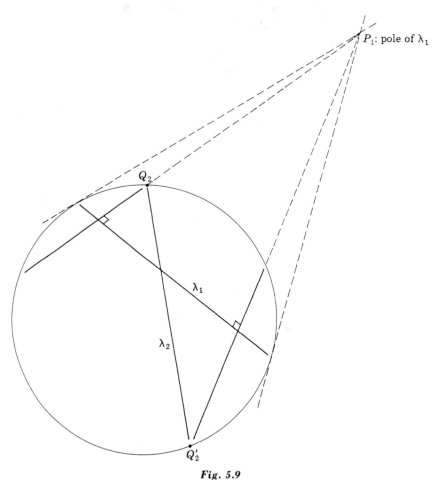

Fig. 5.9

In our study of hyperbolic geometry we also discovered that two
nonintersecting lines always have a unique common perpendicular,
and this too has an interesting interpretation in H_2. In fact, if λ_1
and λ_2 are two lines which are nonintersectors in H_2, but intersect
in E_2 in a point P in the exterior of the metric gauge conic, then the
common perpendicular of λ_1 and λ_2 is the polar line, p, of the inter-

section, P. For λ_1 and λ_2 each pass through the pole of p, namely P. Therefore, each is perpendicular to p (Fig. 5.10). Moreover, any line which is perpendicular to both λ_1 and λ_2 must pass through the pole of λ_1 and the pole of λ_2 and hence is uniquely determined.

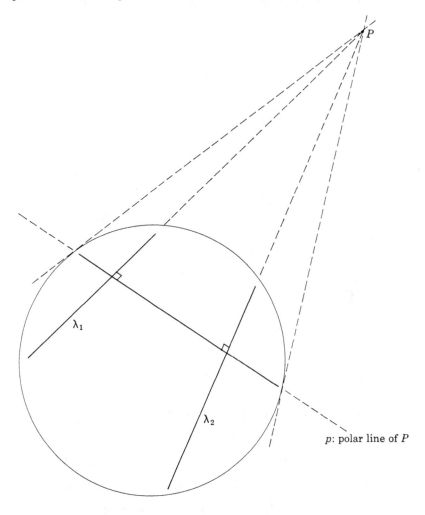

Fig. 5.10

As a final illustration, we observe that in H_2 the famous theorem of Saccheri, which asserts that if three angles of a quadrilateral are right angles, then the fourth angle is acute, is also an immediate consequence of simple polar relations. Specifically, let $ABCD$ be a quadrilateral in H_2 with right angles at A, B, and C; let P be the pole of $\overleftrightarrow{AB}$; let Q be the pole of $\overleftrightarrow{BC}$; let R be the pole of $\overleftrightarrow{AD}$; and let C' be a point, distinct from D, on the ray opposite to $\overrightarrow{DC}$ (Fig. 5.11). Then, since $\overleftrightarrow{AB}$ and $\overleftrightarrow{CD}$ are both perpendicular to $\overleftrightarrow{BC}$, each must pass through the pole, Q, of $\overleftrightarrow{BC}$. Likewise, $\overleftrightarrow{AB}$, being perpendicular to

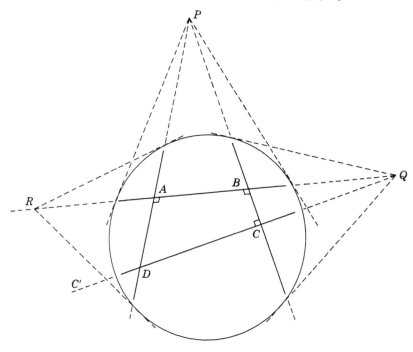

Fig. 5.11

$\overleftrightarrow{AD}$, must pass through the pole, R, of $\overleftrightarrow{AD}$. Now $\overleftrightarrow{CD}$ cannot pass through R, since it already intersects $\overleftrightarrow{AB}$ in Q. Hence $\overleftrightarrow{CD}$ cannot be perpendicular to $\overleftrightarrow{AD}$; that is, $\angle D$ cannot be a right angle. Moreover, since $\overrightarrow{CD}$ cannot intersect $\overline{AR}$, it follows that $\overrightarrow{DC'}$ lies between $\overrightarrow{DR}$ and the ray opposite to $\overrightarrow{DA}$. Hence, by the reasoning that led up to Fig. 5.5, the cosine of $\angle ADC'$ must be negative and $\angle ADC'$ must be obtuse. Therefore its supplement, $\angle ADC$, must be acute, as asserted.

EXERCISES

1. (a) What is the cartesian equation of the common perpendicular in H_2 to $y = 0$ and $x + 4y = 2$?

(b) What is the shortest hyperbolic distance between $y = 0$ and

$$x + 4y = 2$$

2. What are the cartesian equations of the lines of H_2 which are perpendic-
    ular to $2x + 2y = 1$ and parallel to $2x = 1$?

3. In H_2, how many lines, if any, can be drawn parallel to one of two parallel
    lines and perpendicular to the other?

4. In H_2, how many lines, if any, can be drawn parallel to one of two non-
    intersecting lines and perpendicular to the other?

5. Using pole and polar relations, show that in H_2 two parallel lines or two
    intersecting lines cannot have a common perpendicular.

INDEX

The letter *e* after a page number refers to an exercise.

CPSIA information can be obtained
at www.ICGtesting.com
Printed in the USA
LVHW082158300719
625943LV00018B/843/P